T0325114

DYNAMIC FAILURE OF MATERIALS

Theory, Experiments and Numerics

Proceedings of the International Seminar on Dynamic Failure of Materials –
Theory, Experiments and Numerics held in Vienna, Austria, 2–4 January
1991.

DYNAMIC FAILURE OF MATERIALS

Theory, Experiments and Numerics

Edited by

H. P. ROSSMANITH

Technical University of Vienna, Vienna, Austria

and

A. J. ROSAKIS

California Institute of Technology, Pasadena, USA

ELSEVIER APPLIED SCIENCE
LONDON and NEW YORK

ELSEVIER SCIENCE PUBLISHERS LTD
Crown House, Linton Road, Barking, Essex IG11 8JU, England

Sole Distributor in the USA and Canada
ELSEVIER SCIENCE PUBLISHING CO., INC.
655 Avenue of the Americas, New York, NY 10010, USA

WITH 14 TABLES AND 252 ILLUSTRATIONS

© 1991 ELSEVIER SCIENCE PUBLISHERS LTD

British Library Cataloguing in Publication Data

Dynamic failure of materials : theory,
experiments and numerics.
I. Rossmanith, H. P. II. Rosakis, A. P.
620.112

ISBN 1-85166-665-6

Library of Congress CIP data applied for

PREFACE

The rapid pace of current developments in the theoretical, analytical, numerical and experimental fields of dynamic failure of materials called for an international seminar of workshop style aimed at improving the finding and understanding of solutions to the basic physical processes involved in dynamic failure.

The Vienna Seminar DFM-1 was held at the Technical University Vienna in the historic administration building in the city center under the auspices of the university. More than 30 international experts from all over the world followed the invitation to participate at this seminar. High in the list of priorities was the common desire for ample time for discussions after each technical presentation, a fact and a chance frequently made overextensive use of during the seminar. Thus, opportunity was given to the seminar participants to present and expose ideas and results of their original research work, either terminated, ongoing, proposed or conceived and intended, to an international forum of experts for critical discussions, evaluation and appraisal.

The technical program included dynamic failure of polymers and steel, numerical modelling of fracture processes, experimental techniques and analytical/numerical investigation of crack/wave interaction problems. The scope of the contributions stretched from implementation of advanced mathematical techniques in the theoretical developments to most direct applications in various fields of engineering practice.

Papers published in this volume represent revised, updated and expanded versions of the seminar contributions.

Hopefully the new ideas, results and experiences gained at this international seminar will be taken into consideration and put to work by the scientists and engineers who are concerned with the assurance and evaluation of safety and integrity of modern advanced technology.

All participants agreed on the fact that it is mandatory to continue with this type of seminar/workshop in the future to let it become a fruitful, well-received and esteemed tradition.

The local Organizing Committee wishes to sincerely thank the authors for the timely submission of their original manuscripts. The generous help and appreciated advice of many colleagues is kindly acknowledged.

H. P. ROSSMANITH
Technical University Vienna, Austria

CONTENTS

ANALYSIS OF HIGH VELOCITY IMPACT TESTING

JOHN DEAR and HUGH MacGILLIVRAY,
Department of Mechanical Engineering,
Imperial College, Exhibition Road, London SW7 2BX

ABSTRACT

This paper reports on the evaluation of a range of polymeric materials as to their crack resistance to impact loadings. On-specimen sensors are used to study the variations of impact stresses induced in different parts of the specimen for the high impact velocities used. These test data have large oscillatory transient forces present that much influence and sometimes dominate crack initiation and propagation. Under these conditions, the forces acting on the crack initiation site can be significantly different to those seen by sensors placed in the striker or support points for the specimen.

INTRODUCTION

Impact testing is extensively used in research and production of materials and their manufacture into an increasingly wide variety of products. With advances in polymer and other non-metallic materials that provide for lighter, stronger, higher working temperatures, easier-to-fabricate, lower cost and many other improved attributes, so the level of impact testing required varies greatly. A main trend is towards higher impact velocities with a need for more precise measurements. For polymers that have enhanced toughness and other improved dynamic qualities, then, as well as higher impact velocities being needed, better instrumentation of test specimens and more detailed analysis of test results is also becoming very important (1-6). A point is that under more demanding test conditions results can vary greatly for many reasons. For example, with three-point bend testing, the size and nature of the crack initiation zone, the hinge-bend section, side-lip tearing, partial crack arrest and other such features are sensitive to impact test conditions. Also, care is needed in the determination of K_c and G_c when high velocity impact testing is required.

Some factors affecting results are the bounce between the specimen and its supports and with the striker, the dynamic properties of the material, the dimensions of the specimen, the extent the specimen overhangs its supports, symmetry of the three-point bend test and, of course, temperature and other test conditions. Some of these factors can be minimised by using cushioning at the support points. This may be acceptable for routine testing of the same material in production to simplify test results but questionable if used for evaluating a range of different materials or when more precise measurements are required for research or other studies. High-speed photography is used to capture on film the behaviour of the test specimen as it is taken to its failure point of crack initiation and propagation. This is for specimens with and without side-grooves. Optical and scanning electron-microscope studies are made of the fracture surfaces and findings discussed.

EXPERIMENTAL

Equipment

For this study, a drop-weight three-point bend impact tester was used with a falling mass of 65 kg. At a velocity of 5 m s^{-1} this gave an available energy of ca. 0.8 kJ to allow specimens to be tested of thickness B (12mm), width W (40mm) and overall length L (100mm), with a support span S (160mm) and a pre-notch of depth 15mm. Using this size of specimen meant that the 1.5 mm length strain gauges were relatively small making it easier to explore more precisely the strain on different parts of the specimen. At the root of the notch, a razor-sharp pre-crack of depth 5mm was made immediately before the test. The general arrangement is illustrated in Figure 1 which shows the specimen resting on two rollers of diameter 25mm and the impact striker having a tip radius of 10mm. For each specimen fitted with strain gauges, a static load was used to calibrate the outputs from the strain gauges and their amplifiers. The specimen was placed into the same calibration fixture in the two possible orientations as a further check on the gauges and any undue sensitivity to slight misalignments. Given that all these static tests were consistent, then, the gauge was finally calibrated in the static test rig just prior to impact loading.

On-specimen strain gauges

The selected strain gauges were a self-temperature compensated constantan foil with polyimide backing. They were connected in a three-wire 1/4 bridge configuration to amplifiers with a frequency response of 1 dB at 200 kHz and the amplifier outputs were monitored on a 20MHz digital storage scope. The sensor on the striker used a full bridge of semiconductor strain gauges and was monitored in a similar way as the on-specimen gauges. The surface of the specimen where the strain gauges were to be attached was carefully abraded and cleaned taking care not to leave any deep grooves or residue of cleaning materials. An effective and readily available cleaning agent proved to be acetone degreaser. The attachment site was then heat-treated just before bonding of the gauge to the surface with a cyanoacrylate adhesive. A thin glue-line was achieved by using carefully metered amounts of adhesive and quickly applying a firm steady pressure to the bond.

Of particular interest was to monitor the strain as near as possible to the crack initiation site and to relate this to forces monitored at the striker contact point by sensors in the striker. This is to have regard for the yielding of the surface material around the crack tip and also that the fast crack propagation would mostly be influenced by the bulk properties of the specimen material. A position of strain gauge sensor found to be most suitable was W/2 (i.e. 20mm) from the tip of the pre-crack along the length of the specimen as shown in Figure 1. The position of this site was not too critical and it fitted in well with the positioning of strain gauge sensors for other materials including metals.

Of course, once the crack was initiated, then, the strain as seen by this W/2 sensor would be affected by the redistribution of strain within the specimen. Another site that captured better the overall strain within the specimen very similar to that seen by the striker sensor, was that placed in the compression zone of the specimen. This being situated on the same side of the specimen but nearer to the striker. Having regard for the surface effects, the sensor was placed W/8 from the impact edge of the specimen and W/2 away from the crack. This will now be referred to as the W/2xW/8 position.

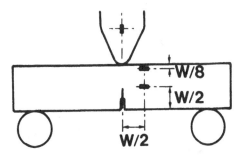

Figure 1. Schematic diagram showing the location sites of strain gauges on the striker and on the specimen in the W/2xW/2 and W/2xW/8 positions.

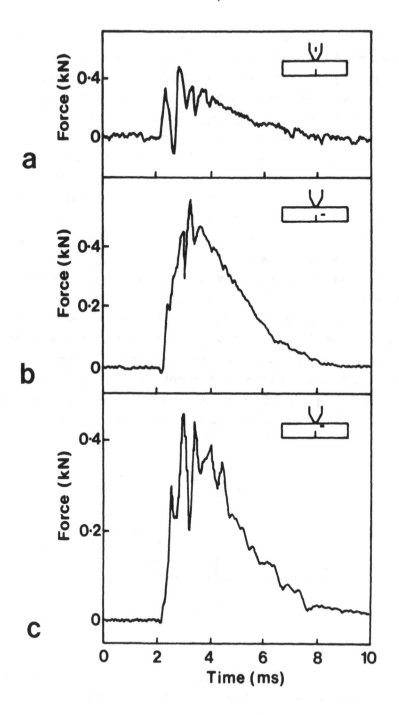

Figure 2. Force-time outputs   (a) Striker sensor output   (b) W/2xW/2 sensor output
(c) W/2xW/8 sensor output.

RESULTS

Figure 2 is a set of force-time test records obtained from fracturing a polyethylene (HDPE) specimen at room temperature for an impact velocity of 2 m s^{-1}. The force-time trace (a) was from the sensor embedded in the striker and traces (b) and (c) were those from two strain gauges attached to the specimen on one side of the crack. One was in a W/2xW/2 position and the other was in a W/2xW/8 position as illustrated in Figure 1. These were found to be the best positions for monitoring the forces acting on the crack tip and those in the compressed part of the specimen. Also for helping to determine the different failure rates for the fast crack in the core of the material, the tearing of the side-lips and the rupturing of the hinge bend material. This was achieved by comparing the W/2xW/2 and W/2xW/8 sensor outputs for specimens with and without side-grooves.The final phase of a specimen failure is the fracture of the hinge material. In some cases, the hinge can survive the impact and this results in additional kinetic energy being given to the fractured specimen as it is driven from its fixture.

Shown in Figure 3 is an optical photograph of a fracture surface of a specimen without side-grooves which shows the size and texture of the fast crack, side tear-lips and hinge bend zones. It was found that the size and texture of these zones varied greatly with impact velocity and specimen temperature.

A reason for deciding upon the chosen size of sample and conditions for impact test was so that better information about the fast crack propagation in the core of the specimen, the tear lips and the hinge failure could be obtained. Figure 4 shows electron micrographs at the same magnification of parts of the initiation, fast crack, side tear-lips and hinge-bend zone to highlight their texture variations. Notable is the much finer micro-ductility of the fast crack surface in the core of the specimen compared with that of the tear lip and the hinge failure surfaces.

To help in the study and analysis of crack and other failure processes in materials under test, high-speed photographic techniques can be used to capture more information than is possible by other means. For the work reported here, then two photographic studies (see Figures 5(a) and 5(b)) were made to observe crack propagation, one being for a specimen tested at room temperature at an impact velocity of 5 m s^{-1} without side-grooves and the other for a specimen with side-grooves. In the former case, of course, what is seen is the well-rounded cleavage point of the tear-lip following on behind the hidden fast crack in the core of the specimen. Comparing the two photographic sequences, without and with side-grooves, provides some information on the effect of tear-lips in the specimen. With side-grooves, the crack visible in the high-speed photographs (Figure 5(b)) can be taken as the fast crack. Without side-grooves, the crack seen by the camera is the side-lip tearing

Figure 3. Optical photograph of fracture surfaces showing initiation zone, arrest lines, tear-lips and hinge bend section.

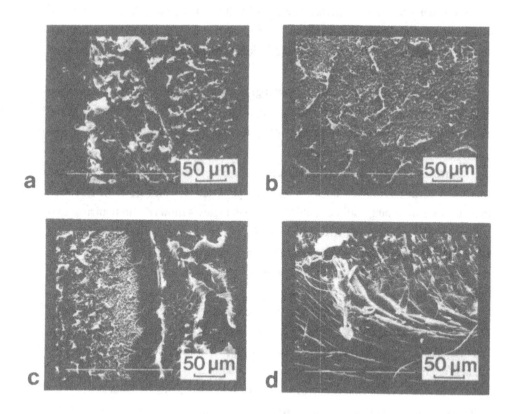

Figure 4. Electron-micrographs of fracture surface (a) initiation zone (b) fast crack area (c) hinge-failure zone (d) tear-lip.

7

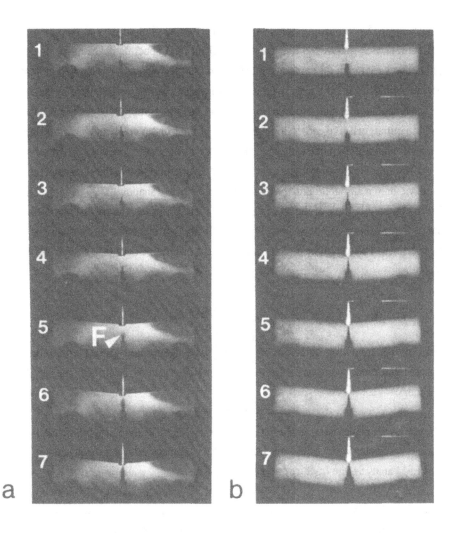

Figure 5. High-speed sequence (interframe time 0.25ms) (a) Non side-grooved specimen
(b) Side-grooved specimen.

(Figure 5(a)). An interesting point when comparing the photographs of specimens with and without side-grooves is that the fast crack in both specimens seems to have been initiated at very much the same overall strain of the specimen and it may well be that the fast crack velocity in both tests is similar but this needs further verification. In the case of the specimen without side-grooves (Figure 5(a)), the photographic sequence has captured well the initial fork-like depression, labelled F, of the tear-lip crack and how this changes to a well-radiused blunt crack-tip as the tear-crack traverses the specimen. To observe this effect, a single illuminating light source was set to one side so that the surface depression of the yielding material around the tear-lips would create a shadow. This, of course, meant that the overall lighting of the specimen is not even. However, the rollers supporting the end of the specimen can be seen and because of the lighting, these throw a shadow onto the specimen. The top of the striker shows up as a bright line because its upper surface is cylindrical.

ANALYSIS

Based on the following equations K_c and G_c were determined using the outputs from sensors in different positions on the specimen and the sensor in the striker. The specimens had a S/W=4 so:

$$K_c = (6P_c Y a^{1/2})/(BW) \qquad (1)$$

where P_c is the peak load at crack initiation, a is the length of the sharp pre-crack, B is the thickness, W is the width of the specimen and Y is a dimensionless geometry factor. From an LEFM analysis, the fracture energy, G_c, is given by:

$$G_c = U_c / (BW\phi) \qquad (2)$$

where U_c is the stored elastic strain energy in the specimen at crack initiation. Table I shows a comparison of K_c and G_c values determined from outputs from the two on-specimen sensors and the striker sensor. Values of K_c and G_c calculated using the striker sensor output are greater than if on-specimen sensor outputs are used and this is more so the higher the impact velocity. The point being that the striker sensor monitors the forces acting on the specimen as a whole from the striker not only to initiate fracture but also to generate dynamic vibrations and to impart kinetic energy to the fractured specimen parts. For some calculations and use of derived values, it is also important to take into account the different sizes of the crack initiation, side-lip tearing and hinge-bend section.

   The values of K_c and G_c in Table I are evaluated using the first peak of the strain gauge force-time curves. Of particular interest is that K_c and G_c values for specimens with and without side-grooves are very similar. This confirms the point made above that the initiation and fast crack propagation can dominate the early failure of the specimen. It is after the fast crack has propagated that the side-lip tearing and hinge failure significantly affect the continuing failure processes as revealed by the force-time traces.

The values of K_c and G_c derived from the on-specimen sensors (K_c^{sp} and G_c^{sp}) have much less scatter than K_c and G_c values derived from the force-time traces of the striker sensor (K_c^{st} and G_c^{st}). Generally, it is easier to identify more precisely the peak load and the time that failure is initiated in the specimen when using an on-specimen sensor. This can be seen by inspection of the traces in Figure 2. The higher the impact velocity, then the stronger and more erratic the oscillatory excursions and the time to failure shorter. It is for these higher impact velocities of the striker that the on-specimen sensors are particularly useful for helping to analyse test results. Also, for calibrating the striker sensor which is then used for routine tests.

Looking in more detail at the reasons for uncertainty in determining the peak load and the elastic energy stored in the specimen prior to its failure, then, the following are some relevant factors. It follows that the higher the impact velocity of the striker then the greater the effect of the initial transient forces on the failure test results. A feature can be that the specimen zone near to the impact site can be significantly deformed by the first contact effects of the impact. This is before the specimen as a whole is taken into a three-point bending configuration. The on-specimen sensors near to the crack site will see the stresses created by this early impact deformation of the specimen. In the case of the tests reported in this paper, it is to be noted that the force-time traces for the on-specimen sensors near to the crack tip first dip before following the build up of three-point bending load in the specimen. This is as the sensors see the early impact stresses. Of course, depending on where the on-specimen sensors are located so observation of the early impact stresses can be different and this is a very useful way of analysing these effects.

TABLE I

Examples of K_c and G_c calculations

Impact velocity (m s^{-1})	K_c^{st} (MPa m$^{1/2}$)	K_c^{sp} (MPa m$^{1/2}$)	G_c^{st} (kJ m^{-2})	G_c^{sp} (kJ m^{-2})
2	2.3	2.1(W/2xW/2) 2.1(W/2xW/8)	3.8	2.7(W/2xW/2) 2.6(W/2xW/8)
5	2.8	2.1 (W/2xW/2) 2.2 (W/2xW/8)	4.6	2.0(W/2xW/2) 2.1(W/2xW/8)
5 (side-grooved)	2.8	2.1(W/2xW/2) 1.9(W/2xW/2)	4.5	1.9(W/2xW/2) 1.7(W/2xW/2)

K_c^{st} and G_c^{st} relate to the sensor on the striker and K_c^{sp} and G_c^{sp} relate to the sensors on the specimen in the W/2xW/2 and W/2xW/8 positions.

CONCLUSIONS

The indications in Table I are quite clear that both K_c and G_c are generally higher when using the output from the striker sensor than when using the output from the on-specimen sensors. A main factor is that the forces seen by the striker sensor relate to those acting on the specimen as a whole not only to initiate fracture but also to generate dynamic vibrations and to impart kinetic energy to the fractured specimen parts whereas the on-specimen sensors tend to see mostly the strain caused by forces focused onto the crack site. There is also the additional point that the transient excursions are less in evidence in the on-specimen sensor traces than those for the striker sensor. This is particularly noticeable for the W/2xW/2 sensor. A further point that is that under strong transient conditions such as in evidence in the research reported upon here, there is the question as to the true quantity of stored energy available in the strained specimen at the point of failure and the extent this can flow into the crack site and generate crack surface. However, for this research the main need was the ability to compare the pre-crack energy stored in the specimen as measured by using the striker sensor and the specimen sensor.

ACKNOWLEDGEMENTS

The authors would like to thank Professor J G Williams for his interest and helpful discussion on this work and one of us (JPD) would very much like to thank the interest and support of BP Chemicals Ltd.

REFERENCES

1. Williams J. G., Fracture Mechanics of Polymers, Ellis Horwood, Chichester (1984).

2. Dear, J.P., High-speed photography of impact effects in three-point bend testing of polymers, J. Appl. Phys. 67 (9) 4304-4312.

3. Dear, J. P., Development of a method for fast crack testing of high-grade polymers. J. Mater. Sci., In press.

4. Dear, J.P., Measurement of the toughness and dynamic properties of polymers for pipes and similar products. KVIV Pipeline Technology, Ostend, Belgium (October, 1990).

5. Dear, J.P., Impact and other testing of high-grade polymers. III European Polymer Federation Symposium on Polymeric Materials, Sorrento, Italy, (October,1990).

6. Dear, J.P.,Validation of impact testing of high-grade polymers. 9th International Conference on Experimental Mechanics, Copenhagen, Denmark (August, 1990).

CLEAVAGE FRACTURE IN STEEL SPECIMENS UNDER SHORT STRESS PULSE LOADING

Hiroomi Homma, Yasuhiro Kanto and Kohji Tanaka
Department of Energy Engineering
Toyohashi University of Technology
Tempaku-cho Toyohashi 441 Japan

ABSTRACT

Cracks in SM50A, steel for welding structure chilled to -40 and -160 C were loaded by various stress intensity pulses with durations of 20, 40, and 80 μs to generate experimental data of critical stress intensity levels for crack instability. Fracture surfaces were observed by a scanning electron microscope to examine cleavage nucleation origins ahead of the crack tips. The obtained experimental results were discussed from the minimum time criterion and the dislocation dynamics.

The critical stress intensity obtained by the minimum time criterion was decreased as the temperature was lowered, while the minimum time had the maximum at -80 C. The cleavage nucleation origin approached the crack tip as the temperature decreased. Finally, the temperature dependence of the minimum time was interpreted by the dislocation dynamics.

INTRODUCTION

Crack instability is strongly affected by loading rate and test temperature especially for body-centered cubic metallic materials as steel. ASTM standard E399 provides a test method for fracture toughness measurement under rapid loading. However, there is a rstriction of loading rate so that quasi-static stress field would take place in a specimen.

For rapid loading under which the stress field near a crack tip is significantly affected by the inertia, we lack reliable crack instability criteria and experimental data. The crack instability criteria proposed heretofore are classified to two types. One of them is based on the quasi-static concept that a crack become unstable immediately when a fracture parameter evaluated by full dynamic analysis equals or exceeds the critical value[1,2]. The other takes into account of time effect that the crack instability is brought about when the fracture parameter equals or exceeds the critical value for a certain period[3-5]. The period is called "minimum time" or "incubation time". In the previous work, Homma et al[6] obtained values of the minimum time for steel for welding structures, SM50A(JIS) at -40 C and -80 C.

The aim of this work is to generate experimental data of crack instability at lower temperature than -80C and to examine the minimum time or the incubation time for cleavage fracture in steel at the low temperature.

EXPERIMENTAL METHOD

Material and Specimens

Material used in the experiment was steel for welding structures, SM50A(Japanese Industrial Standards) and its chemical composition and mechanical properties are indicated in Table 1. Three kinds of specimen configurations were prepared as shown in Fig.1. They had the same width, but different length. A single-edged notch was machined in each specimen to initiate a 3 to 5 mm long pre-fatigue crack from the notch root. When a short pulsative load was applied at the center of back surface of the specimen without any supports, the specimen was bent and swung back due to the inertia effect. This loading is called one point bending[7]. A period of this movement depends on specimen configuration. The specimen geometries shown in Fig.1 were used to generate stress intensity histories with three different periods. In the longest specimen, 80μs-duration stress intensity history was generated, in the intermediate specimen, 40μs-duration stress intensity history, and in the smallest specimen, 20μs-duration stress intensity history.

Test temperatures were -120 C and -160 C which are lower than the transition temperature of the steel so that the whole fracture surface would be covered by cleavage.

TABEL 1
Chemical composition of SM50A

Material	C	Si	Mn	P	S	Fe
SM50A	<0.2	<0.55	<1.5	<0.04	<0.04	Re.

TABLE 2
Mechanical properties of SM50A

Material	Yield Point	Tensile Strength	Elongation	Strain n.e	Young's Modulus
Unit	(MPa)	(MPa)	(%)	(n)	(MPa)
SM50A	422	588	31	0.25	206000

TABLE 3
Calculated dynamic yield strength and plastic zone size

Temp (C)	Dynamic yield strength (MPa)	Plastic zone size (μm)	Cleavage origin (μm)	Elastic stress at the origin (GPa)
-40	695	124	165	1.06
-80	790	56	106	1.03
-120	936	33	22	2.00
-160	1184	19	20	2.01

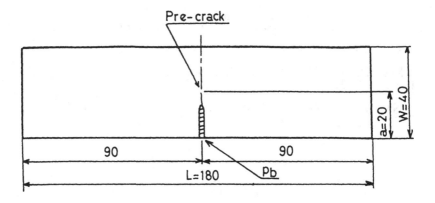

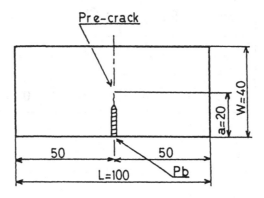

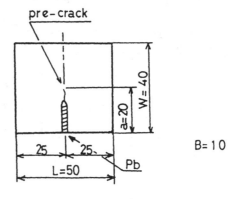

Figure 1. Specimen geometry and dimensions

Dynamic Loading Device and Measurement of Dynamic Stress Intensity Histories

The loading device is shown in Fig.2. This consists of a launching device, and a load transfer rod. A specimen was placed contacting with the round end of the load transfer rod. The launching device is made up of a cylindrical barrel, 2000 mm long by 20 mm inner diameter, a reservoir for nitrogen gas, and a solenoid valve. A projectile, 20 mm long by 20 mm diameter, was positioned at the end of the barrel, and the reservoir was filled with nitrogen gas to the desired pressure. The projectile was launched by opening the solenoid valve. The pressurized gas flew into the barrel at a high speed and pushed the projectile to accelerate it against the load transfer rod. Impact velocity was controlled by the gas pressure.

Dynamic stress intensity history experienced by a crack was measured by a strain gage mounted 7 mm far from the crack tip according to the method developed by Dally and Sanford[8].

RESULTS

Critical Stress Intensity for Crack Growth

The maximum value of dynamic stress intensity pulse measured by a strain gage mounted near a crack tip is plotted against the pulse duration as a parameter of crack growth event in Figs.3 (a) and (b) for the both test temperatures. The open circle means that the crack did not grow under the stress intensity pulse and the solid circle means that the crack grew unde the pulse. A critical value of dynamic stress intensity for the crack initiation was defined as the middle of the adjacent solid and open circles. The critical value is shown as a function of the stress intensity pulse duration in Fig.4 including the previous results at −40 C and −80 C[6]. It is indicated that the critical values for 40 and 80 μs durations are almost same and those are much smaller than the value for 20 μs duration. It is also seen that there is temperature dependence of the critical value.

Observation of Cleavage Fracture

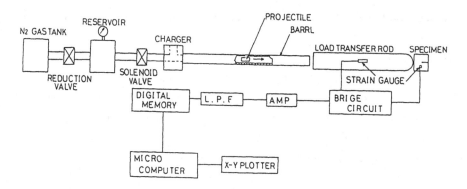

Figure 2. Loading device

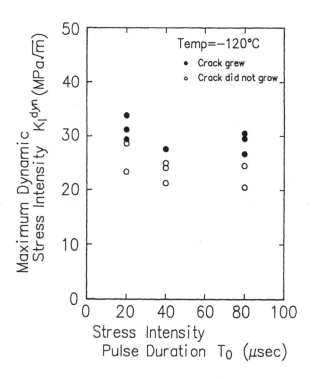

Figure 3(a). Maximum dynamic stress intensity (-120 C)

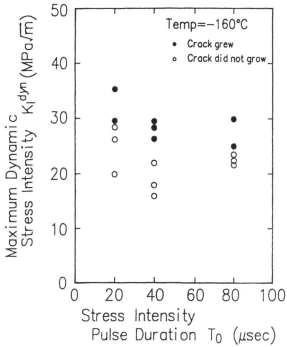

Figure 3(b). Maximum dynamic stress intensity(-160 C)

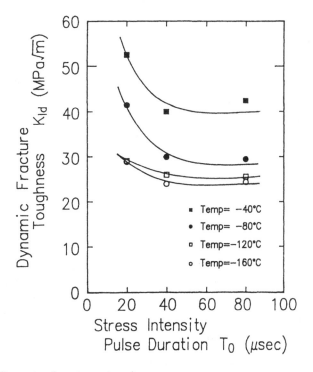

Figure 4. Dynamic fracture toughness

Fracture surface near the initial crack tip was observed by a scanning electron microscope. Microscopic fractographs for the temperatures of -120 C and -160C are shown in Fig.5. For all the experimental conditions, the whole of fracture surface was covered by the cleavage.

The cleavage was always nucleated ahead of the initial crack tip. The nucleation origin can be detected from flow directions of river patterns. In the previous work[6], first, a chevron pattern was found on a photograph of low magnified fracture surface. Its sharp edge indicates a macro-scopic fracture origin. Then, several fractographs near the origin were taken with high magnification (x400). Observing the flow directions of the river patterns, all the rivers were traced to their original source. The source was defined as the nucleation origin of the cleavage. However, the chevron pattern could not be detected on the low magnified photograph of the fracture surface at the test temperature lower than -120 C. Then, the nucleation origin was found by the direct observation of river patterns in the vicinity of the initial crack tip. The origin was a flat and smooth facet, and its direction almost equaled the macro-scopic fracture surface direction.

The distance between the initial crack tip and the nucleation origin were measured and is shown as a function of the duration of the stress intensity pulse in Fig.6 including the previous results[6]. It is seen that the nucleation origin is almost independent of the pulse duration while it is significantly dependent on the test temperature higher than -120 C. The nucleation distance is shown as a function of the test temperature in Fig.7.

DISCUSSION

The critical maximum value of dynamic stress intensity pulse steeply increased at the duration of 20 μs as shown in Fig.3. The steep increase is contrary to the effect of loading rate on the dynamic fracture toughness that the higher loading rate enhances yield strength of the material at the crack tip to lower the dynamic fracture toughness since the loading rate is higher in the stress intensity pulse with the 20 μs duration than in those with 40 and 80 μs durations. It is well explain from the minimum time or

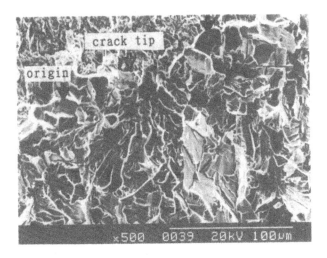

(a) −120 C T_c = 20 microsec

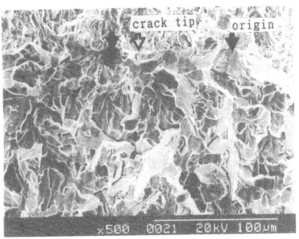

(b) −160 C T_0 = 80 microsec

Figure 5. Fracture surface near the initial crack tip

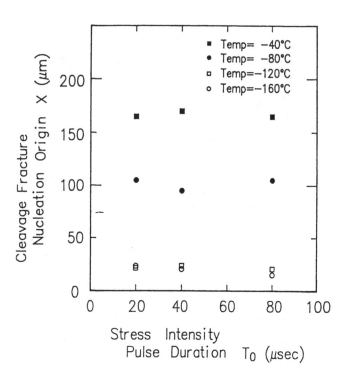

Figure 6. Cleavage nucleation origin as a function of pulse duration

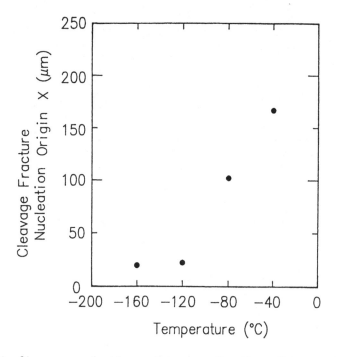

Figure 7. Cleavage nucleation origin as a function of temperature

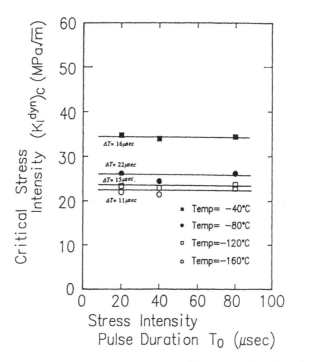

Figure 8. Critical stress intensity as a function of pulse duration

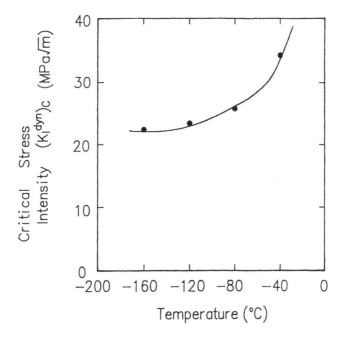

Figure 9. Critical stress intensity as a function of temperature

the incubation time criterion[3-5]. Estimation of the minimum time was tried. The details of the method is described in [9]. Several minimum times were assumed as the trial. The critical stress intensity values were determined for those times using the time histories of the stress intensity pulses. When deviation of the critical stress intensity values is the smallest if a certain time is taken as the minimum time, the time is defined as the minimum time. The critical stress intensity determined by the minimum time criterion is shown as a function of the pulse duration in Fig.8. In the figure, the estimated minimum times are also indicated. In Fig.9, the critical stress intensity is plotted as a function of the test temperature. It is seen that the temperature effect is significant in the range of higher than -120 C.

The nucleation origin of the cleavage will be considered in association with the plastic zone. Dynamic yield strength of the material was estimated by the equation given in ASTM standard E399 Annex 7. The dynamic yield strength was indicated in Table 3 for 80 μs duration pulse. It can be considered that the small scale yielding and plane strain conditions prevail near the initial crack tip at the crack initiation. The plastic zone size was calculated by the following equation,

$$r_p = (K/\sigma_{yd})^2/6\pi \qquad\qquad (1)$$

The calculated results are listed in Table 3. Comparing these results with the nucleation origin of the cleavage shown in Fig.7, it is seen that the cleavage fracture was initiated 1.06 to 2 times of the plastic zone size far from the initial crack tip. However, since average grain diameter of the used steel is around 50 μm, only one or three grains will exist in the plastic zone. Therefore, the plastic zone size given by Eq.(1) based on the continuum mechanics may not correspond to the real one. The cleavage was nucleated one to three grains ahead of the crack tip. Especially, the cleavage fracture at the temperature of lower than -120 C was nucleated in the grain at the crack tip of which the cleavage plane was approximately parallel to the macro-scopic fracture surface. When the crack was loaded by the critical stress intensity, the elastic stress at the cleavage origin can be calculated by

$$\sigma = K / \sqrt{2\pi r} \qquad\qquad (2)$$

The calculated stress is also indicated in Table 3. It is seen that the elastic stresses for -120 C and -160 C are equal to each other and are twice as large as those for -40 C and -80C which also equal each other.

The minimum time or the incubation time obtained by the experiment will be discussed by the dislocation dynamics. Hahn et al[10] reviewed the proposed several dislocation models for the cleavage fracture and explained their experimental data using a dislocation model. In all the dislocation models, the cleavage fracture is initiated by the pile-up of dislocations against an obstacle such as the grain boundary and inclusions.

In this experiment, dislocations are emanated from the crack tip and pile up at the nucleation origin. Emanation rate and moving velocity of a dislocation will be important factors to consider the minimum time in framework of the dislocation dynamics.

According to the dislocation dynamics, the dislocation velocity v is given by

$$v = v^* \exp\{-D/\sigma\} \qquad\qquad (2)$$

where v^* is the terminal velocity, D is the drag stress, and σ is the applied stress. If the drag stress increases as the temperature decreases like the yield strength, the dislocation velocity decreases with the temperature.

The emanating rate of dislocation from the crack tip was analysed by Yokobori et al[11] assuming that stress field around a moving dislocation can be expressed by that for a stationary dislocation. According to the result, the time required to emanate the number N of the dislocations is given as

$$t = \beta [N(\frac{\tau^0}{\tau^*})^3]^{1/4} \qquad\qquad (3)$$

where β is a constant, τ^* is the applied stress and τ^0 is shear stress necessary to move an isolated dislocation at a velocity of 1 cm/s.

We consider that temperature dependence of τ^0 is similar to that of yield strength as the first approximation. The time t becomes longer as the temperature is lowered. The number of the piled-up dislocations to initiate the cleavage fracture may depend on the elastic stress calculated by Eq.(2). The number of the dislocations emanated from the crack tip to nucleate the cleavage fracture will become smaller as the temperature is lowered. Therefore, the minimum time decreases with the temperature. However, the experimental result shows that the estimated minimum time have the maximum value at -80 C. In order to explain the experimental result more completely, we must need to examine the dislocation behavior in more detail taking into account of traveling time from the crack tip to the leading edge of piling up as well as emanating rate.

CONCLUSIONS

Experiments on crack instability under short pulse loading were made at low temperatures of -120 and -160 C using steel for welding structure SM50A so that cleavage fracture would predominantly take place at the crack initiation. The following conclusions were obtained:

1. The maximum stress intensity experienced by a crack at the crack instability was much larger for 20 μs-duration stress intensity pulse than for 40 and 80 μs-duration stress intensity pulses. This was well explained by the minimum time criterion.

2. The obtained minimum times were 15 μs and 11 μs for the test temperatures of -120 and -160 C, respectively.

3. The cleavage fracture nucleation origins were around 20 μm far from the initial crack for the both temperatures while those were around 170 and 100 μm for -40 and -80 C respectively.

4. The minimum time was interpreted as the time required to generate and pile up the number of the dislocations which is large enough to bring about the cleavage fracture ahead of the obstacle.

REFERENCES

1. Achenbach,J.D. and Brock,L.M.,in Dynamic Crack Propagation, G.C.Sih,Ed.,
    Noordhoff, The Netherlands, 1973, p.529
2. Lehnigk,S.H. in Dynamic Crack Propagation, G.C. Sih, Ed., Noordhoff, The
    Netherlands, 1973, p.333
3. Kalthoff,J.F. and Shockey,D.A., Journal of Applied Physics, Vol.43,No.3,
    1977,p.986
4. Homma,H., Shockey,D.A., and Murayama,Y., Journal of the Mechanics and
    Physics of Solids, Vol.31,No.3, 1983,p.261
5. Kalthoff,J.F., Engineering Fracture Mechanics, Vol.23, No.1, 1986,
    p.289
6. Homma,H., Kanto,Y and Tanaka,K.,To appear on ASTM STP for the
    proceedings of symposium on "Rapid Load Fracture Testing" April 22-27
    1990, San Francisco USA
7. Bohme,W. and Kalthoff,J.F., International Journal of Fracture, Vol.20,
    No.1, 1982,p.139
8. Dally,J.W. and Sanford,R.J., Experimental Mechanics, Vol.27, No.4, 1987
    p.381
9. Homma,H., Shockey,D.A., and Hada,S., in Fracture Mechanics:Seventeenth
    Volume, ASTM STP 906, American Society for Testing and Materials,
    Philadelphia, 1986,p.683
10. Hahn,G.T., Averbach,B.L., Owen,W.S., and Cohen.M., in Fracture,
    Averbach et al Ed., The M.I.T. Press, USA, 1959, p.91
11. Yokobori,T.Jr., Yokobori,T., and Kamei,A., "Computer Simulation of
    Dynamics of Dislocation Group under Constant Applied Stress Rate",
    Transaction of Japan Society for Mechanical Engineers, Vol.42,No.358,
    1976, p.1652

CRACK PROPAGATION AND ARREST STUDY AT STRESS PULSE LOADING

R. CLOS, U. SCHREPPEL, U. ZENCKER, T. RAHMEL
Magdeburg University of Technology, Department of Technical
Physics, Universitätsplatz 2, 3040 Magdeburg, FRG

K. KLENK, U. MAYER
Stuttgart University, Federal Materials Testing Lab.,
Pfaffenwaldring 32, 7000 Stuttgart 80, FRG

ABSTRACT

A method for the investigation of crack initiation, propagation and arrest under short pulse loading conditions is given. Experimentally a modified version of the Hopkinson-bar technique has been used in connection with suitable methods for measuring the instantaneous crack length. The main advantage of the Hopkinson-bar technique, the use of nearly plane waves, is maintained here. For the interpretation of the experimental data in terms of fracture mechanics a numerical elastodynamic analysis is used. By comparison with analytical solutions the applicability of the numerical method has been verified. Preliminary results obtained for a structural steel and PMMA are presented.

INTRODUCTION

The investigation of the behaviour of cracks under highly transient loading conditions is important for the safety assessment of dynamically loaded structures as well as for fundamental research in the field of dynamic fracture. In such investigations usually the phase of crack start is considered [1-6]. Besides the determination of a dynamic initiation fracture toughness K_{Id} [1,3,6], which is of practical interest, the question of the validity of the crack initiation criterion $K_I(t=t_f) = K_{Id}$ (K_I-stress intensity factor, t-time, t_f-crack

start time) at high crack tip loading rates $\dot{K}$ is investigated
[2,4,5,6]. It seems that this criterion should be completed by
an additional time criterion (minimum time criterion [7,8]) at
extremely high $\dot{K}$ (or equivalently short crack start time t_f),
but it is open how long the relevant times (incubation times)
are [4,9,10].

Another point of interrest in dynamic fracture is the
question of the uniqueness of a crack propagation toughness
K_{ID} - crack velocity relationship [11-13], which is also related to crack arrest. Most investigations in this field were
carried out under quasistatic loading conditions [13,14]. Only
few crack propagation/arrest experiments has been done under
transient loading conditions [15-17]. In the following an experimental method is proposed, which allows to study crack initiation, propagation and arrest under shorttime stress pulse
loading. For the interpretation of the experiments in terms of
fracture mechanics a combined experimental-computational approach has been used. Preliminary results for structural steel
and PMMA are given. As pointed out by Kanninen et al. [14] one
practical advantage of stress pulse loading is, that a large
amount of energy is delivered to the crack tip region in a
short time. So in principle in connection with an embrittled
starter material crack propagation and arrest in high
toughness materials may be investigated using relatively small
specimens.

EXPERIMENTAL METHOD

The loading setup is shown schematically in Fig. 1. It consists essentially of a gas gun, a long input bar (circular
cross section, diameter 20 mm, length about 1 m) and the fatigue precracked specimen. The specimen is pressed slightly
against the bar in such a way, that its right end can be considered as a free end in the investigated time range. By the
impact of the projectile a pressure pulse of short duration is
generated in the input bar. It propagates through the bar and

the specimen. At the free end of the specimen it is reflected as a tensile pulse, which interacts with the crack and can cause crack propagation depending on its amplitude. The pulse

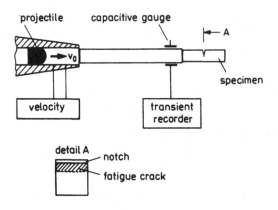

Figure 1. Pulse loading setup schematically

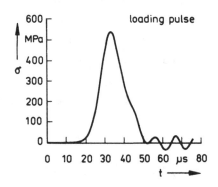

Figure 2. Typical loading pulse

is measured with a coaxial capacitive gauge and recorded with a digital oscilloscope (time resolution usually 0.2 μs). A typical loading pulse is shown in Fig. 2. The duration of the main part is τ_0 = 28 μs. The specimen has nearly the same length as the pulse L = $c_0 \tau_0$ (c_0-bar velocity). The trailing

oscillations are small and result from geometrical dispersion.
It can be shown, that the pulse has a nearly homogeneous distribution of the (time dependent) axial stress over the cross
section and shear stresses may be neglected under the chosen
experimental conditions . So nearly pure mode I crack opening
occurs and furthermore the boundary conditions for the numerical analysis become relatively simple [18].

The calculated normalized stress intensity factor for
resting cracks of different length is given in Fig. 3. σ_0 denotes the pulse amplitude. In the calculation the material
parameters of steel were used. The dashed line is the normalized pulse (main part). It is important, that the K-factor
decreases after passing a maximum value. In principle this is
the reason, why a crack which has been started for example in
the point A (Fig. 3), can arrest again in such experiments.

The investigated materials are the structural steel H 52
(ferritic-pearlitic microstructure, specimen size: 17.7 x 17.7
x 130 cmm) and the polymeric material PMMA (specimen size: 10
x 30 x 100 cmm, pulse duration about 55 µs here).

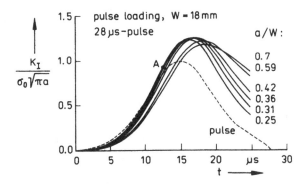

Figure 3. Normalized stress intensity factor versus time for
different crack length

The main difficulty in the experiments is the measurement
of the crack length versus time. This has been done for PMMA

with a strip technique, which is shown in Fig. 4. As strips
evaporated silver layers has been used. The time to rupture of
each strip was measured by separate time counters, which were

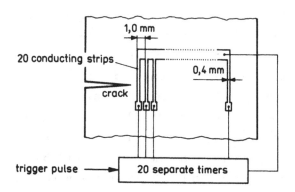

Figure 4. Measurement of the crack length, PMMA

triggered together with the digital oscilloscope (Fig. 1). A
definite delay time between the pulse record and the counters
was realized. While this method works for PMMA, there are some
problems in using it for steel. For this reason in the experiments with steel the crack tip position was recorded by means
of a high speed camera, Imacon 790, with an interframe time of
2.05 μs. In order to avoid greater differences between the instantaneous crack length on the surface and in the interior of
the specimen, the steel experiments were made at a relatively
low temperature of -80°C. From earlier investigations it was
known, that brittle behaviour should occur at this temperature
[5].

NUMERICAL ANALYSIS

A twodimensional numerical analysis was made for isotropic
linear-elastic material behaviour and assuming plane strain
conditions. The equations of motion under initial and boundary

conditions, which correspond closely to the experimental conditions, were solved by means of a finite difference method using explicit time integration. For handling running cracks a moving grid technique was used similar to the one applied by Ravichandran and Clifton [19]. Details of the analysis will be published [20].

The mode I stress intensity factor was calculated in each time step from the numerical solution using the relationship

$$J_1' = \frac{\dot{a}^2 \, \beta_1}{2 \, \mu \, c_2^2 \, R(\dot{a})} \, K_I^2 \qquad (1)$$

between the J_1'-integral and K_I, where

$$\beta_{1/2}^2 = 1 - \dot{a}/c_{1/2}^2, \quad R(\dot{a}) = 4 \, \beta_1 \, \beta_2 - (1 + \beta_2^2)^2$$

and μ is the shear modulus. The path independent domain integral in the notation given by Nishioka and Atluri [21] is

$$J_1' = \lim_{\varepsilon \to 0} \int_{\Gamma + \Gamma_c} \left[(W + T) \, n_1 - t_i \, u_{i,1} \right] dS$$

$$+ \int_{V - V_\ell} \left[\rho \, \ddot{u}_i \, u_{i,1} - \rho \, \dot{u}_i \, \dot{u}_{i,1} \right] dV, \qquad (2)$$

other integrals of this type [22] were also considered.

Alternatively K_I was calculated from the crack opening displacements using the known analytical form of the near crack tip field [21]. The agreement of both methods was reasonable.

Some results of test calculations, in which a semiinfinite crack was simulated, are given. It is wellknown, that for a semiinfinite crack in an infinite body in mode I holds [23,24]

$$K_I(t,\dot{a}) = k(\dot{a}) \, K_I(t,0), \qquad (3)$$

where $K_I(t,\dot{a})$ is the stress intensity factor for the moving crack, $K_I(t,0)$ is the K-factor of the resting crack at the same time and k(ȧ) is a function of the instantaneous crack velocity

$$k(\dot{a}) = \frac{1 - \dot{a}/c_R}{\sqrt{1 - \dot{a}/c_2}} \exp\left\{ \frac{1}{\pi} \int_{\kappa}^{1} \arctan\left[\frac{(x^2 - 0.5)^2}{x^2 \sqrt{(x^2 - \kappa^2)(1 - x^2)}} \right] \frac{dx}{x - c_2/\dot{a}} \right\}, \quad (4)$$

with $\kappa = c_2/c_1$, c_R is the Rayleigh wave velocity. k(ȧ) is the same as derived by Freund [23] but in a slightly different analytical form. Because the exponential term deviates not much from unity for all velocities the preexponential term may be used as a good approximation.

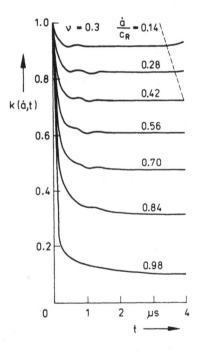

Figure 5. k(ȧ,t) for different crack velocities

Of course while for a semiinfinite crack eqn (3) is an exact result for arbitrary crack tip motion, which holds also in a finite body in the moment of crack start, the region of

K-dominance may become extremely small at rapid changes of the crack velocity and/or the stress intensity factor. From this in a real solid, where e.g. a finite process zone exists, fundamental problems arise.

Numerically the K-factor for a moving and a resting crack can be calculated. The ratio

$$k(\dot{a},t) = K_I(t,\dot{a}) / K_I(t,0)$$

should be equal to the velocity function eqn (4) at least after some time. Results are given in Fig. 5. In the calculation

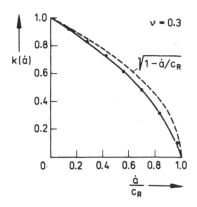

Figure 6. Velocity function, numerically and analytically

the crack surface was loaded by constant pressure at the time zero, at the same time the crack started with a given velocity. The domain was chosen such that waves from the boundary did not reach the crack tip during a certain time. Really $k(\dot{a},t)$ becomes constant after some time, which is the longer the higher the velocity jumps (the dashed line indicates the influence of waves from the boundary). From these constant values Fig. 6 is found. The points denote the numerical values, the full line is the theoretical velocity function

from eqn (4) and the dashed line shows an approximation sometimes used in literature.

The next figures show the K-factor for a semiinfinite
crack, whose surfaces is loaded by a pressure σ_o over its
whole length at the time zero, at a later time the crack
starts to move with constant velocity. The theoretical $K_I(t,0)$
in eqn (3) is in this case

$$K_I(t,0) = 4\ \sigma_o\ [\kappa\ (1\ -\ \kappa^2)\ /\pi]^{1/2}\ (c_2\ t)^{1/2}.$$

The full line shows in each case the analytical curve. The numerical solution oscillates immediately after the velocity
jump. Such a behaviour has been found also with other numerical methods [19]. If there is no velocity jump the numerical
and the analytical solution agree well for all times (Fig. 9).

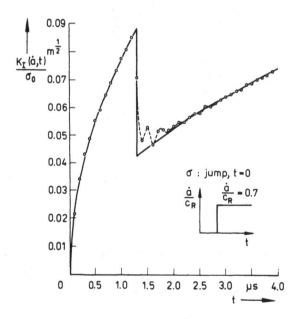

Figure 7. K-factor for a semiinfinite crack loaded by a step
function pressure, velocity jump 0.7 c_R

In the experiments no velocity jumps were detected and the crack velocities were lower than in these test calculations. So it can be concluded, that the numerical model may be used.

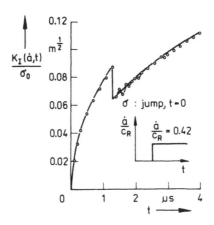

Figure 8. As figure 7, velocity jump 0.42 c_R

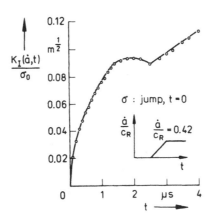

Figure 9. As figure 7, smooth velocity change

Important input data for such calculations are the loading pulse and the measured crack length versus time curve.

RESULTS

In Fig. 10 the experimental data (pulse, crack length) for a steel specimen as well as the stress intensity factor calculated from these data are shown. The points of crack start and arrest are marked on the K_I versus time curve. The crack velocity is derived from a(t) by differentiation. The maximum velocity is about 750 m/s.

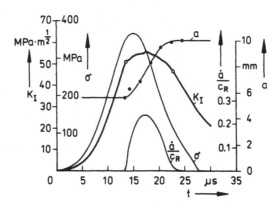

Figure 10. Results for the steel H 52, temperature -80°C

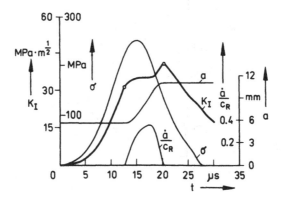

Figure 11. Results for the steel H 52, temperature -80°C

Fig.11 shows the results of an experiment with a loading pulse of different amplitude. The fatigue crack length in both cases was about 6 mm. The maximum crack velocity is about 1000 m/s here. Contrary to Fig. 10 the arrest value of the K-factor is higher than the initiation value, the reason for is the sharply decreasing crack velocity during the last stage of the propagation. On the other hand the difference of both values is not very large. There is no reason to overrate it because little inaccuracies of the crack length data can influence the velocity curve appreciably, especially during the start and arrest phase. Moreover due to the time resolution of 2.05 μs the start and arrest points are somewhat unsure.

The global statement, which can be made for the steel is, that the stress intensity factor is nearly constant during initiation, propagation and arrest. This is consistent with the literature in this velocity range [13].

A comparison of both experiments shows that the initiation fracture toughness in Fig. 11 is about 35 MPam$^{1/2}$ but in Fig. 10 about 50 MPam$^{1/2}$. From other experiments we know, that the initiation toughness (determined at $\dot{K}$ = 2 10^6 MPam$^{1/2}$/s) at the test temperature is about 25 MPam$^{1/2}$ [5], which is near to the experiment Fig. 11. An inspection of the specimen used in Fig. 10 revealed that there were very small shear lips on the fracture surface and the arrest length in the middle of the specimen was much longer than on the surface. Obviously the crack has been started on the surface later than in the middle of the specimen, from which an apparently higher initiation fracture toughness results. In the specimen of Fig. 11 no shear lips were detected and it is assumed, that the measurement of the crack length on the surface of the specimen was applicable in this case. It should be noted, that the conditions for the determination of valid (static) fracture toughness values [25] were fulfilled in both cases, so the differences between Figs. 10 and 11 are somewhat surprisingly. Nevertheless it indicates, that further experiments should be done with side grooved specimens even at low temperatures. Furthermore should be remarked, that in experiments, of which the results were not presented here, with increasing pulse amplitude an increasing initiation toughness was found. While

the effect is smaller than in the known Kalthoff-experiments
[4], is seems that it is not due to experimental errors.
Although this fact could be explained by the concept of an incubation time [7,8], this says nothing about the origin of
such a time and other explanations are possible too.

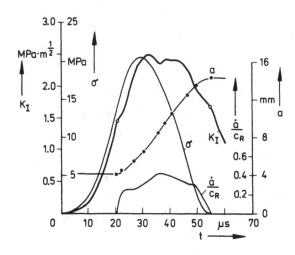

Figure 12. Experimental and numerical results for PMMA

A typical experiment on PMMA is given in Fig. 12. The initiation value of the K-factor is about 1.5 MPam$^{1/2}$, which
agrees with initiation values reported in literature [26], the
arrest toughness is only slightly higher. During the propagation phase there is a pronounced increase of the K-factor.
This result follows from the used purely elastic analysis.
Considering the extremely localized damage zone in PMMA material models similar to the strip yield model recently analysed
by Freund and Lee [27] could be more appropriate.

On the base of the presented preliminary results no defi-
nitive statement about the concept of a propagation fracture
toughness can be made. But if the K-factor during the propagation is identified with a fracture toughness it can be concluded, that this quantity is nearly independent on the crack
velocity for the investigated structural steel while it in-

creases strongly with the velocity for PMMA. This holds in the investigated velocity range up to about 0.4 c_R. These findings agree essentially with results given in literature for crack propagation initiated under quasistatic loading conditions [13,26].

REFERENCES

1. Costin, L.S., Duffy, J. and Freund, L.B., Fracture initiation in metals under stress wave loading conditions. In ASTM STP 627, Philadelphia, 1977, pp. 301-18.

2. Homma, H., Shockey, D.A. and Muryama, Y., Response of cracks in structural materials to short pulse loads. J. Mech. Phys. Solids, 1983, 31, 261-79.

3. Stroppe, H., Schreppel, U., Clos, R. and Schmidt, R., Experimentelle Methode zur Bestimmung der Bruchzähigkeit von Werkstoffen bei dynamischer Belastung. Wiss. Z. Techn. Hochsch. Magdeburg, 1984, 28, 51-5.

4. Kalthoff, J.F., Fracture behavior under high rates of loading. Engng. Frac. Mech., 1986, 23, 289-98.

5. Stroppe, H., Schreppel, U. and Clos,R., Dynamic fracture of steel at short loading times. In Impact Loading and Dynamic Behaviour of Materials, eds. C.Y. Chiem, H.-D. Kunze and L.W. Meyer, DGM Informationsgesellschaft-Verlag, Oberursel, 1988, pp. 161-7.

6. Giovanola, J.H., Crack initiation and extension in steel for short loading times. J. Physique, 1985, 46, 171-8.

7. Kalthoff, J.F. and Shockey, D.A., Instability of cracks under impulsive loads. J. Appl. Phys., 1977, 48, 986-93.

8. Shockey, D., Kalthoff, J.F., Homma, H. and Erlich, D.C., Response of cracks to short pulse loads. In Workshop on Dynamic Fracture, eds. W.G. Knauss, K. Ravi-Chandar and A.J. Rosakis, Cal. Inst. Tech., Pasadena, 1983, pp. 57-71.

9. Curran, D.R., Incubation times and rate-dependent toughness for macrocrack initiation. Paper on The American Physical Society Conf. on Shock Compression of Condensed Matter, Albuquerque, New Mexico, August 1989.

10. Homma, H., Kanto, Y. and Tanaka, K., Cleavage fracture in steel specimens under short stress pulse loading. Paper on Int. Seminar on Dynamic Failure of Materials, Vienna, January 1991.

11. Dally, J.W., Fourney, W.L. and Irwin, G.R., On the uniqueness of the stress intensity factor - crack velocity relationship. Int. J. Frac., 1985, 27, 159-68.

12. Knauss, W.G. and Ravi-Chandar, K., Fundamental considerations in dynamic fracture. Engng. Frac. Mech., 1986, 23, 9-20.

13. Kanninen, M.F. and Popelar, C.H., Advanced Fracture Mechanics, Oxford University Press, New York, Clarendon Press, Oxford, 1985.

14. Kanninen, M.F., Hudak, S.J., Couque, H.R., Dexter, R.J. and O'Donoghue, P.E., Viscoplastic-dynamic crack propagation: Experimental and analysis research for crack arrest application in engineering structures. Int. J. Frac., 1990, 42, 239-60.

15. Giovanola, J.H., Klopp, R.W. and Simons, J.W., Effect of shear lips on dynamic crack propagation. In Dynamic Fracture, eds. H. Homma and Y. Kanto, Chuo Technical Drawing Co. Ltd., Toyohashi, 1989, pp. 93-102.

16. Oda, J., Shibahara, M. and Yasuda, K., On crack extension phenomena of cracked plate subjected to a tensile pulse loading. In Dynamic Fracture, eds. H. Homma and Y. Kanto, Chuo Technical Drawing Co. Ltd., Toyohashi, 1989, pp. 224-35.

17. Couque, H., Hudak, S.J. and Lindholm, K.S., On the use of coupled pressure bars to measure dynamic initiation and propagation toughness of pressure vessel steel. J. Physique, 1988, 49, 347-53.

18. Clos, R., Berechnung des Spannungsintensitätsfaktors eines spannungswellenbelasteten axialsymmetrischen Risses mittels eines Differenzenverfahrens. <u>Wiss. Z. Techn. Univ. Magdeburg,</u> 1988, 32, pp. 1-8.

19. Ravichandran, G. and Clifton, R.J., Dynamic fracture under plane wave loading, <u>Int. J. Frac.,</u> 1989, 40, 157-201.

20. Zencker, U. and Clos, R., Numerische elastodynamische Analyse bewegter Risse. <u>Wiss. Z. Techn. Univ. Magdeburg,</u> (to be published).

21. Nishioka, T. and Atluri, S.N., Path-independent integrals, energy release rates, and general solutions of near-tip fields in mixed-mode dynamic fracture mechanics. <u>Engng. Frac. Mech.,</u> 1983, 18, 1-22.

22. Nishioka, T. and Atluri, S.N., A numerical study of the use of path independent integrals in elasto-dynamic crack propagation. <u>Engng. Frac. Mech.,</u> 1983, 18, 23-33.

23. Freund, L.B., Crack propagation in an elastic solid. Part 2: Non-uniform rate of extension. <u>J. Mech. Phys. Solids,</u> 1983, 1972, 20, 141-52.

24. Kostrov, B.V., On the crack propagation with variable velocity. <u>Int. J. Frac.,</u> 1975, 11, 47-56.

25. E 399-74 Standard Test Method for Plane-Strain Fracture Toughness of Metallic Materials. 1976 Annual Book of ASTM Standards, ASTM, Philadelphia, 1976, Part 10, 471-90.

26. Takahashi, K. and Arakawa, K., A parameter influential in dynamic fracture. In <u>Dynamic Fracture,</u> eds. H. Homma and Y. Kanto, Chuo Technical Drawing Co. Ltd., Toyohashi, 1989, pp. 8-19.

27. Freund, L.B. and Lee, Y.L., Observations on high strain rate crack growth based on a strip yield model, <u>Int. J. Frac.,</u> 1990, 42, 261-76.

DYNAMIC CRACK PROPAGATION AND CRACK ARREST IN A STRUCTURAL STEEL :
COMPARISON BETWEEN ISOTHERMAL AND THERMAL SHOCK TESTS

Marc DI FANT*, André FONTAINE** and André PINEAU*
* Centre des Matériaux-Ecole des Mines,
BP.87 - 91003 EVRY Cedex (FRANCE) URA CNRS N°866.
** IRSID, 185, rue du Président Roosevelt,
78104 SAINT-GERMAIN EN LAYE Cedex (FRANCE)

ABSTRACT

The dynamic fracture and crack arrest behaviour of a C-Mn-Nb steel was investigated using two widely different experimental techniques. The first technique which follows the ASTM recommandations, is based on isothermal tests carried out on two specimen geometries with different in-plane dimensions : the "Compact Crack Arrest" geometry and the "Reduced Dynamic Effect" geometry. These specimens were instrumented in order to measure the crack velocity. The second technique, which is more original, is based on a thermal shock transient loading applied to thin discs containing at their external periphery a fatigue precrack. These specimens were cooled down and then heated up by an induction coil set in the center of the inner diameter to develop very severe temperature gradients which produce a tensile hoop stress at the external surface. A crack is initiated from the fatigue precrack which propagates very rapidly ($\simeq$100 µs) over a distance of a few centimeters and then stops in the hot material. The measured radial temperature distribution is used as the input for finite element calculations of the stress intensity factor.

Both types of experiments are analyzed using a static approach to determine the stress intensity factor at crack arrest, K_{1A}. Similar results showing a significant shift in the brittle-ductile transition temperature are obtained from both techniques, in spite of their large differences which are discussed.

INTRODUCTION

The integrity of mechanical structures is usually insured by preventing the onset of unstable crack initiation, which is characterized by the fracture toughness, K_{1c}. However, in structures requiring a high level of integrity and safety, both engineers and researchers are very much concerned with the concept of crack arrest, which is usually interpreted in terms of the crack arrest fracture toughness, K_{1A}.

In steels the crack arrest behaviour is usually characterized by using isothermal mechanical tests carried out on double cantilever beam (DCB) type specimens in which crack initiation and crack arrest are investigated under displacement control conditions [1-3]. These tests are interpreted using a static approach, as described in ASTM standards [4], although there are still many discussions in the literature to know whether a dynamic analysis should be preferred (see Eg.[5,6]).

In a number of circumstances, the crack arrest phenomenon is partly related to thermal shocks. A good example is provided by the nuclear power industry in which large-scale experiments have been performed to investigate the fracture behaviour of a flawed shell of a pressurized water reactor under the severe thermal shock corresponding to a postulated loss of cooling agent (see e.g. [7-9]). In these experiments crack arrest was observed to take place after crack initiation and rapid crack growth. These large-scale experiments are very expensive. Therefore there is a growing need for the development of a small-sized crack arrest experiment, which was one of the main issue of the present investigation. This experiment has already been described in detail elsewhere [10]. In the present paper the comparison between both methods of testing is made.

MATERIAL

The material investigated is a C-Mn-Nb steel (E36) used in the offshore industry. The composition was, in weight percent : C = 0.17, Mn = 1.44, Si = 0.27, Nb = 0.033, P = 0.011, S = 0.010 and Al = 0.005. This material was taken from a thin ($\simeq$ 20 mm) plate containing a mixed microstructure of ferrite and pearlite aligned along the rolling direction, with a grain size about 7 μm (Fig.1). The crack was perpendicular to the rolling direction.

Tension and compression tests were performed at temperatures ranging from -196°C to R.T. and at two loading rates, ie $\simeq 10^{-3} s^{-1}$, (quasi-static conventional tensile tests), and $1.1 \ 10^{3} s^{-1}$, (Hopkinson bar compression tests).In the latter case the yield strength was defined for a plastic strain of 2 percent. The results are given in Fig.2.

Fracture toughness measurements were carried out in order to determine the static brittle-ductile transition curve. These measurements was made on CT type or 3 points bend specimens with a thickness of 17.5 mm. The results are shown in Fig.3 where the square symbols refer to valid measurements of K_{1c} according to the ASTM E399-83 standard, the lozinge symbols indicate non valid measurements of K_{1c} according to the same standard, while the triangle symbols refer to the initiation of slow stable ductile crack growth. In this case the fracture toughness was inferred from J_{1c} measurements according to the ASTM E813-87 standard and the fracture toughness was measured as : $K_{Jc} = \left[EJ_{1c}/(1-\nu^2) \right]^{1/2}$,where E is the Young modulus and ν the Poisson ratio (ν = 0.30).These tests show that the static transition temperature is approximately -140°C.

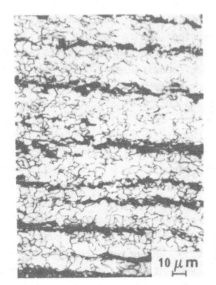

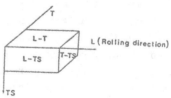

Fig.1 : Microstructure of the E36 steel observed on the T-TS surface.

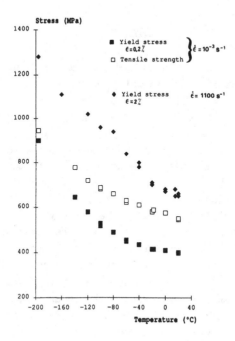

Fig.2 : Static yield stress and tensile strength ($\epsilon^\bullet$ = 10^{-3}s^{-1}) and dynamic yield stress ($\epsilon^\bullet$ = 1100 s^{-1}) as a function of temperature.

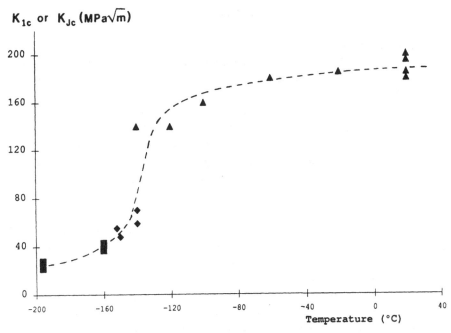

Fig.3 : Static fracture toughness as a function of temperature.

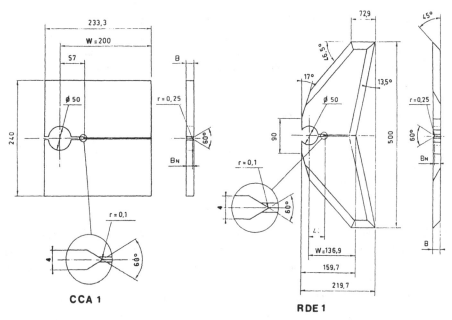

Fig.4 : Specimen geometries for isothermal tests.

ISOTHERMAL TESTS

Experimental Procedures

Two types of specimens were tested, as shown in Fig.4, i.e. the "Compact Crack Arrest" (CCA) geometry which is recommanded by the ASTM and the "Reduced Dynamic Effect" (RDE) specimen introduced by Kalthoff et al [3,11,12]. The geometries CCA1 and CCA2, RDE1 and RDE2 have homethetic in-plane dimensions, (W = 200 mm and 150 mm for CCA1 and CCA2; W = 136.9 mm and 102.7 mm for RDE1 and RDE2, respectively). These specimens contained on one face, a side groove with a depth between 1.5 mm and 2.8 mm. The other face was polished for metallographical observations and for the fitting of crack velocity measurement gages. The notch was embrittled by spot welding and subsequently machined by spark erosion.

These specimens were tested under displacement control by forcing a wedge into two split pins, (Fig.5). In these tests the compressive load applied to the wedge is increased progressively until unstable crack extension occurs. The crack mouth opening displacement, δ is measured with an extensometer located at a distance of 16.7 mm from the back face of the specimens. For CCA1 specimens this corresponds to a distance of 0.25 W from the load line axis. Crack mouth opening displacement (CMOD) was recorded both at crack initiation (δ_o), and after crack arrest (δ_a). Crack velocity measurements were carried out with special gages made of twenty conductive wires connected to a high speed memory oscilloscope, (Fig.6). The precise measurement of the crack length at arrest was made after the tests by the examination of the fracture surface after heat tinting or fatigue cracking (Fig.7). SEM observations showed that the fracture surface corresponding to unstable crack growth was essentially covered by cleavage facets.

Results

14 experiments were performed. In table 1 we have indicated the initial crack length, a_o measured from the load line axis, the crack extension distance, Δa occurring during unstable crack propagation, the values of the CMOD measured at the onset of unstable crack growth, $2\delta_o$ and those determined after crack arrest, $2\delta_a$.

The results of crack velocity measurements for a typical test (ND11; -45°C) are reported in Fig.8, where it is observed that the crack arrest phenomenon occurs abruptly over a short distance of less than 5 mm. This figure shows also that the crack velocity is constant over a large distance ($\simeq 30$ mm). The measured crack velocities were found to be included between 400 m/s and 800 m/s.

In Table 1, the initial values of the stress intensity factor at crack initiation K_{Io} and those of K_{IA} were calculated by using the ASTM method, ie : $K = 2E\delta/\sqrt{W} \cdot f(a/W) \cdot \sqrt{B/B_N}$, where 2δ is the CMOD ($2\delta_o$ or $2\delta_a$), E is the Young modulus (200.000 MPa), B is the initial specimen thickness, while B_N is the specimen thickness at crack plane. The functions $f(a/W)$ were calculated by finite element method. The results are reported in Fig.9

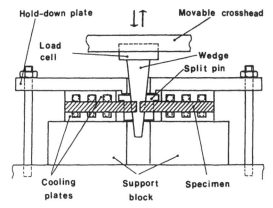

Fig.5 : Schematic illustration of the loading arrangement used for isothermal tests.

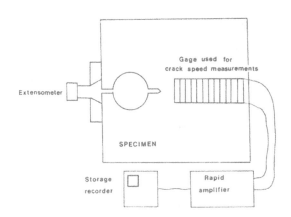

Fig.6 : Schematic illustration of the crack velocity instrumentation for isothermal tests.

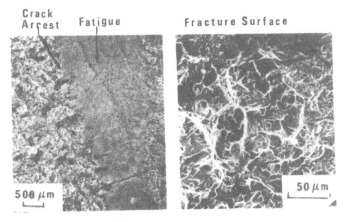

Fig.7 : SEM observations showing the method used to identify the fracture mode at crack arrest.

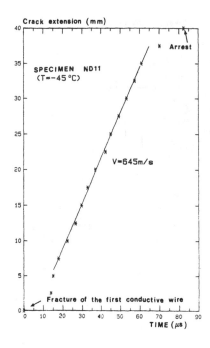

Fig.8 : Example of a crack extension versus time measurement.

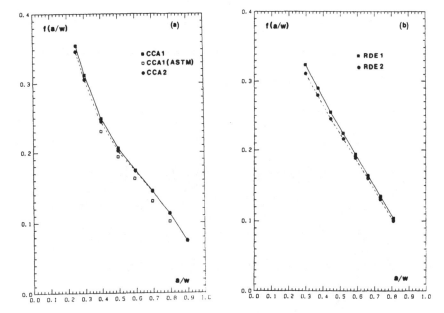

Fig.9 : Variations of the functions f(a/w) for various specimen geometries, a) CCA1 and CCA2 specimens; b) RDE1 and RDE2 specimens.

Specimen	Type	T (°C)	a_o (mm)	Δa (mm)	$2\delta_o$ (mm)	$2\delta_a$ (mm)	K_{1o} (MPa√m)	K_{1A} (MPa√m)
ND4	CCA1	-100	98.5	69.5	1.02	1.05	112	53.1
ND1	CCA1	-100	79.2	60.5	0.83	0.87	105.6	63.9
ND19	CCA1	-100	84.7	57	0.77	0.92	92.9	66.1
A2	CCA1	- 75	57.2	41.3	0.54	0.64	84.3	64.4
ND6	CCA1	- 58	80	65	1.24	1.38	155.5	95.5
ND14	CCA1	- 45	86.7	80	1.34	1.38	173.1	75.2
ND17	CCA1	- 45	86.7	86	1.06	1.12	174.4	66.8
ND9	CCA2	-105	98.3	41.7	0.98	1.08	87.1	38.4
ND11	CCA2	- 45	76.5	56	1.33	1.35	150.3	61.5
ND25	EDR1	-100	54	38.2	0.64	0.69	102.6	63.1
ND24	EDR1	- 62	39	29.7	0.60	0.62	120.8	85.8
ND26	EDR1	- 45	53.2	49.5	1.08	1.17	173.4	85.8
ND27	EDR2	-100	36.4	47	0.59	0.67	115.3	44.4
ND29	EDR2	- 55	25.1	39.9	0.59	0.60	138.6	69.9

Table 1 : Results of isothermal crack arrest tests.

Specimen	Δt (s)	a_o (mm)	T_o (°C)	a_a (mm)	T_a (°C)	K_{1C} (MPa√m)	K_{1A} (MPa√m)
C1	6	14.15	-196	39.05	-65	22	58.3
C2	7	13.40	-165	35.92	-48	30	69

Δt = Time during which the power was supplied before crack initiation; a_o = Initial crack length; T_o = Temperature at which crack initiation occurred; a_o = Crack length at crack arrest; T_a = Temperature of crack arrest.

Table 2 : Results of thermal shock experiments.

where a good agreement is observed for the CCA1 geometry between the ASTM expression and those derived from our numerical calculations (Fig.9a).

THERMAL SHOCK EXPERIMENTS

Experimental Procedures

The specimens used for the thermal shock tests are hollow discs which are 20 mm thick and with an outer and an inner diameter of 150 mm and 50 mm, respectively (Fig.10). A notch is machined from the outside surface and is sharpened by a fatigue precrack. Holes of 3 mm in diameter are drilled through the wall thickness of the discs to receive thermocouples which are spot welded at various depths. This arrangement was adopted in order to measure the radial thermal gradient and to control the axisymmetry of the temperature distribution. Full details are given elsewhere [10,13,14].

An induction heating device was developed to introduce large thermal gradients through the wall thickness of the discs. In these experiments the induction coil is a key component which was designed to introduce only radial thermal gradients and no gradient through the specimen thickness [13].

Each specimen is instrumented to give direct measurements of CMOD and radial temperature profiles. It is then placed around the induction coil, as shown in Fig.11. This assembly is placed in a tank filled with liquid nitrogen. Once isothermal conditions at -196°C are reached, the coolant is drained and the specimen slowly air-warmed up to the test temperature. Then a thermal shock is applied by manual activation of the induction power supply. This results in an abrupt heating of the inner surface of the disc and in the development of severe thermal gradients through the wall thickness. This temperature distribution produces compressive hoop stresses at the inner part of the specimen and tensile hoop stresses at the outer part. Once the fracture toughness at crack initiation is reached, a crack run-arrest event takes place in approximately 100 μs. This crack jump, which is instantaneous on the time scale of the temperature transient, is detected by an audible noise and, more importantly, by a sudden increase in CMOD.

Results and analysis

A number of tests have been carried out, including tests on thicker specimens [10,14]. The results of two experiments (C_1 and C_2) are described here. The radial thermal gradients measured at crack initiation are shown in Fig.12, while more details are reported in Table 2. It is observed that thermal gradients as large as 550°C can be developed through the wall thickness in a few seconds. SEM observations of the fracture surfaces showed that, in both specimens, crack propagation and crack arrest occurred in a predominantly cleavage mode.

The interpretation of these experiments requires thermoelastic and

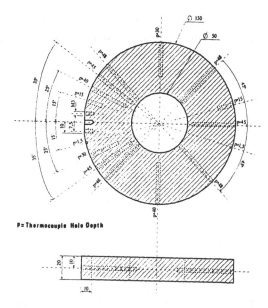

P = Thermocouple Hole Depth

Fig.10 : Specimen geometry used for the thermal shock experiments.

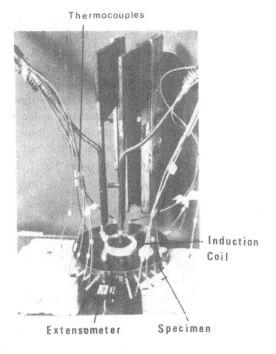

Thermocouples

Induction
Coil

Extensometer   Specimen

Fig.11 : Instrumented specimen placed around the inductor.

thermoplastic 2D FEM analysis which was performed using the ZEBULON code developed in the authors laboratory. The temperature profiles shown in Fig.12 were introduced directly as loading conditions. Cracks of different lengths were simulated and, for each specimen which was assumed to be under plane stress conditions, the K factor was computed using the virtual crack extension technique introduced by Parks [15]. These calculations were purely static. The results corresponding to Test C1 are reported in Fig.13. This figure shows how the fracture toughness, K_{1c} or K_{1A} can be determined, knowing the initial crack length and measuring the crack length at arrest. The values of K_{Ic} and K_{IA} determined on both tests are reported in Table 2. The values of K_{1c} derived from these tests are very similar to those determined from tests carried out on conventional CT type specimens (Fig.3).

DISCUSSION

Both methods used to investigate crack arrest phenomenon can now be compared.

(1) Specimen preparation

The specimen geometry used for thermal shock experiments is quite simple and can easily be machined compared to the specimens used in the isothermal tests, which, in particular, require the machining of a notch by spark erosion in a spot weld. The thermal shock specimens could also be tested only with a notch instead of a fatigue precrack which necessitates the use of a rather powerful machine. Typically the specimens used in the present study were precracked with a servohydraulic machine under a cyclic load of approximately $2\ 10^5$N, and at a frequency of 10 Hz. The geometry of these specimens could be modified to some extent to reduce the fatigue precracking load by increasing the ratio between the inner and the outer diameter of the discs.

(2) Testing procedures

It is difficult to compare the testing equipments used in both types of experiments. The comparison can only be made on the various operations which have to be performed before running an experiment.

It might appear that the thermal shock specimens necessitate a longer preparation, in particular to set the thermocouples. However it should be kept in mind that a smaller number of thermocouples, lower than 15, is sufficient provided that the heating conditions leading to a purely axisymmetric temperature distribution with no thermal gradient through the specimen thickness, which is necessary for a 2D analysis, are obtained. The cooling arrangements are very similar in both cases. In the present study crack velocity measurements were made only on the isothermal specimens.

Electrical noise due to induction heating might appear to be a limitation for the triggering of the high speed memory oscilloscope. However it is believed that this difficulty could easily be overcomed.

The boundary "mechanical" conditions can be much more easily controlled in the thermal shock experiments. It is extremely difficult to control the displacement of the wedge used in the ASTM specimens during crack propagation. In the method proposed by the ASTM it is assumed that the test is carried out under a constant displacement. This condition is not strictly verified since the measurements of the distance between the split pins showed a supplementary wedge penetration during crack propagation [13]. This is partly responsible for the difference of the CMOD measured before crack initiation and after crack arrest, (Table 1).

(3) Interpretation of crack arrest experiments

The results of the 16 experiments corresponding to both isothermal and thermal shock tests are shown in Fig.14. At first glance this figure shows no significant influence of either the specimen geometry for the isothermal specimens or the testing procedures, when the relative large scatter in the results is considered. Fig.14 shows also an interesting behaviour since the K_{1A} values are much lower than the fracture toughness at crack initiation, K_{1C}. Thus, in this material, like in other steels, there is a drastic shift of the transition temperature, of the order of 80°C to 100°C, between static tests at crack initiation and crack arrest phenomenon.

The fact that both methods might lead to similar results is not trivial when one considers the largely different conditions, ie. (i) an initial notch in an embrittled material with the isothermal tests compared to a fatigue precrack in the base material for the thermal shock experiments and (ii) widely different K histories, (increasing and decreasing K in the ASTM specimens and thermal shock specimens, respectively). In the present study we have tested relatively thin (15 to 20 mm) specimens. This rises the question of plane strain or plane stress conditions. In the ASTM procedure, for a given displacement, the analysis involves the factor $(1-\nu^2)$ between plane strain and plane stress conditions, that is only a difference of about 10%. In the analysis of our isothermal tests we assumed plane strain conditions.For thermal shock experiments we assumed that the specimens were under plane stress conditions. If we had assumed plane strain conditions, with an elastic analysis, larger K values would have been obtained by a factor of $1/(1-\nu)$, that is about 30%. The appropriate conditions for thermal shock specimens, especially for thicker discs, would be generalized plane strain conditions, ie the axial strain, ϵ_{zz} in each cross section is independent of the position of that section. Thicker specimens which fulfilled this condition have actually been tested [10,14].

All the experiments have been interpreted in terms of a static approach because it is the simplest one. However elastodynamic numerical calculations have also been made on the ASTM specimens with ABAQUS finite

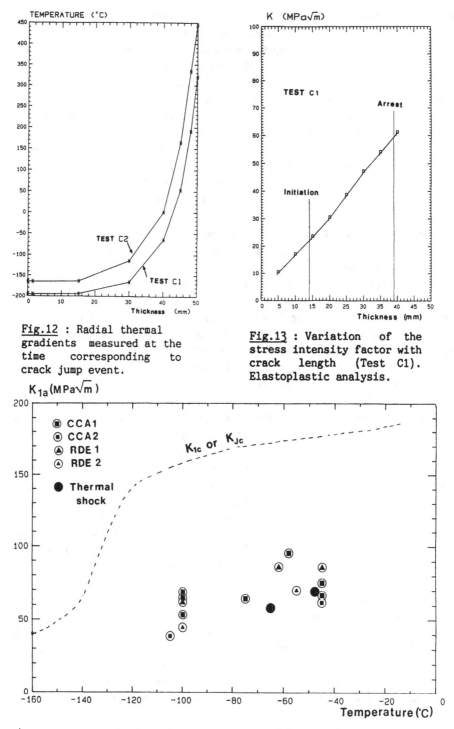

Fig.12 : Radial thermal gradients measured at the time corresponding to crack jump event.

Fig.13 : Variation of the stress intensity factor with crack length (Test C1). Elastoplastic analysis.

Fig.14 : Variation of the static fracture toughness at crack arrest with temperature. Comparison with the fracture toughness at crack initiation.

element code [13]. These calculations showed that the static K factor was lower than the calculated dynamic one, K_{1d}, as suggested by Kalthoff et al [3], and as also shown recently in another numerical study [16]. The conclusion according to which the static approach is conservative applies to ASTM test specimens but not necessarily to other loading and geometrical conditions. In these specimens the difference between static and dynamic analysis is mainly associated with wave reflections taking place at the free surfaces. This is the reason why the RDE specimens were proposed [12]. It is felt that this effect should also largely be reduced in thermal shock round shaped specimens.

CONCLUSIONS

1. The fracture toughness at crack arrest of a C-Mn-Nb steel is significantly lower than the fracture toughness at crack initiation even when the failure mode corresponds to brittle cleavage fracture.
2. A flexible thermal shock experiment applied to small-sized specimens has been described in some detail.
3. It is shown that, in spite of widely different conditions, the ASTM method for determining crack arrest fracture toughness and a novel technique based on thermal shock lead to consistent results.

REFERENCES

1. Crosley, P.B., Fourney, W.L., Hahn, G.T., Hoagland, R.G., Irwin, G.R. and Ripling, E.J. Final report on cooperative test program on crack arrest toughness measurement NUREG/CR-3261 University of Maryland, College Park, MD, 1983.

2. Rosenfield, A.R., Mincer, P.N., Marschallo, C.W. and Markworth, A.J. Recent advances in crack arrest technology Fracture Mechanics : Fifteenth Symposium ASTM STP 833 (Ed. R.J. Sanford) American Society for Testing and Materials, Philadelphia, PA,1984, pp.149-164.

3. Kalthoff, J.F., Beinert, J., Winkler, S. and Klemm, W. Experimental analysis of dynamic effects in different crack arrest test specimens Crack Arrest Methodology and Applications ASTM STP 711 (Eds. G.T. Hahn and M.F. Kanninen) American Society for Testing and Materials, Philadelphia, PA, 1980, pp.109-127.

4. ASTM E1221-88 Standard test metod for determining plane-strain crack arrest fracture toughness, K_{1A} of ferritic steels, 1988, pp.844-859.

5. Kanninen, M.F., Popelar, C. and Gehlen, P.C. Dynamic analysis of crack arrest in contoured and straight-sided DCB-test specimens for wedge loading and machine loading conditions, Fast Fracture and Crack Arrest, ASTM STP 627, 1977, pp.19-39.

6. Kalthoff, J.F. Interpretation of crack arrest fracture tougness measured with varius steels, Proceeding of the 7th International Conference on Fracture, Advances in fracture research, Houston, Eds. K. Salama, K. Ravi-Chandar, 1989, pp.705-714.

7. Bryan, R.J., Bass, B.R., Bolt, S.E., Bryson, J.W., Edmonds, D.P., McCulloch, R.W., Merkle, J.G., Nanstad, R.K., Robinson, G.C., Thoms, K.R. and Whitman, G.D. Pressurized-thermal-shock test of 6-in-thick pressure vessels. NUREG/CR-4106 Oak Ridge National Laboratory, Oak Ridge, TN, 1965.

8. Kussmaul, K, Föhl, J. and Roos, E. Some conclusions with regard to the safety assessment of cracked components drawn from the research program "integrity of components" (FKS II) at present state 12. MPA Seminar MPA, Universität Stuttgart, 1986.

9. Sollogoub, P. and Pellissier-Tanon, A. Essai de fissuration d'une virole par choc thermique à l'azote liquide. DT 82 1185 Framatome, Paris, 1982.

10. Di Fant, M., Genty, A. and Pineau, A. Thermal-shock-induced crack arrest of two low-alloy steels. High Temperature Technology, 1990, 8, pp.105-114.

11. Kalthoff, J.F. On the measurement of dynamic fracture tougness - A review of recent work, Int. J. of Fract., (1985), Vol.27, pp.277-298.

12. Beinert, J. and Kalthoff, J.F. The development of a crack arrest test specimen with reduced dynamic effects, Application of Fracture Mechanics to Materials and Structures, Eds. G.C. Sih, E. Sommer, W. Oal, 1983, pp.493-511.

13. Di Fant, M. Mécanique et mécanismes de la rupture dynamique dans un acier ferritique peu allié. Thesis, Ecole des Mines, Paris, 1990.

14. Genty, A. Etude de la rupture par choc thermique d'un acier 15MND6, Thesis Ecole des Mines, Paris 13 Décember 1989.

15. Parks, D.M. A stiffness derivative finite element technique for determination of crack tip stress intensity factor. Int. J. of Fracture, 1974, 10, pp.487-502.

16. Dahlberg, L. Crack arrest. Additional safety against catastrophic fracture. Final Report of the NKA Project MAT 550, 1989, Nordisk Kontakdorgan for atomenergispergsmal.

ON THE PRACTICAL SIGNIFICANCE OF CRACK ARREST

THOMAS VARGA and GÜNTHER SCHNEEWEISS
Institute for Testing and Research in Materials Technology
(TVFA)
and
Department for Material Science and Testing
at the Technical University of Vienna
A - 1040 Wien, Karlsplatz 13

INTRODUCTION

In case crack initiation and propagation in a component or structure cannot be excluded, the safety against fracture depends on the arrest of the running crack. The practical significance of crack arrest will be demonstrated concerning ships, airplanes, pipelines, penstocks and mentioned for components of power plants. The measurement of the Crack Arrest Fracture Toughness K_{Ia} by testing three point bend specimens will be discussed. Tests and results concerning a fully killed weldable fine grain steel TStE 355 will be reported.

METHODS TO OBTAIN CRACK ARREST

Ships
After introducing the series production of ships by welding whithout subsequent stress relief heat treatment (Liberty ships, T2 tankers), severe cracking of these occured [1]. In Fig.1 some main types of crack arrestors are shown according to a study of the U.S. Ship Structure Committee [2].
One solution to attain crack arrest is by welding a strip of highly ductile steel between two of the usual hull plates (inserted type of crack arrestor). In the case of ships a fully killed Al-treated fine grain steel served as crack arrestor [3]. Those areas in ships which used to be equipped by high fracture toughness materials are described in [2, 4].
Another possibility is to attach patches (or stiffeners) to the plate by welding, riveting or adhesive bonding perpendicular to the expected crack path (patch type crack arrestor). When a crack runs into this region, the patches

will be subjected to tension. The resulting compressive reaction force suppresses the crack driving force in the cracked plate and the crack arrests.

Reduction of constraint can be achieved by reducing the plate thickness on the bottom of a wide groove. Because of reduced triaxiality a significant increase in toughness should arrest the running cleavage crack (ditch type crack arrestor). Contrary to this, an increase in thickness was believed to reduce stresses and stress intensities to such a low level, that even cleavage cracks would arrest. A similar effect was observed, when cracks did not run through ribs (stiffener type crack arrestors).

The all-welded structure of a ship makes crack arrestors very essential. In case of a discontinous riveted hull structure, a crack hardly continues to propagate over a riveted seam. Therefore, the simplest type of crack arrestor is to use riveted seams at the vital portions of welded structures (riveted seam type crack arrestor).

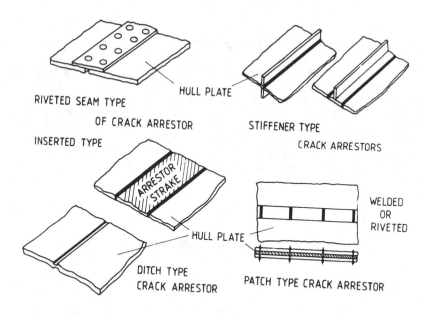

Figure 1. Different types of crack arrestors.

The geometry of crack arrestors of the riveted seam type is shown in Fig.2a with a longitudinal slot in the hull or in the deck plate [4]. The version according to Fig.2b [4] contains arrestor holes as a means to limit crack propagation without riveted cover plates. The crack may hereby originate in the hull plate or in the bilge keel.

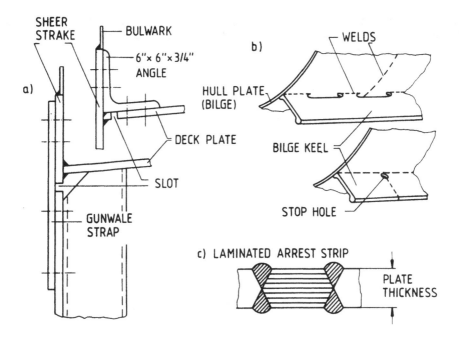

Figure 2. Examples for the geometry of crack arrestors

On the circumference of a hole the stress intensity is lower than at the tip of a crack [5]. Near to the hole the stress intensity factor, K, is very much increased at the tip of a crack approaching the hole [6]. On fatigue crack propagation the effects in sum will therefore become small in case the crack passes through the hole [7]. K will be reduced, if the crack path lies between the holes [8]. Therefore crack arrest will be attained by the coalescence of the crack and the hole or if the crack extends between the holes [5].
The crack arrestor in Fig.2c, consisting of thin lamellae, increases fracture energy (and deformation capacity) significantly [9].

Airplanes
Airplane wings consist mainly of the load carrying main profile spars, the ribs and the outer skin [10]. In Fig.3 this is shown according to [11]. The fuselage represents a light weight pressure vessel, realized as a combination of shape-forming circumferential and longitudinal framework and the skin again [10], see Fig.4 [11]. In both cases there are longitudinal stringers, which are important in respect to crack arrest.
The riveted skin-stringer design of many aircraft structures is basically a crack-arrest structure. Aircraft are presently designed to arrest a two-bay crack; i.e., a crack

originating at a stringer is to be arrested at the two adjacent stringers (MIL-A-83444, 'Airplane Damage Tolerance Design Requirement' [2]).

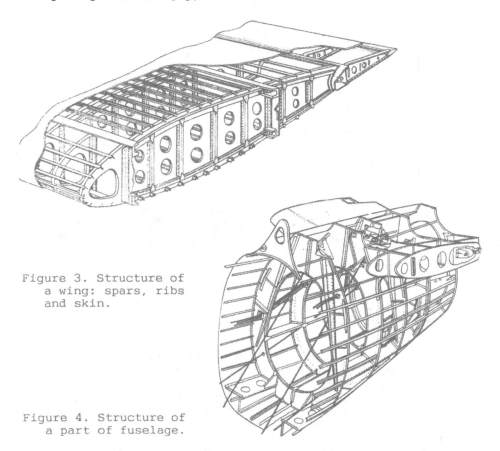

Figure 3. Structure of a wing: spars, ribs and skin.

Figure 4. Structure of a part of fuselage.

Tests were carried out by Harpur [12] to check the effect of crack stopper strips, which were either riveted to, bonded to, or machined integrally with the basic sheet. In order of decreasing strength for these specimens is bonding, integral machining and riveting. However, the riveted crack arrestor permits a much greater crack length before failure than the integral and hence allows more time for the crack to be spotted. The order of fail-safe effectiveness is therefore bonding, riveting and integral machining and this confirms the similar result obtained on box beam tests [13-15].

Fatigue cracks are often stopped by drilling a hole at the crack tip. The stress intensity factor may be substantially reduced by this means. Still more favorable conditions may be introduced by expanding the hole to introduce compressive stresses; this is applied also to increase the chance of crack arrest at holes [16,17]. Fatigue crack propagation and start of unstable cracks from crack tips is much less probable if also residual compressive stresses are introduced by a steel ball pressed in the material at the crack tip [17,18].

Gas pipelines

Some patented proposals for crack arrestors in pipelines are shown in Fig.5 [19]. The first catastrophic failures in pipelines [20,21] were most probable of the cleavage type. Two arresting methods, the application of compressive stresses or forces on one hand like in examples 1 and 4, Fig. 5, or on the other hand the replacement of a pipe section of usual quality by a pipe section of high toughness, examples 2 and 3 (partly), seem to be practicable.

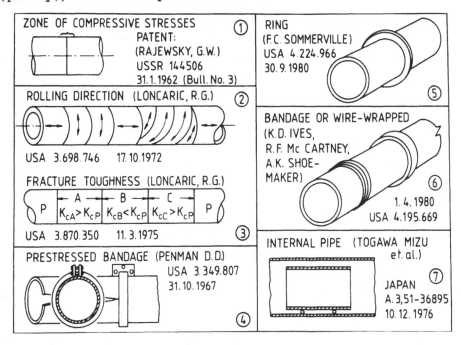

Figure 5. Examples of crack arrestors in line pipe according
to patents

In pipeline construction the principal rolling direction is parallel to the longitudinal axis of the pipe. The resistance against a running crack is generally higher if the rolling direction is perpendicular or inclined to the pipe axis. The patent according to example 2 is based on this effect. Example 3 shows the insertion of some pipe sections of differing toughness K_c.

Cleavage cracks are not always arrested by means used to arrest shear cracks. Wire wrapping for instance was not effective in arresting cleavage cracks in lit. [22]. The velocity of a running cleavage crack is up to 600 m/s [22] and is therefore slower than the decompression wave in fluids (water 1400 m/s, diesel fuel 1500 m/s approx.), but more fastly propagating than the decompression wave in natural gas (methane 400 m/s approx.). The driving force for a cleavage crack is therefore originating mainly from the stored elastic energy [23] (but not in case of ductile fracture [24]). A running cleavage crack may therefore get arrested, if the elastic energy becomes sufficiently low.

Line pipe steels suffer nowadays shear fractures only [25]. The toughness necessary for crack arrest was calculated [26] in a way that the results of all published full scale burst tests could be predicted to an accuracy of approximately ± 10%.The necessary parameters have been directly determined from Charpy impact and from Battelle DWTT [27] energy values. In some older references the opinion is expressed, that higher fracture toughness of the material is not always sufficient to guarantee arrest of a running shear crack [28].

Shear crack propagation usually occurs in a longitudinal direction along the top of the pipeline [29,30]. During running of the crack, the shape of the pipe in advance of the crack tip becomes oval, due to the flap opening behind the crack tip. As the crack approaches, the longitudinal tensile strain in a band on the top of the pipe continously increases to a maximum value. This maximum value is attained at a distance of about two pipe diameters from the crack tip. The longitudinal tensile strain in this "tongue" is 1 to 2% or more. Behind the crack tip, the gas pressure is acting on the flaps over a distance up to three pipe diameters. The pressure is resulting in a force, acting on the mass of the pipe wall which causes an acceleration of the flap. This force acts both normal to the crack surface as a crack-opening force and as a radial force driving the flaps outward. Shear fracture velocities observed lie between 60 and 250 m/s [32]. Maximum velocities in tests were up to 300 m/s [28,33].

Constraint of the flaps behind the propagating crack tip by the use of wires, straps, or rings around the pipe is therefore an effective means for arresting cracks. The crack arrestors 5 and 6 in Fig.5 employ this concept. In a full-scale test, wire wrapping resulted in crack arrest [28].

An alternate means of restraining the flap opening is to provide periodically large masses along the pipe length which must be accelerated by the pressure acting on the flaps behind the propagating crack [28]. According to [34], fracture toughness of an uncovered pipe has to amount to more than twice as much as the fracture toughness needed for crack arrest in a buried pipe. For a further example, concrete anchor blocks require significant masses to be accelerated, if the concrete is prevented from spalling off the pipe during ovaling ahead of the running crack.

At last, a possibility to arrest a longitudinal propagating shear crack is to initiate a circumferential crack in advance of the longitudinal crack. This effect is due to high longitudinal strain levels at stress concentrations such as girth welds.

Additional crack arrestors according to [33] are given in Fig.6. The example in Fig.6c belongs to a patent [35]. A similar idea is presented in [36].

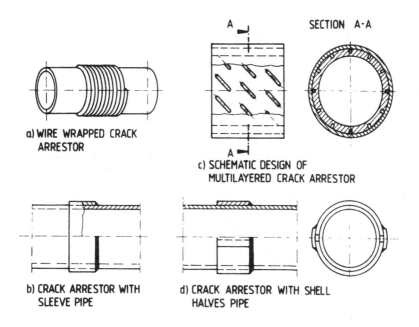

Figure 6. Crack arrestors in gas pipelines.

In [19] specimens were tested, built up of four layers of plates formed by hot rolling. The two inner layers were interrupted by slots. The crack reached the through-slots in the inner layers and stopped in all the layers of the specimen with or without a slot.

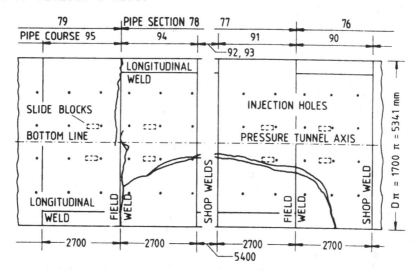

Figure 7. Cracks in the penstock after damage, hydraulic power plant Gerlos.

Penstocks

Pressurized tubes, filled by water, usually do not exhibit very long cracks. Therefore no crack arrestors are known. Crack arrest, however, should be investigated in penstocks. As an example, the cracking of the Gerlos penstock is shown in Fig.7. Arrest of longitudinal cracks was observed either at higher toughness tube sections or at a circumferential weld.

Power Plant Components

The same or similar considerations may be applied for design against unstable crack propagation in power plant components. Unstable cracks may be initiated by local thermoshock. Arrest may become in such cases an essential feature for safety against fracture. Due to this fact ASTM Code Section XI includes crack arrest as a main item to handle pressurized thermal shock of the reactor pressure vessel (RPV).

The first standardized crack arrest testing method for steel was not established before 1989 [ASTM E 1221] using still relatively large and not easy to produce specimens.

DEVELOPMENT OF AN ALTERNATIVE TEST METHOD

The arrest of cleavage cracks was observed during testing of large CT specimens of RPV steel in a servohydraulic machine. Cleavage jumps have been observed not only in strain controlled tests, but also during force control. Even impact test specimens exhibit arrest phenomena [38].

Therefore the development of a testing method was envisaged, which is characterized as follows:

- simple geometry of the specimen
- high starting speed of the cleavage crack
- both crack mouth opening displacement (CMOD) and force at arrest are to be measured to deduce fracture arrest toughness

In consequence, three point bend specimen geometry with a brittle crack starter weld was chosen [39].

The nominal size of the side-grooved three point bend test specimens employed is: height W=20 mm, width B=10 mm, net width B_N=7,5 mm. The main notch is of 45° opening angle, the depth is 3 (or 4) mm. The notch radius was kept constant, r = 0,25 mm. The side groove was of the Charpy V notch type, with a depth of 1,25 mm. The steel used in this work represents a Si and Al killed weldable normalized fine grain type with the designation TStE 355 according to DIN 17102. This steel was also investigated in conjunction with the martensitic stainless steel X5 CrNi 13 4 for heterogenous welded joints [40].

Chemical composition was C=0,16; Si=0,41; Mn=1,3; P=0,014 and S=0,016 mass percent. The specimens were machined in two layers taken from 30 mm thick plate, the notch being perpendicular to the surface. The machined specimens were 5 mm deep grooved; the groove was subsequently filled by crack starter weld metal FOX DUR 500 with minimum dilution.

Testing was done in a device for measuring crack mouth opening displacement, CMOD. The device was placed into the frame of a 250 kN servomechanic testing machine Schenck-Trebel. The anvil was fixed onto the machine bed, whereas the loading rod was attached to the crosshead. The device was inside a cooling chamber Cryoson. The loading rate was 2 mm/min.

The measuring system is shown in Fig.8: The load was measured by strain gauges attached to the loading rod, the loading point displacement by inductive means, the CMOD by a clip gauge of rather high stiffness.

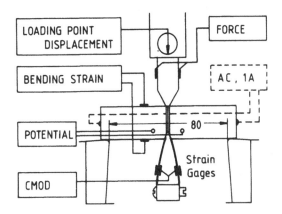

Figure 8. Three point bend specimen and scheme of measuring equipment for arrest testing.

Further the potential during testing was mesured between the end faces of the specimens. Due to the magnetic polarity of the specimens a transient change in potential could be observed during loading and especially cracking. This was indicated by potential recording between contact points fixed on the specimen sides in a distance of 5 mm besides the tip of the notch. Alternatively, a crack length measuring device Matelec CGM 5 was used, as indicated in Fig.8. The measuring points were attached to the specimen side, in a distance of 10 mm from the notch midline and 5 mm from the edge of the specimen.

For registration a 14-channel magnetic tape recorder AMPEX PR 2230, band width 80 kilo-cycles per second (Wideband Group I acc. to IRIG) at 120 ips was used. By playing back at 1/64 of the above velocity it was possible to reproduce the signals of cracking and crack arrest using a high velocity graphic recorder Gould ES 1000.

All amplifiers were of the 100 kilo-cycles per second measuring capacity (-3 dB). Therefore even high frequency signals could be evaluated.

Fig.9 shows for specimen 28 the CMOD, the load point displacement (LPD) and the force over time. Three crack jumps can be distinguished best regarding the force history. The fracture surface of this specimen can be seen in Fig.10. The first cleavage jump ended at the transition between crack starter weld and base metal. The jumps end at narrow deformation bands, characterized by fibrous, greyish appearance. Testing was discontinued after the last crack jump. Therefore no ductile failure band is visible.

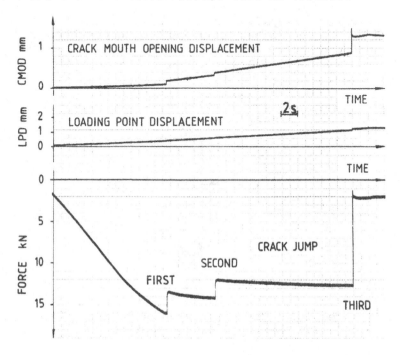

Figure 9. Crack mouth opening displacement, loading point displacement and force over time, specimen 28.

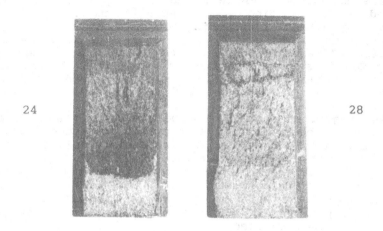

Figure 10. Fracture surfaces of the specimens 24 and 28.

After testing the specimen was chilled in liquid nitrogen and broken up. The last fracture surface is more flat than the foregoing; the length of the last jump can therefore be measured easily. No heat tinting was used in case of multiple crack jumps because the differences between cleavage and ductile fracture surface became less pronounced, the only well distinguishable being the last arrest. Heat tinting was used, however, where only a single crack jump occured (see Fig.10, specimen 24).

As an example, force over time during crack jump is shown for the specimens 24, 151, 23 and 19, see Fig.11, using an enlarged time scale. In the lower part of the figure, CMOD over time for the specimens 24 and 23 can be seen. Especially strong oscillations are visible in graphs belonging to specimens exhibiting large crack jumps. Small jumps, however, correspond to small oscillations.

The evaluation was based on the static crack arrest concept [41]. This concept assumes, that all dynamic effects which start at the crack tip and are reflected from the edges of the specimen do not influence the running crack. To calculate the crack arrest toughness K_{Ia}, the force P_a, measured shortly after arrest and the crack length a_a, are used.

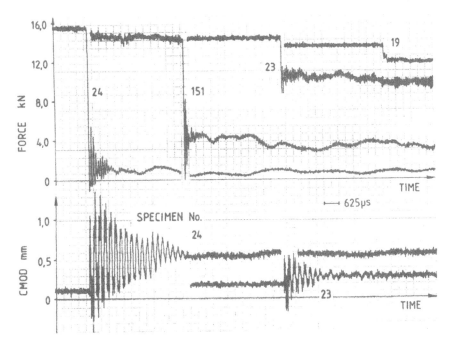

Figure 11. Force and crack mouth opening displacement of specimens 24, 151, 23 and 19 in function of time.

Starting from the stress intensity factor which is

developing at the tip of a crack of the length 2a situated in a large plate, K_{Ia} may be calculated by the folllowing equation:

$$K_{Ia} = \sigma \cdot \sqrt{\pi \cdot a_a} \cdot f_{(x)} = \frac{3 \cdot P_a \cdot S}{2 \cdot B_N \cdot W^2} \cdot \sqrt{\pi \cdot a_a} \cdot f_{(x)}$$

For the ratio span S to height W of the specimen which amounts to 4 for the actual geometry, the geometry function $f_{(x)}$ becomes:

$$f_{(x)} = \frac{1,99 - x \cdot (1 - x) \cdot (2,15 - 3,93x + 2,7x^2)}{\sqrt{\pi} \cdot (1 + 2x) \cdot \sqrt{(1 - x)^3}}$$

with error limits of ±0,5% for $x = a_a/W \le 1,0$ [42]. a_a represents the mean length of the crack according to ASTM E 399 i.e. the mean value from those values measured at $B_N/4$, $B_N/2$ and $3B_N/4$ (1/4, 1/2 and 3/4 of the net section width). The K_{Ia}-values, as calculated by using the above mentioned formulae, are depicted in Fig.12. K_{Ia}-values for short cracks are included, too. K_{Ia} seems to be dependent from the relative crack length.

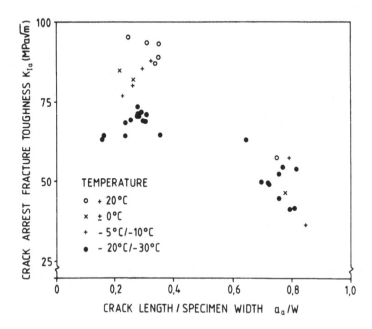

Figure 12. Crack arrest toughness (K_{Ia}) values based on arrest force and crack length measurements on the fine grain weldable steel TStE 355 in function of crack length

DISCUSSION AND CONCLUSIONS

If crack initiation and propagation can not be excluded, than crack arrest has to be proven to demonstrate safety against fracture. Crack arrest may be achieved by design and by applying appropriate, tough steel locally or by the combination of both.

As an additional test to measure Crack Arrest Toughness K_{Ia}, three point bend specimens with a brittle crack starter of $10 \times 20 \times 92$ mm^3, seem to be suitable. As examples K_{Ia} values measured on a Si and Al killed weldable fine grain steel StE 355 were presented. The K_{Ia}-values are decreasing with increasing a/w ratio. This effect, however, may be also depending on the steel examined: former investigations on steel A 533 grade B cl.1 did not show this ratio dependence [39].

Because no dynamic effects are taken into account, same limitations than for ASTM E 1221 apply.

REFERENCES

1. Boyd, G.M.: Fracture Design Practices for Ship Structures. In: Fracture, An Advanced Treatise, Vol. 5, Fracture Design of Structures, ed. H. Liebowitz. Academic Press, New York and London, 1969, pp. 383-470.

2. Kanninen, M., Mills, E., Hahn, G., Marschall, C., Broek, D., Coyle, A., Masubushi, K., Itoga, K.: Final Technical Report on Project SR-226, Hull Crack Arrester Systems: A Study of Ship Hull Crack Arrester Systems. Ship Structure Committee, SSC-265 (1977).

3. Lloyds Register of Shipping: Extracts from the Rules for the Construction and Classification of Steel Ships. Metric Edition No.4, Chapter P, Section 2, pp. 452-457. London, 1967.

4. Final Report of a Board of Investigation: The Design and Methods of Construction of Welded Steel Merchant Vessels, 15 July 1946. Government Printing Office, Washington, 1947.

5. Sumi, Y., Ohashi, K,. Emura, H.: On Crack Arrestability by a Circular Hole based on Computational Crack Path Prediction. Naval Arch. and Ocean Engng. 25, 1989, pp. 173-180.

6. Isida, M.: On the determination of stress intensity factors for some common structural problems. Eng. Fract. Mech. 2, 1970, pp. 61-79.

7. Broek, D.: Elementary Engineering Fracture Mechanics. Sijthoff & Noordhoff, Alphen aan den Rijn, The Netherlands, 1978.

8. Yoshiki, M., Kanazawa, T.: Studies on the Brittle Fracture Problems in Japan. 60th Anniv. Ser., Vol.13. Soc. Naval Arch. Japan, Tokyo, 1967.

9. Bluhm, J.I.: Fracture Arrest. In: Fracture, An Advanced Treatise, Vol.5, Fracture Design of Structures, ed. H. Liebowitz. Academic Press, New York and London, 1968, pp. 1-63.

10. Chapel, C.E., Bent, R.D.: Aircraft Basic Science. McGraw-Hill Book Comp., Inc., New York, Toronto, London, 1953, Sections V and VII.

11. Neville, L.E.: Aircraft Designer's Data Book. McGraw-Hill Book Company, Inc., New York, Toronto, London, 1950.

12. Harpur, N.F.: Fail-Safe Structural Design. J. Royal Aeronautical Society 62, 1958, May, pp. 363-376.

13. Leybold, H.A.: Residual static strength of aluminium alloy box beams containing fatigue cracks in the tension covers. NASA TN-D-796, April 1961.

14. Hertel, H: Ermüdungsfestigkeit der Konstruktionen. Springer-Verlag, Berlin-Heidelberg-New York, 1969.

15. Hardrath, H.F., Leybold, H.A., Landers, C.B., Hauschild, L.W.: Fatigue-Crack Propagation in Aluminium-Alloy Box Beams. National Advisory Committee for Aeronautics, Techn. Note 3856, Washington, August 1956.

16. De Rijk, P.: Empirical investigation on some methods for stopping the growth of fatigue cracks. Nat. Aerospace Inst. Amsterdam, Rept. TR 70021, 1970.

17. Van Leeuwen, H.P. et al.: The repair of fatigue cracks in low-alloy steel sheet. Nat. Aerospace Inst. Amsterdam, Rept. TR 70029, 1970.

18. Eggwirtz, S.: Review of some Swedish investigations on fatigue during the period 1967-1969. Swedish Aerospace Inst. FFA Rept. TN-HE-1270, 1969.

19. Paton, B.E., Biletzkiy, S.M., Barvinko, Y.P., Pivovarskiy, N.B.: Mechanism of Limiting of the Ductile Fracture Propagation in the Wall of the Rolled Quasi-Laminated Crack Arrestor. Int. Seminar on Application of Rolled Quasi-Laminated Crack Arrestors in Gas Transmission Pipelines, October 22-23, 1985, Kiev, USSR. E.O. Paton Electric Welding Institute of the Ukrainian S.S.R. Academy of Sciences, Kiev, 1985, pp. 7-12.

20. Anon.: Rupture Will Not Delay Transwestern's Line. Oil and Gas J. 58, 1960, No.17 (April 25), p.105.

21. Stránský, J.: Brittle Fractures in Piping. DOC. Commission IIW-XI-202-68. Welding Research Institut, Bratislava, 1968.

22. Fearnehough, G.D.: Fracture propagation in gas pipelines - Relevance to submarine lines. 3R international 15, 1976, H.9, S. 509-515.

23. Fearnehough, G.D., Jude, D.W., Weiner, R.T.: The Arrest of Brittle Fracture in Pipelines. Conference on Practical Application of Fracture Mechanics to Pressure Vessel Technology. London, May 1971. Proc. Instn. Mech. Engrs. C42/71 pp. 156-162, London, 1971.

24. Poynton, W.A., Shannon, R.W.E., Fearnehough, G.D.: The Design and Aplication of Shear Fracture Propagation Studies. Trans. ASME, Ser. H: J. of Engineering Materials and Technology 96, 1974, No.4 (October), pp. 323-329.

25. Coors, P.Ph.C., Fearnehough, G.D., Koch, F.O., Kügler, J., Venzi, S., Vogt, G.H.: Sicherheit gegen Rißausbreitung in Großrohren - eine europäische Gemeinschaftsuntersuchung. 3R international 18, 1979, H.6, S. 380-386.

26. Priest, A.H., Holmes, B.: The Characterisation of Crack Arrest Toughness in Gas Transmission Pipelines in Terms of Shear Fracture Propagation Energy, Final Report. Commmission of the European Communities, Report Eur 11176 EN, 1988.

27. Standard Method for Drop-Weight Tear Tests of Ferritic Steels. ASTM Designation: E 436-74 (Reapproved 1986).

28. Shoemaker, A.K., McCartney, R.F., Ives, K.D.: Mechanical Crack-Arrestors Concepts for Line-Pipe Applications. Conference on Materials Engineering in the Arctic, St. Jovite, Quebec, 1976, pp. 298-305.

29. Ives, K.D., McCartney, R.F., Shoemaker, A.K.: Method of Arresting Crack Propagation in Line Pipe Characterized by Ductile Fracture. U.S. Patent 4.195.669, April 1, 1980.

30. Loncaric, R.G.: Crack Arrester System. U.S. Patent 3.870.350, March 11, 1975.

31. Ives, K.D., Shoemaker, A.K., McCartney, R.F.: Pipe Deformation During a Running Shear Fracture in Line Pipe. Trans. ASME, Ser.H: J. of Engineering Materials and Technology 96, 1974, No.4 (October), pp. 309-317.

32. Fearnehough, G.D.:Fracture propagation in gas pipelines - Relevance to submarine lines. 3R international 15, 1976, H.9, pp. 509-515.

33. Wiedenhoff, W.W., Vogt, G.: Fracture behaviour of line-pipe - Recent developments. 3R international 26, 1987, H.8, pp. 522-527.

34. Shoemaker, A.K., McCartney, R.F.: Displacement Consideration for a Ductile Propagating Fracture in Line Pipe. Trans. ASME, Ser.H: J. of Engineering Materials and Technology 96, 1974, No.4 (October), pp. 319-322.

35. Wiedenhoff, W., Gärtner, A.W., Vogt, G., Weisgerber, C.: Stahlrohr als Rißstopper für Gasleitung. Patentschrift DE 3.437.564 C2, 19.2.1987.

36. Paton, B.E., Medovar, B.I., Saenko, V.Y., US, I., Krendeleva, A.I., Pivovarskiy, N.B., Astafyev, N.A.: Rolled Quasi-Laminated Crack Arrestors of Gass Transmission Pipelines. Int. Seminar on Application of Rolled Quasi-Laminated Crack Arrestors in Gas Transmission Pipelines, October 22-23, 1985, Kiev, USSR. E.O. Paton Electric Welding Institute of the Ukrainian S.S.R. Academy of Sciences, Kiev, 1985, pp. 14-26.

37. Neuhauser, E.: Die Ursache der Rohrbrüche im Druckschacht des Gerloskraftwerkes. Österr. Wasserwirtschaft 2, 1950, H.8/9, S. 185-211.

38. Konkoly, T., Straube, H., Varga, T.: Investigations on MAG Weld Metal for Critical Valuation of Fracture Mechanics Properties. Int. Conf. on Fracture, New Delhi, Nov. 1984. Proc. pp. 1137-1143.

39. Schneeweiß, G., Varga, T.: Erfahrungen mit Dreipunkt-Biegeproben zur Ermittlung der Rißauffangzähigkeit K_{Ia} an niedriglegiertem Stahl. Swiss Materials 1, 1989, Nr.6, S. 29-37.

40. Varga, T., Straube, H., Strigl, M.: Investigations on Ductility and Fracture of Compound Steel Weldments. Proc. 7th European Conference on Fracture. Failure Analysis - Theory and Practice. Ed. E. Czoboly, Budapest 1988, Vol. II, pp. 821-837.

41. Ripling, E.J., Crosley, P.B., Wiersma, S.J.: A review of static crack arrest concepts. Eng. Fract. Mech. 23, 1986, No.1, pp. 21-33.

42. Srawley, J.E.: Wide Range Stress Intensity Factor Expressions for ASTM E 399 Standard Fracture Toughness Specimens. Int. J. Fract. Mech. 12, 1976, No.6, pp.475-476.

INFLUENCE OF GRAIN SIZE ON TRIAXIAL DYNAMIC BEHAVIOR OF ALUMINA

Jacques CAGNOUX and Antonio COSCULLUELA
Centre d'Etudes de Gramat
46500 GRAMAT, FRANCE

ABSTRACT

The aim of this paper is a preliminary synthesis of the triaxial dynamic behavior of alumina. Stress-wave experiments were selected to avoid the effect of the surface flaws on the response of the material. Plate impacts, bar impacts and detonation of sphere-shaped H.E. were performed to explore the whole space of stresses. The phenomenological analysis shows two types of failure : i) a ductile failure and ii) a brittle failure. Ductile failure is due to movements of dislocations and leads to a compaction of the alumina, without damaging the fine-grained material. Brittle failure is due to coalescence of in-material cracks. The intrinsic curve of failure states is positioned in the space of stresses for fine-grained and for coarse-grained materials respectively.

INTRODUCTION

The behavior of ceramics under shock loading has become a topic of great importance, particularly due to the improvement of heavy armor. Heavy ceramic armor employs tiles that are typically several times thicker than the impactor diameter. The velocity of the impactor is about 1500 m/s for kinetic-energy projectiles and about 8000 m/s for shaped-charge jets. The first stage of impact is a shock phase. The deformation associated with shock loading is one-dimensional strain. Shock stresses (Hugoniot stresses) in the ceramic are higher than the Hugoniot elastic limit (HEL). Some microstructural modifications and some damage may be induced in the ceramic by these Hugoniot stresses. After the initial shock passage, tensile stresses (hoop stresses) develop in the ceramic due to the diverging of the flow behind the shock wave. Damaging of the ceramic is also induced by these tensile stresses ahead of the projectile. The problems we face are the following :

- The description of the modifications and of the damage induced in the unpenetrated ceramic by the passage of the stress-wave.

- The micromechanical interpretation of these modifications and damage.

These problems involve a phenomenological analysis of the ceramic behavior in the whole space of stresses. The aim of this paper is a preliminary synthesis of the triaxial dynamic behavior of alumina, the material used in our research. The intrinsic response of alumina is studied using stress-wave experiments, to avoid the effect of the surface flaws. Our experimental strategy is described in the next section. Then, the behavior are described, for a fine-grained alumina and for a coarse-grained alumina, respectively. The final section is a discussion of the influence of grain size on triaxial dynamic behavior of alumina.

EXPERIMENTAL STRATEGY

Our strategy was as follows :

- To identify the behaviors of the material in interpolating results of experiments involving simple deformation processes. These experiments were impacts performed with the Demeter gas gun at the Centre d'Etudes de Gramat [1].

- To validate these behaviors by performing experiments to induce the propagation of a spherical dilatational wave. The experiments consisted in the detonation of a sphere-shaped H.E. in contact with ceramic [2].

The identification experiments were :

- Plate-impact experiments, inducing the propagation of a dilatational wave into the alumina. The strain is one-dimensional, and the strain rate is a few 10^5/s. These experiments were used to obtain results in tri-compression and tri-tension (spall experiments).

- Experiments involving impacts of a plate onto an alumina long rod, inducing the propagation of a longitudinal wave into the rod. The stress is uniaxial [3] and the strain rate is a few 10^3/s. These experiments were used to obtain results in uniaxial stress (compression and tension).

For each type of experiments, including the validation experiments, the Hugoniot stress or the particle-velocity history was measured using piezo-resistive gages or a VISAR-velocity interferometer. Furthermore, in order to identify the micromechanisms, soft-recovery experiments were performed using samples and momentum traps designed to avoid the influence of lateral unloading waves [4, 5].

PHENOMENOLOGICAL ANALYSIS OF THE DYNAMIC BEHAVIOR
OF A FINE-GRAINED ALUMINA

Material

We used alumina T 299, manufactured by CICE[*]. The grain size is 2-7 μm. The purity is 99,7% Al_2O_3. The density is 3.86 g/cm³ for long rods, 3.91 g/cm³ for plates, and 3.86 g/cm³ for samples tested in spherical geometry.

---

[*] CICE - 63, Rue Beaumarchais - B.P. N° 69 - 93104 MONTREUIL CEDEX FRANCE

Identification of the behavior in bi-tension-compression

This state of stress is defined by $\sigma_2 = \sigma_3$ as tensile stresses and σ_1 as compressive stress. The behavior of alumina for this state of stress was interpolated between results obtained in uniaxial compression and those obtained in uniaxial tension [3].

Optical observations of the soft-recovered rods reveal two networks of cracks (Figure 1). The first one consists of in-material radial cracks oriented parallel to the direction of the compression (Figure 1b). These cracks result from the uniaxial compressive stress induced by the propagation of the longitudinal wave in the long rod, and occur for a uniaxial stress of 2 GPa and more. They probably originate from local tensile stresses due to the heterogeneity of the material. The failure in uniaxial compression results from the coalescence of these in-material cracks. The second network of cracks consists of in-material transversal cracks (Figure 1c), normal to the axis of the rod. These cracks result from the tensile uniaxial stresses induced by the reflection of the longitudinal wave on the extremity of the rod. Their nucleation threshold is close to the tensile strength of the rod [6].

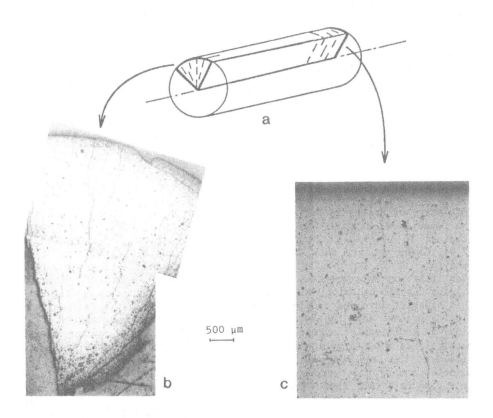

Figure 1. In-material cracks (b) and transversal cracks (c) in soft-recovered long rods of alumina T299.

No microplastic activity was detected by TEM into the soft-recovered rods.

Both types of cracks exhibit essentially transgranular fracture. In the case of radial cracks, however it is intergranular in places where the grain size is lower than 3 μm.

On Figure 2, we have attempted an preliminary synthesis of results previously described. Study of the damage micromechanisms leads to the definition of an intrinsic curve of nucleation thresholds. This is shown on Figure 2. As the above discussion shows, its position is not yet precisely defined. The brittle failure of the material results from the coalescence of in-material cracks. The intrinsic curve of failure states has been drawn on Figure 2, interpolating ultimate strengths measured in uniaxial compression and in uniaxial tension.

Identification of the behavior in tri-tension

This state of stress is defined by $\sigma_2 = \sigma_3$ and σ_1 as tensile stresses. The behavior of alumina for this state of stress was interpolated between results obtained in uniaxial tensile stress (impacts on long rod), and those obtained in tri-tension (spall experiments), though the strain rates of these two deformation processes are different.

Optical and SEM observations show that spall failure results from the coalescence of in-material cracks, about 60°-inclined to the direction of the maximum principal tensile stress σ_1 (Figure 3). These cracks are essentially transgranular but in places intergranular, and numerous cleavage steps appear. This type of fracture does not differ from the one described for the bi-tension-compression state of stress. In tri-tension, the damage consists of cracks with their plane depending on the state of stress. Figure 2 shows the intrinsic curve of failure states in tri-tension drawn with an uniaxial tensile strength of 0.45 GPa ± 0.05 GPa and a spall strength of 0.4 GPa ± 0.07 GPa.

Identification in tri-compression

This state of stress is defined by $\sigma_2 = \sigma_3$ and σ_1 as compressive stresses. The behavior of alumina for this state of stress was deduced from results obtained in uniaxial stress [3] and those obtained in uniaxial strain [7,8]. These results show a brittle-ductile transition ; i) for brittle behavior, it is not possible to find the crack orientation by means of identification experiments. Data can be obtained by spherical geometry ; ii) for ductile behavior, the material densifies without damage (Figure 4). This behavior involves closing of pores due to microplastic deformation of alumina grains. When the material achieves the theoretical density of alumina, an additional compression provokes the plastic flow of the densified material, as a continuation of the microplastic deformation of grains initiated at the beginning of the compaction. The relative positions of the Hugoniot and of the hydrostatic curves of Lucalox (taken as the reference densified material) show a longitudinal deviatoric stress of 5 GPa, constant for stresses higher than the HEL [9]. Furthermore, Howitt and Kelsey [10] showed that there are sufficient slip systems to satisfy the Von Misès criterion during plastic deformation of alumina. These results are reported on Figure 5.

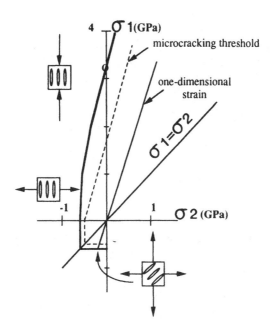

Figure 2. Positions of the curve of nucleation threshold and of the curve of failure states, for alumina T299 in bi-tension-compression.

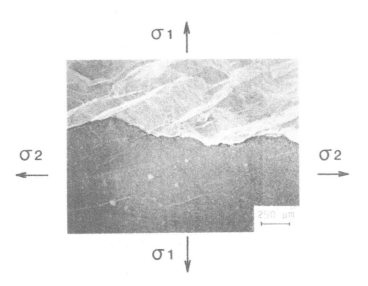

Figure 3. Spalling of alumina T299. Photograph showing in-material cracks 60°-inclined to the direction of the maximum principal tensile stress σ_1 ($\sigma_2/\sigma_1 \simeq 0.3$).

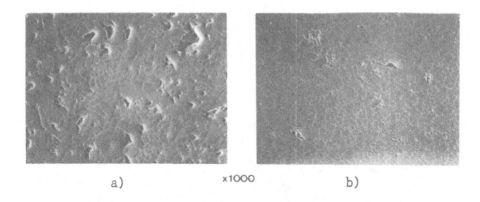

a)   ×1000   b)

Figure 4. SEM photographs of alumina T299 - a) initial material ; b) after 1.5 x HEL loading.

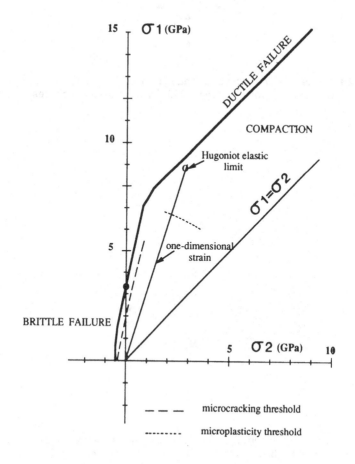

Figure 5. Positions of the curve of nucleation threshold and of the curve of failure states, for alumina T299 in tri-compression.

In uniaxial strain,TEM observations show a microplastic activity which increases with the Hugoniot stress, from a microplastic threshold lower than the HEL [8]. The HEL (8.5 GPa for the alumina T299) is a threshold from which the coupling between microplastic deformation of grains and the microscopic behavior of the material becomes effective. On Figure 5, the uniaxial strain path shows that the compaction of the alumina T299 is achieved at a Hugoniot stress of about 8.8 GPa. Alumina T299 is then quasi densified at its HEL. This result, shown on Figure 5, is in agreement with the density 3.95 g/cm³ measured by Longy and Cagnoux [8] on a 7.4 GPa shocked sample.

Validation in spherical geometry

The validation was performed using Cagnoux's results [5]. Observations of soft-recovered samples confirm the two basic phenomena defined above : i) in a zone near the explosive, the alumina is not fragmented, and exhibits a higher density than initially. A numerical simulation of this experiment, using a one-dimensional lagrangian code [11] showed that this zone was submitted to a one-dimensional strain loading with a longitudinal stress higher than the HEL, and the unloading was effected with a tri-compression state of stress ; ii) beyond this unfragmented zone, the alumina is fragmented. The numerical simulation showed that this zone was submitted to a tri-compression state of stress during the loading, and to a bi-tension-compression state of stress during the unloading. In this zone, cracks are normal to the direction of principal tensile stresses (hoop stresses) which develop during the diverging of the spherical flow. These cracks are transgranular and exhibit numerous cleavage steps. Observations show that the brittle-ductile transition is close to the σ_1 axis on Figure 5. It may be interpreted as the threshold of confinement necessary and sufficient to inhibit the local tensile stresses which initiate cracks under a macroscopic compressive state of stress.

PHENOMENOLOGICAL ANALYSIS OF THE DYNAMIC BEHAVIOR
OF A COARSE-GRAINED ALUMINA

Material

We used alumina AL23, manufactured by DEGUSSA[*]. The grain size is 20-70 µm. The purity is 99.7% AL_2O_3. The density is 3.92 g/cm³ for long rods, 3.82 g/cm³ for plates, and 3.85 g/cm³ for samples tested in spherical geometry.

Identification of the behavior in bi-tension-compression

The observations of soft-recovered samples lead to the same description of damage in the alumina AL23 as that described for alumina T299.

---

[*]DEGUSSA - 37, Avenue Marceau - 92400 COURBEVOIE, FRANCE.

Identification of the behavior in tri-tension

In uniaxial tensile stress, damage has been identified as in-material transgranular cracks.

In uniaxial strain (spalling), SEM observations show that spall fracture results from the coalescence of in-material cracks, normal to the direction of the maximum principal tensile stress σ_1. These cracks are intergranular, and there is in places a process zone of transgranular damage of one grain size thick.

In tri-tension, there is a transition in the damage mechanisms. However, in both cases, cracks are normal to the maximum principal tensile stress.

Identification of the behavior in tri-compression

In uniaxial compressive stress, the fracture of alumina AL23 results from the coalescence of in-material transgranular cracks, parallel to the direction of compression.

In uniaxial strain, in-material intergranular cracks appear from 0.8 times the HEL (Figure 6). The orientation of these cracks has not been measured because they are randomly oriented. This damage is not sufficient to reduce the spall strength (spall strength of AL23 is 0.5 GPa ± 0.07 GPa after a Hugoniot stress of 1.5 times the HEL).

In tri-compression, there is a transition in the damage mechanisms. According to the analysis of the behavior of the alumina T299, we suggest the location of the threshold of the two types of damage as shown on Figure 7. The HEL of alumina AL 23 (6.3 GPa) appears as the result of a sudden increasing of the microplastic activity. The ductile behavior of alumina AL23 is the same as the one described for alumina T299. However, for this material the respective locations of the failure curve and the uniaxial-strain path show that the compaction is not achieved at the HEL. The cracked nature of the recovered AL23 sample prevents any verification of this deduction.

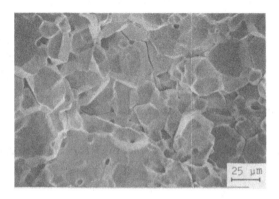

Figure 6. Intergranular cracks in alumina AL23 after 0.8 x HEL one-dimensional strain loading.

Validation in spherical geometry

This validation was performed using Cagnoux's results [5]. Observations of soft-recovered samples show : i) a microfragmented zone near the explosive. Cracks are intergranular and numerous twins are seen. The intergranular nature of cracks in this zone accords with observations on uniaxial-strained material. These results show that cracks initiated during shock loading propagate and coalesce during divergent unloading. Twins appear during unloading, but it seems an independent phenomenon if we consider the intergranular nature of the fracture. In fact these twins show the type of microplastic activity developed between the uniaxial-strain path and the brittle failure curve. On Figure 7, the brittle failure occuring in the upper region of the tri-compression results from the coalescence of intergranular cracks ; ii) a macrofragmented zone, distant from the explosive. It is at an identical distance from the explosive as the same zone in the T299 sample. Transgranular cracks appear, with the same description as those of alumina T299. On Figure 7, the brittle failure occuring in bi-traction-compression and for weak confinement results from the coalescence of these transgranular cracks. For alumina AL23 there is also a threshold of confinement necessary and sufficient to inhibit the local tensile stresses which initiate transgranular in-material cracks.

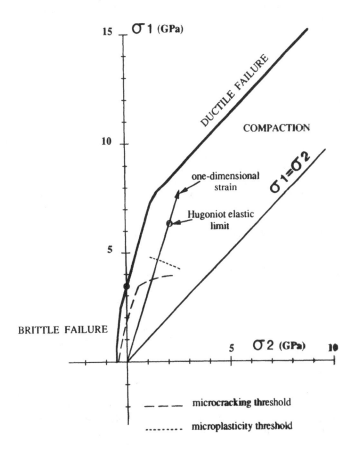

Figure 7. Positions of the curve of nucleation threshold and of the curve of failure states, for alumina AL23 in tri-compression.

DISCUSSION AND CONCLUSION

Bi-traction-compression

In the two studied aluminas, the damage consists of transgranular in-material cracks. The cracks propagate normally to the direction of the maximum principal tensile stresses.

In uniaxial compression, nucleation of cracks may result from local-tensile stresses due to the heterogeneity of the material. Then cracks propagate parallel to the direction of the compression (Figure 1b). Furthermore, comparing quasi static and dynamic ($10^3 s^{-1}$) data, Brar [12] did not observe any influence of strain rate in the uniaxial compressive strength of alumina.

Tri-traction

In the fine-grained alumina T299, the damage results from the propagation of transgranular cracks as a continuation of the bi-traction-compression state of stress. The orientation of cracks depends on the two tensile stresses (Figure 3). In the coarse-grained alumina AL23, the damage is due to two mechanisms : i) propagation of transgranular cracks for uniaxial tensile stress, and ii) propagation of intergranular cracks for the uniaxial strain. The orientation of cracks remains normal to the maximum principal tensile stress.

Tri-compression

For fine-grained alumina, damage resulting from the propagation of transgranular cracks is confined to loading paths close to the uniaxial-stress loading path. Weak confinement inhibits this damage which is probably nucleated by local tensile stresses. There is a brittle-ductile transition. The plastic flow of the densified alumina is preceded by a compaction of the porous material resulting from microplastic activity (slips) inside the grains. The increase of grain size has two consequences :

- The initiation of intergranular cracks (Figure 6). This occurs jointly with the microplastic activity. TEM observations show that these two phenomena are not linked. Intergranular cracks may be due to the conjunction of two factors : i) residual stresses, significant for grain size larger than 20 μm [13] and ii) a weakening of grain joints due to a relatively larger thickness of the glassy phase in coarse-grained alumina.

- A reduction of the microplasticity threshold. In particular the HEL is related to the grain size, as it is shown in the Petch-diagram of Figure 8. The HEL is not rate-dependent [14].

These results imply that the shock loading of a divergent-stress wave does not provoke damage in pure fine-grained alumina, and provokes intergranular damage in pure coarse-grained alumina. During the unloading of the divergent-stress wave, these intergranular cracks coalesce, leading to the failure of the coarse-grained material. In fine-grained alumina, the failure takes place in bi-tension-compression by coalescence of transgranular cracks nucleated by local tensile stresses.

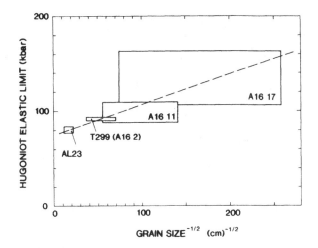

Figure 8. Petch diagram of Hugoniot elastic limits of pure dense alumina. Error bands are due to the distribution of grain size, and to the extrapolation of zero porosity respectively.

REFERENCES

1. Chartagnac, P. and Jimenez, B., Description of the Demeter Gas Gun and of the Associated Data Acquisition Channels (in Fr.). Rev. Phys. Appl. (France), 1984, 19, pp. 609-18.

2. Cagnoux, J., Déformation et ruine d'un verre pyrex soumis à un choc intense : étude expérimentale et modélisation du comportement, Thèse de Doctorat d'Etat, UER-ENSMA, Poitiers, 1985.

3. Cosculluela, A., Cagnoux, J. and Collombet, F., Compression uniaxiale de l'alumine : influence de la vitesse de déformation et de la taille des grains. Accepted for publication in the Proceeding of 1991 DYMAT Conference, Strasbourg, France. Edited by les Editions de Physique, France, 1991.

4. Longy, F. and Cagnoux, J., Plate Impact Recovery Experiments of Ceramics. In Proceeding of the American Physical Society Topical Conference, Shock Compression of Condensed Matter, Albuquerque, 1989. Edited by S.C. Schmidt, J.N. Johnson and L.W. Davison, North Holland, 1990, pp. 441-444.

5. Cagnoux, J., Spherical waves in pure alumina ; effects of grain size on flow and fracture. In Proceeding of the American Physical Society Topical Conference, Shock Compression of Condensed Matter, Albuquerque, 1989. Edited by S.C. Schmidt, J.N. Johnson and L.W. Davison, North Holland 1990, pp. 445-448.

6. Deobald, L.R., Taya, M., Kobayashi, A.S. and Yoon, H.S., Spall Resistance of Alumina. Technical Report N° UWA/DME/TR.89/1, Department of Mechanical Engineering University of Washington, by the Office of Naval Research Contract n° 00014-87-K-0326, 1989.

7. Longy, F. and Cagnoux, J., Macro and Micromechanical Aspects of Shock Loading of Alumina. In Proceeding of the International Conference on Impact Loading and Dynamic Behavior of Materials, Bremen, FRG, 1987. Edited by C.Y. Chiem, H.D. Kunze and L.W. Meyer, DGM Informationsgeselleshaft mbH, Oberursels, FRG, 1988, pp. 1001-1008.

8. Longy, F. and Cagnoux, J., Plasticity and Microcracking in Shock-loaded Alumina. J. Am. Ceram. Soc., 1989, 72, pp. 971-79.

9. Ahrens, T.J., Gust, W.H., Boyce E.B., J. Appl. Phys., 1968, 39, p. 4610.

10. Howitt, D.G. and Kelsey, P.V.,Dislocation Microstructure of Shock-loaded Alumina. In Proceeding of the International Conference on Impact Loading and Dynamic Behavior of Materials, Bremen, 1987. Edited by C.Y. Chiem, H.D. Kunze and L.W. Meyer, DGM Informationsgesellshaft mbH, Oberursels, FRG, 1988, pp. 249-56.

11. Cagnoux, J., non published work.

12. Brar, N.S., Bless, S.J. and Rosenberg, Z., Brittle Failure of ceramic rods under Dynamic Compression. In Proceeding of the 1988 DYMAT Conference, Ajaccio, France. Edited by les Editions de Physique, France, 1988, pp. 607-612.

13. Evans, A.G. and Fu, Y., The Mechanical Behavior of alumina a model anisotropic brittle Solid. In Fracture in Ceramic Materials, Edited by A.G. Evans, Noyes Publications, USA, 1984.

14. Cagnoux, J. and Longy, F., Is the Dynamic Strength of Alumina Rate-Dependent ? In Proceeding of the American Physical Society Topical Conference, Shock Waves in Condensed Matter, Monterey, 1987. Edited by S.C. Schmidt and N.C. Holmes, North Holland, 1989, pp. 293-96.

HIGH SPEED DOUBLE TORSION TESTING OF PIPE GRADE POLYETHYLENES:- EXPERIMENT AND ANALYSIS

M.A. WHEEL & P.S. LEEVERS
Department of Mechanical Engineering
Imperial College
Exhibition Road, London, SW7 2BX, UK

ABSTRACT

Rapid Crack Propagation (RCP) has been induced in two pipe grade polyethylene (PE) materials using an impact version of the conventional double torsion test. Striker speed, crack velocity and support load were measured during each test. A subsequent linear elastic analysis of each experiment produced scattered dynamic fracture resistance, G_D, values. A torsional impedance method was used to measure shear modulus values appropriate to the fracture test. The data obtained using this method implied that the materials exhibit nonlinear elasticity. A nonlinear analysis of the fracture test data reduced the scatter in range of G_D values. A clear difference between the resistance of the two materials to RCP is now seen.

1. INTRODUCTION

Rapid Crack Propagation (RCP) has been identified as an unlikely but potentially catastrophic failure mode in polyethylene (PE) gas and water pipeline systems. Full scale field tests conducted by British Gas [1] have shown that once a high speed fracture, with velocity å of up to 350m/s, has been initiated in a section of pipe, it has the potential to propagate for a considerable distance, perhaps hundreds of meters. The section of pipe lying ahead of the crack tip only partially depressurises, so energy is available within the pipe wall to sustain the crack growth. The field tests have also shown that there is a well defined critical pipeline pressure, p_C, below which RCP cannot be maintained and crack arrest occurs shortly after initiation.

Therefore, these full scale tests are used to establish safe operating conditions for PE pipes, but they are extremely expensive to perform. Yayla and Leevers [2] have developed an economical small scale experiment that tests short sections of pipe and also identifies p_C values. While the information obtained from this test is extremely valuable to the pipeline producer and the gas and water utilities, the material manufacturers are interested in comparing the performance of candidate PE materials before committing themselves to the expense of extruding pipe sections.

In order to serve the requirements of the manufacturers the High Speed Double Torsion (HSDT) Test has been developed to induce RCP in small test samples of PE and measure the resulting crack velocities. A post-mortem analysis of the test determines the dynamic fracture resistance, G_D, of each candidate PE material tested, which then allows the performance of the various grades to be compared.

2. THE HIGH SPEED DOUBLE TORSION TEST:- EXPERIMENTAL DETAILS

The test sample consists of a 6 or 9mm thick plate measuring 100×200mm, which is supported at four points in the test rig, as shown in Figure 1. A 1mm deep 'V' groove is machined on the underside of the sample along the major axis and then a prenotch of length a_0=40mm is machined part way along this axis. A 1.5mm deep razor blade slit is scored along the axis on the upper surface of the specimen. The sample is impacted normally close to the notched end, which causes the material lying either side of the notch to be deformed as two rectangular

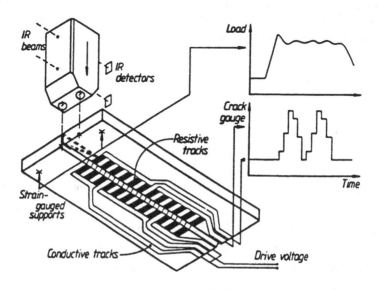

Fig. 1 The High Speed Double Torsion Test

sectioned torsion beams rotating in opposite senses. The resulting torsional disturbance travels along each half of the specimen and loads the notch tip, causing a crack to initiate at some time t_f after the impact event. This crack then propagates axially at speeds of up to 240m/s in 6mm thick samples. The load history is measured by piezo-electric load cells under the load plane supports while the crack length is monitored in 10mm increments during the test, using a pattern of conductive and resistive strips painted onto the lower surface of the sample. The striker velocity is found by measuring the time of flight between two infra-red detectors just prior to it impacting the sample. It is assumed that the striker velocity remains constant throughout the test.

Both a high density and a medium density grade PE have been tested using the HSDT technique. The majority of tests were performed at 0°C for two reasons:-

i) This is the likely minimum operating temperature of buried PE pipes.

ii) The limited number of tests performed at 23°C all exhibited permanent deformation of the sample after testing, suggesting that plastic flow had occurred. Any elastic analysis would be invalid.

Initially, 6mm thick samples of each material were tested using striker speeds in the range of 14-31m/s. A few tests were performed using striker speeds between 7-14m/s. In tests where the striker speed was greater than 14m/s continuous constant velocity crack growth throughout most of the sample length was observed at velocities in the range of 125-260m/s. In tests where the striker speed was less than 14m/s several arrest markings were frequently seen on the fracture surface. Subsequently, 9mm thick samples of the medium density grade were impacted using striker speeds of 14-31m/s. Crack velocities in the range of 260-350m/s were observed.

3. A LINEAR ELASTIC MODEL OF THE HSDT TEST.

Each half of the sample can be modelled as a torsion beam that is unrestrained behind the crack tip (x<a) but restrained by a torsional elastic foundation ahead of the crack tip (x>a). The restoring torque applied by this foundation at any particular section is then proportional to the rotation, $\phi(x,t)$, of that section at that time. Kanninen [3] developed a similar model of the double cantilever beam sample. The model for the HSDT test is:-

$$\mu K\phi_{xx} - \rho I\phi_{tt} - H(x-a)\sigma\phi = 0 \qquad (1)$$

where H is the Heaviside step function:-

$$H(x-a) = 0, \quad x < a \qquad H(x-a) = 1, \quad x > a \qquad (2)$$

The torsional rigidity of the beam section is μK, the polar moment of inertia of the section per unit length is ρI, σ is the torsional stiffness per unit length of the foundation and ρ is the mass density. The shear modulus, μ, was found using ultrasonic methods which give an estimate of the modulus under high strain rate, low strain loading conditions. The elastic properties of the two grades tested are given in Table 1.

Table 1 Material Properties

Material	Ultrasonically Measured Shear Modulus, GPa	Density, kG/m^3
HDPE 1	1.12	960
MDPE 3	1.09	950

The torsional wave speed, C_T {$=(\mu K/\rho I)^{1/2}$}, is typically 250m/s for a 6mm thick sample and 350m/s for a 9mm thick specimen. The initial conditions are:-

$$\phi(x,0) = 0 \qquad \phi_t(x,0) = 0 \qquad (3)$$

which state that the sample is undeformed and at rest prior to loading. Since the striker velocity is assumed to remain constant during the test, the design of the striker and support then ensures that the load plane rotation rate, $\phi_t(0,t)$, remains almost constant throughout, so the load plane boundary condition can be expressed as :-

$$\phi(0,t) = \phi_t(0,t) t \qquad (4)$$

Crack growth is simulated by allowing the foundation boundary to translate at the crack tip speed after fracture initiation, that is, by setting:-

$$a = a_0 + H(t - t_f) \text{ å } (t - t_f) \qquad (5)$$

in equation (1), which can then be solved numerically by integrating along the characteristics [4], allowing the sample deformation to be determined as a function of time.

In order to check the validity of the model represented by equations (1) to (5) HSDT samples were marked and instrumented with the normal crack length gauge and the tests filmed using a high speed camera. Figure 2 shows a typical sequence of frames viewing the underside of the sample, with the crack propagating towards the camera. The initial rotation of the load plane and propagation of the torsional disturbance down each half of the test piece can clearly be seen. Figure 3 compares the rotation profiles predicted by the model at three particular times, to those actually measured from the photographs. The measured profiles indicate that the rotation varies linearly behind the crack tip and decays exponentially in front. The model simulates these features accurately.

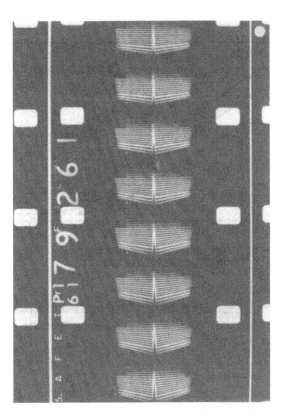

Fig.2 Deformation of an HSDT Sample Viewed From the Underside
With the Crack Propagating Towards the High Speed Film Camera
(Film Speed 8000 Pictures per Second)

G_D can then be determined at each timestep in the integration procedure by calculating the imbalance between the rate at which external work is being supplied to the sample, dU_W/dt, and the rates of increase of kinetic energy, dU_K/dt, and strain energy, dU_S/dt, within the specimen :-

$$G_D = \frac{1}{B_C \mathring{a}} \left\{ \frac{dU_W}{dt} - \frac{dU_S}{dt} - \frac{dU_K}{dt} \right\} \tag{6}$$

where B_C is the width of the fracture surface.

Section Rotation (Degrees)

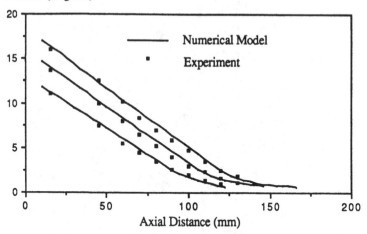

Fig.3 Comparison Between Computed and Experimentally
Measured Rotation Profiles Along the Specimen Axis

Fracture Resistance (kJ/m^2)

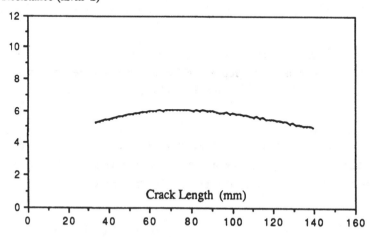

Fig.4 Variation in Fracture Resistance with Crack Length
for a Typical Test (Linear Elastic Model)

Figure 4 shows the computed variation in G_D with crack length for a typical test. The main feature of this Figure is that G_D remains approximately constant as the crack length increases and hence a mean value of G_D can be determined for each test. However, as Figure 5 shows, when mean G_D values are computed using experimental data obtained from a series of tests performed on the high density PE grade and are plotted as a function of crack velocity, considerable scatter in the resistance results is seen. A similar spread is observed when the resistance results for the medium density material are plotted against crack velocity, as shown in Figure 6. This Figure also suggests that the lower bound G_D values may be thickness dependent. However, this Figure does show that for each sample thickness tested, no crack velocities greater than the appropriate torsional wave speed were recorded. Clearly C_T dictates the limiting crack speed in the double torsion test geometry. The scatter, seen in each Figure, is systematic: at a given crack velocity, a higher G_D value is yielded by a test performed at a higher impact speed. Another inconsistent feature of the model is illustrated by Figure 7, which shows that the predicted applied load is much greater than that actually measured during a typical test, suggesting that inappropriate modulus values are being used in the simulations.

4. DETERMINATION OF SHEAR MODULUS VALUES APPROPRIATE TO THE HSDT TEST

Previously, Truss [5] has shown that pipe grade PE materials exhibit nonlinear behaviour when twisted statically under an imposed hydrostatic pressure and subjected to shear strains of up to 5%. In order to determine shear modulus values appropriate to the HSDT test a technique that makes use of the HSDT geometry has been developed. Two independent rectangular sectioned torsion beams are clamped together at one end, to produce a sample with the same dimensions as the fracture test piece as illustrated in Figure 8. These samples are then loaded by impact using a range of striker speeds in the HSDT test rig, with the striker twisting each beam at its unrestrained end. Figure 9 shows the assumed deformation of the sample during the test. The front of the disturbance propagates at a torsional wave speed that depends on the ultrasonically measured low strain shear modulus, μ_0, because the region ahead of the disturbance is unstrained. The torque applied to each beam, T, can be determined from the mean measured load, while the uniform twist induced in the disturbed region, ϕ_x, is given by :-

$$\phi_x = \frac{\phi_t(0,t)}{C_T} \quad , \qquad C_T = \sqrt{\frac{\mu_0 K}{\rho I}} \qquad (7)$$

The twist can be varied simply by changing the impact speed. The effective shear stress, τ_{EFF}, and maximum shear strain, γ_{MAX}, in any rectangular section are defined by :-

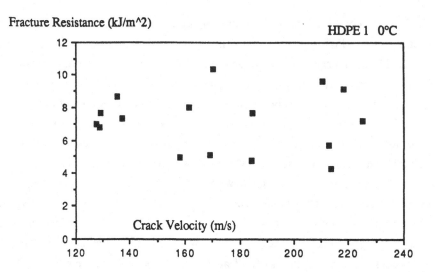

Fig.5 Variation in Fracture Resistance with Crack Velocity
for High Density PE Grade (Linear Elastic Model)

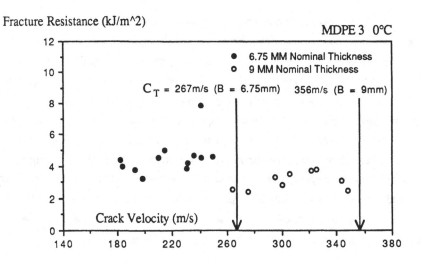

Fig.6 Variation in Fracture Resistance with Crack Velocity
for Medium Density PE Grade (Linear Elastic Model)

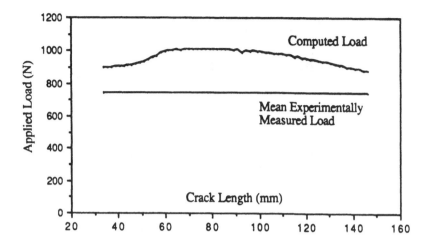

Fig.7 Comparison Between the Load Computed Using the Linear Elastic Model and
the Mean Experimentally Measured Load for a Typical Test

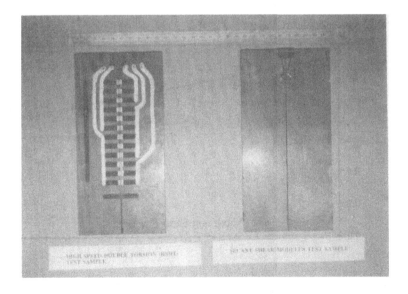

Fig.8 The Modulus Test Piece and Fracture Test Sample

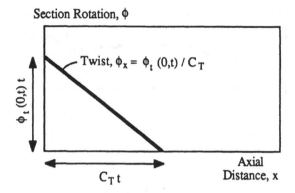

Fig.9 Assumed Deformation in the Modulus Test Piece

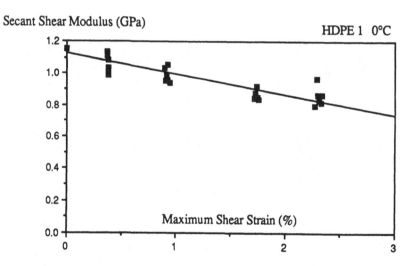

Fig.10 Variation in Secant Shear Modulus with Shear Strain
for High Density PE Grade

$$\tau_{EFF} = T\frac{B}{K} , \qquad \gamma_{MAX} = B\phi_x \qquad (8)$$

where B is the sample thickness. The shear strain induced in the disturbed region thus depends on the striker velocity. An 'effective' secant shear modulus, μ_{SEC}, can then be defined as :-

$$\mu_{SEC} = \frac{\tau_{EFF}}{\gamma_{MAX}} = \frac{T}{K\phi_x} = \frac{TB}{K\gamma_{MAX}} \qquad (9)$$

This allows μ_{SEC} to be found and plotted as a function of the shear strain. Figure 10 shows how μ_{SEC} varies with shear strain for the high density PE grade. The ultrasonically measured shear modulus value is also plotted at zero strain. A linear approximation:-

$$\mu_{SEC} = \mu_0 - m\gamma_{MAX} = \mu_0 - mB\phi_x \qquad (10)$$

is then used to represent the decay in secant shear modulus. Clearly the material is exhibiting nonlinear behaviour under impact loading conditions, because the effective shear modulus falls as the applied shear strain is increased. A simple analysis gives the 'effective' tangent shear modulus, μ_{TAN}, defined by $d\tau_{EFF}/d\gamma_{MAX}$, as:-

$$\mu_{TAN} = \mu_0 - 2m\gamma_{MAX} = \mu_0 - 2mB\phi_x \qquad (11)$$

The decay constants, m, for the high density and medium density materials were found to be 13 and 11GN per unit shear strain respectively.

5. A NON LINEAR ELASTIC MODEL OF THE HSDT TEST

Two improvements can be introduced simultaneously into the existing model. Firstly the variation in 'effective' tangent shear modulus with strain given by equation (11) can be included. The second enhancement is to consider the suppression of free warping of the rectangular beam sections, coupled with the appearance of axial stresses, in the region where the twist varies [6]. The model is now described by the nonlinear equation:-

$$2(1+\nu)\mu_{TAN}\lambda^2 I\phi_{xxxx} - \mu_{TAN}K\phi_{xx} + \rho I\phi_{tt} + H(x-a)\sigma\phi = 0 \quad (12)$$

where ν is Poisson's ratio and $\lambda/B \approx 0.3$ for the test piece. The second inclusion produces an effective stiffening of the beams ahead of the crack tip causing the loading wave to disperse in this region. One additional boundary condition stating that the axial stresses are zero must be applied at the load plane so:-

$$\phi_{xx}(0,t) = 0 \qquad (13)$$

while at the far end of the sample two conditions which state that both the axial and shear stresses are zero :-

$$\phi_{xx}(L,t) = 0, \qquad \mu_{TAN} K\phi_x - 2(1+v)\mu_{TAN} \lambda^2 I\phi_{xxx} = 0 \qquad (14)$$

must be applied. Equation (12) has been solved by employing an explicit finite difference scheme in which the values of the nonlinear coefficients are computed at the start of each timestep. These are then used to compute the rotations at the next timestep.

Figure 11 compares typical rotation profiles of the beams predicted by the linear and nonlinear models. Like the linear model, the nonlinear model predicts a linear variation in rotation behind the crack tip, in accordance with the photographic results. However, in the nonlinear simulation the wavefront now disperses in the region in front of the crack tip and decays in a sinusoidal manner. Figure 12 shows that the load variation for a typical test, computed by the nonlinear model, is now in much better agreement with the mean measured load.

Reanalysing the experimental data using the nonlinear model produces the G_D data shown in Figures 13 and 14 for the high and medium density grade materials respectively. These Figures show considerably less scatter in the range of G_D values at any given crack speed than do Figures 5 and 6. Additionally, both Figures 13 and 14 suggest that G_D is almost independent of crack velocity, while Figure 14 also indicates that the lower bound G_D values are now thickness independent. The average G_D values for the high and medium density materials are 4.75 and $2.60kJ/m^2$ respectively. Obviously, on the basis of resistance to RCP, the high density material is more suited to the pipeline application.

CONCLUSIONS

The HSDT test has successfully generated RCP in both high and medium density pipe grade PE materials. Crack velocities up to the geometry dependent torsional wave speed have been measured. Initially, a numerical model that assumes linear elastic material behaviour was used to compute fracture resistance values. This data showed considerable scatter in the range of G_D values at a given crack speed. A technique that determines shear modulus data appropriate to the HSDT has also been developed. The data obtained from this technique indicate that the shear moduli fall as the applied shear strain is increased. The numerical model has been modified to include this nonlinear elastic behaviour. The recomputed resistance data shows significantly less scatter and indicates that the high density material performs better than the medium density grade. The average G_D values for the two materials correlate well with small scale pipe test p_C data, obtained for pipe sections in which plane strain conditions were promoted by introducing an internal axial slit along the whole section length.

Section Rotation (Degrees)

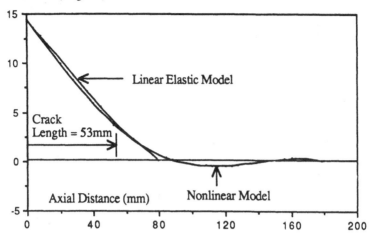

Fig.11 Comparison Between the Rotation Profiles Calculated Using the
Linear Elastic and Nonlinear Models

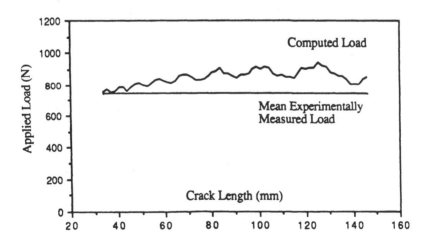

Fig.12 Comparison Between the Load Computed Using the Nonlinear Model and
the Mean Experimentally Measured Load for a Typical Test

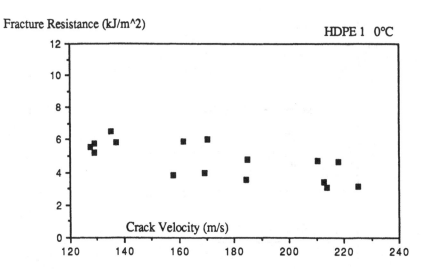

Fig.13 Variation in Fracture Resistance with Crack Velocity
for High Density PE Grade (Nonlinear Model)

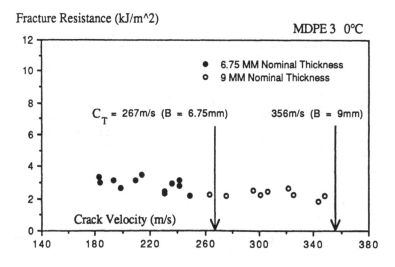

Fig.14 Variation in Fracture Resistance with Crack Velocity
for Medium Density PE Grade (Nonlinear Model)

99

REFERENCES

[1] J.M. GREIG and L. EWING, Proc. of 'Plastic Pipes V', York, UK, (PRI, London), 8-10 Sept. 1982
[2] P.S. LEEVERS and P. YAYLA, This Conference
[3] M.F. KANNINEN, Int. J. of Fracture, Vol.10 No.3, 1974
[4] W.F. AMES, Numerical Methods for Partial Differential Equations, (Nelson, 1969)
[5] R.W. TRUSS, R.A. DUCKETT and I.M. WARD, J. Mat. Sci. 19, pp 413-422, 1984
[6] J.M. GERE, J. App. Mech., Vol. 21, pp 381-387, Dec. 1954

DYNAMIC FRACTURE TOUGHNESS OF FILLED POLYMERS USING A GLOBAL ENERGETIC CRITERION

G PLUVINAGE* AND R NEVIERE**

* LABORATOIRE DE FIABILITE MECANIQUE
UNIVERSITE DE METZ (FRANCE)
** SNPE VERT LE PETIT (FRANCE)

ABSTRACT

Dynamic fracture toughness of a filled polymer and a polyurethan was determined using a global energetic criterion from Andrews. Results are obtained for different conditions of temperature and loading rate. Dynamic loading of specimens was obtained using tensile Split Hopkinson Pressure Bars. All results are presented on a master curve using the principle of time-temperature equivalence.

INTRODUCTION :

Because of their viscoelastic behaviour, polymers exhibit a strong dependence of their physical and mechanical properties such as strain rate, temperature or environmental effects. Solid propellant, a hard particle filled polyurethan lies in this range. This material exhibits non-linear viscoelastic properties due to decohesion of particles from polymeric substrate leading to volume variation. Few results on fracture analysis of solid propellant are available in literature. Due to the complexity of the material and his constitutive equation, only non-linear fracture mechanic (N.L.F.M) can be used for this problem. The fracture toughness of such a material exhibits a strong time-temperature dependence. The difficulties mentioned previously can be over - come if we use a global energetic approach as fracture criterion.

A typical global energetic approach consist in using the non-linear strain energy release rate. This quantity was first defined by Liebowitz and Eftis /1/ and for constant displacement given by equation /1/ :

$$\widetilde{G} = \frac{1}{B}\frac{\partial U_{nl}}{\partial a} \qquad\qquad /1/$$

Where B is the thickness of the sample, U_{nl} the non-linear stored energy. Andrews /2/ suggests that this non linear strain energy release rate is a simple function of the uniform remote strain energy density W^*_0.

$$\widetilde{G} = f\,(W^*_0) \qquad\qquad /2/$$

This suggestion has some analogy with the elastic stress intensity factor formula because the stress intensity factor is function of the global remote stress σ_g :

$$K = f\,(\sigma_g) \qquad\qquad /3/$$

We must remember that the elastic fracture process is sensitive to the global stress, and the fully plastic fracture process to the nominal stress σ_N. For any kind of fracture, the truth is probably intermediate. In Andrew's criterion called the generalised fracture theory, we use the assumption that for any kind of fracture the fracture process is governed by the uniform stress and strain state.

In this paper, the generalised fracture theory has been used to measure the fracture toughness of polymers. This theory was precised by introducing a non-linear geometrical factor using a tensile Split Hopkinson Pressure Bar and a classic tension machine. The influence of loading rate on linear critical energy release rate $\widetilde{G}$ was determined on propellant and polyurethan.

II. GENERALISED FRACTURE THEORY :

II.1 Basic theory :

We consider that the strain energy density W^* is only disturbed in an area near the crack tip. In other places, the strain energy distribution is uniform and equal to W^*_0. In the region of the strain energy gradient, W^* is function of coordinates, the level of strain and the uniform strain energy density.

This function may be written :

$$W^* = W^*_o \cdot f(x, y, \varepsilon_g)$$ /4/

In order to introduce a crack length independence, non-dimensional coordinates X and Y were used :

$$X = \frac{x}{a}; Y = \frac{Y}{a}$$ /5/

The differentiation of W* with respect to crack length gives for any point P :

$$\frac{dW^*}{da}(P) = -g(X, Y, W^*_o) W^*_o / a$$ /6/

and summation over all points P is written as :

$$\frac{dU_{nl}}{da} = \sum_P \frac{dW^*(P)}{da} \cdot dV^*$$ /7/

Where d V* is the volume increment :

$$dV^* = B\,dx \cdot dy = B\,a^2\,dX \cdot dY$$ /8/

$$-\frac{dU_{nl}}{da} = -W^*_o \Big/ a \int_{v^*} B \cdot a^2 \cdot g(X, Y, W^*_o) \cdot dX\,dY$$ /9/

It is shown that the newly created area is dA = 2 B.da :

$$-\frac{dU_{nl}}{dA} = W^*_o \cdot a \cdot k(W^*_o)$$ /10/

For critical conditions which appear when $W^*_o = W^*_{o, c}$:

$$\tilde{G} = \tilde{G}_c = k(W^*_{o, c}) \cdot a \cdot W^*_{o, c}$$ /11/

Function $k(W^*_o)$ is an indicator of the degree of non-linearity of the load displacement diagram and in the elastic case k is equal to 3,14.

$$\tilde{G} = G = \pi\,(\sigma_g / 2E) \cdot a = \pi \cdot W_o^* \cdot a$$ /12/

In other cases k increases rapidly, reaches a maximum value when the process zone invades the total ligament and then decreases (figure 1).

II.2 Geometrical correction factor :

The basic expression is given only for an infinite plate. In the case of a specimen of finite size, a geometry correction $F_{W^*_0}$ must be introduced :

$$\tilde{G} = k \ (W^*_o) \cdot W^*_o \cdot a \cdot F_{W^*_o} \ (a/w) \qquad\qquad /13/$$

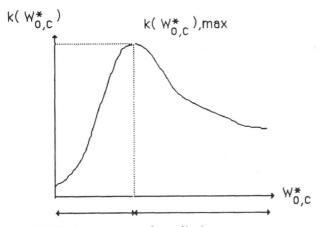

FIGURE 1: SCHEMATIC DIAGRAM OF THE EVOLUTION OF THE PARAMETER K

Andrews /2/ proposed a simple formula for this correction :

$$F_{W^*_o} = \frac{B}{B - a} \qquad\qquad /14/$$

Naït-Abdellaziz and Al /3/ suggested to use the elastic geometrical correction for the stress intensity factor F_σ with :

$$F_{W^*_o} = \left[F_\sigma \left(a/L \right) \right]^2 \qquad\qquad /15/$$

which gives the following expressions :

* For CRB sample (L is the sample length) :

$$F_{W^*_o} = (1{,}25 \ (a/2)^2 + 0{,}44 \ (a^3/3L) - 2{,}64 \left(a^4/4L^2 \right) + 3{,}84 \frac{a^5}{5L^3} + 2{,}21 \left(a^6/6L^4 \right)$$

$$- 4{,}63 \left(a^7 / 7L^5 \right) + 3{,}72 \left(a^8 / 8L^6 \right) \qquad \text{/16.a/}$$

* For axisymetric sample (R is the specimen radius) :

$$F_{W^*_0} = \left[1/(1 - (a/R)) \cdot \left[0{,}8 + 4 \,(a/R) \, / \, 1 - a/R \right] \right]^{-1} \quad \text{/16.b/}$$

II.3 Experimental determination :

The experimental determination of the fracture toughness with the Andrews criterion is carried out as follows :

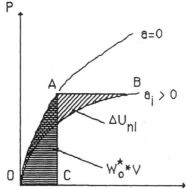

FIGURE N°2: DEFINITION OF THE QUANTITIES ΔU_{nl} AND W^*_0

*The variation of internal energy ΔU_{nl} is defined with the reference of an uncracked specimen (a = 0) and a given value of the uniform strain energy density W^*_0. This value is equal to area OAC divided by specimen volume V^* (Figure 2).

ΔU_{nl} is represented by area OAB and is in relation with the energy for a crack extension from a_0 to a_c.

* Relation /10/ can be integrated with respect to the crack area :

$$- \Delta U_{nl} = k \cdot W^*_0 \int a \cdot dA \qquad \text{/17/}$$

Since dA = 2 B da

$$- \Delta U_{nl} = k \cdot W^*_0 \, B \cdot a^2 \qquad \text{/18/}$$

ΔU_{nl} is proportional to the square of the crack length. By plotting ΔU_{nl} versus a^2, we obtain a graph of slope s (figure 3 .a). Parameter k is then obtained from the following relationship :

$$k = \frac{s}{B. W^*_o}$$ /19/

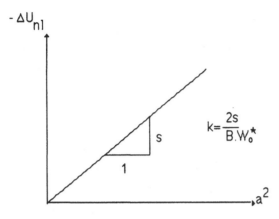

FIGURE 3a : CURVE - ΔU_{nl} VERSUS a^2

The product K . $W^*_{o,c}$ is plotted versus 1/a. The slope of the curve gives the G_c value figure (3b).

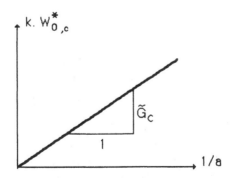

FIGURE 3B : PRINCIPLE OF FRACTURE TOUGHNESS DETERMINATION

III. **EXExPERIMENTAL INVESTIGATIONS** :

III.1) Materials and specimens :

Two materials were tested ,an inert propellant where calcium sulfate is associated with polyurethan and the same polyurethan alone.

The influence of strain rate and temperature on fracture toughness of these polymers was investigated using two different experimental devices. Axisymetric samples of diameter R=10 mm for static tests and 9 mm for dynamic tests were used. Height h is 15 mm for static and 10 mm for dynamic tests. An annular slot was made with a razor blade at mid height.

III.2) Static loading :

A conventional Instron tension machine giving two extreme crosshead speeds namely, static speed (0.5 mm/mn) and intermediate speed (500 mm/mn). Specimens are cemented, using a two component epoxy glue on metallic supports. The evolution of the traction forces on the specimen are measured through a dynamometer connected to an electrical amplifier allowing load recording. Displacements of clamps (loading points) are measured using an electromagnetic extensometer. The corresponding load-displacement curves are recorded on a graphic plotter for static tests or on a numerical oscilloscope for intermediate speed tests.

III.3 Dynamic loading :

Tests under dynamic loading conditions are carried out using a split Hoplinson Tensile Bars device. Our tensile testing device is issued from Kolsky's. As can be seen in figure 4, the device is essentially composed of two long cylindrical rods of aluminium alloy, respectively called incident and transmitted. The specimen is cemented between the two bars and loaded by a tensile stress wave generated by the impact of a tubular projectile on an anvil screwed on the opposite end of the incident bar. The impact of the projectile produces an elastic compression wave (noted σ_c) in the anvil, which, by reflection on the free end, will produced the needed tensile wave (σ_{tr}) in the incident rod. When reaching the input rod-sample boundary, wave $\sigma_I = \sigma_{tr}$

splits in the reflected and transmitted components, respectively called σ_R and σ_T. Time dependent stress, strain and strain rate curves of the tested material are then obtained from σ_I, σ_R, σ_T recordings.

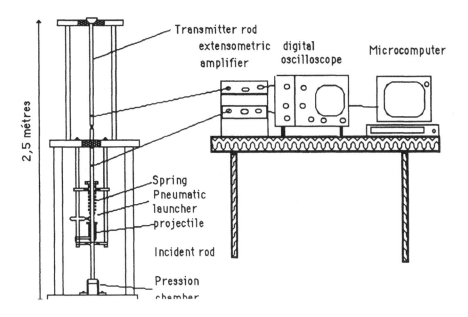

FIGURE 4 : TENSILE SPLIT HOPKINSON PRESSURE BAR DEVICE FOR TEST ON POLYMERS.

The elastic deformations, related to σ_I, σ_R, σ_T are measured through the medium of semi-conductor strain gages glued on the bars and connected to form Wheastone bridges. Using a dynamic voltage amplifier, stress variations versus time are recorded on a numerical oscilloscope and then saved in micro computer memories ulteriorly allowing fracture mechanics analysis.

IV. **EXPERIMENTAL RESULTS AND DISCUSSION** :

IV.1) Example of fracture toughness determination :

The following example concerns only a polyrethan broken under static loading and using an axysimetric sample. In this case the value of ΔU_{nl} is measured at constant load. For the axisymetric sample :

$$-dU_{nl}/dA = k\,(W^*_o)\,.\,W^*_o\,.\,a^2/2\,.\,\pi\,(R\text{-}a) \qquad\qquad /20/$$

$$\text{and}\; -\Delta U_{nl} = k\,(W^*_o)\,.\,W^*_o\,.\,a^3/3 \qquad\qquad /24/$$

$$\text{with}\; -dA=2\pi\,(R\text{-}a)\,.\,da \qquad\qquad /25/$$

When ΔU is plotted versus a^3, a satisfactory linear regression is obtained. Then $k\,(W^*_o)$ is plotted versus the W^*_o values for several crack sizes (a = 2.0, 3.0, 4.5, 7.0 mm).

The limited number of samples introduces some discontinuities, but we can see that value k increases when W^*_o increases and then decreases as mentioned previously. Finally G_c was obtained by measuring the slope of the

$$k\,(W^*_o)\,.\,W^*_\infty =1/F\!\left(\frac{2\,\pi\,(R\text{-}a)}{a^2}\right) \qquad\qquad /26/$$

curve :

where F(a,r) is a geometrical calibration curve which will be discussed later. The slope of the curve leads to the value of G_c equalling 370 J/m^2.

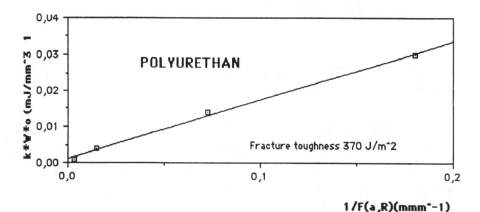

FIGURE 5 : EXAMPLE OF FRACTURE TOUGHNESS DETERMINATION.

V.2) Influence of strain rate and temperature on fracture toughness of inert propellant and polyurethan :

The toughness of these materials was measured for 4 temperatures: (20°C, - 20°C, - 50°C, - 70°C) and five loading rates, for static rates (0.5, 10, 5, 50 mm/m) and one dynamic.

The viscoelastic character of these polymers leads to an increase fracture toughnesss with the loading rate as it can be seen on figure 6. At 20 °C , a linear relationship with the logarithm of the loading rate was found (G= 362+93 * log ε). It is interesting to notice that a similar relationship was found for the polyurethan (G= 164+93 * log ε). The fact that the slope of the curve is similar clearly indicates that the viscoelastic properties of this polymer govern the viscoelasticity of the propellant. The inert material which is used to filled the propellant has a nonloading rate sensitive effect on the fracture toughness.

INERT PROPELLANT

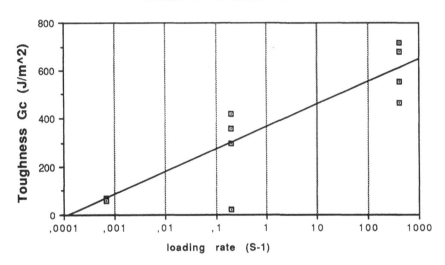

FIGURE 6 : INFLUENCE OF LOADING RATE ON THE FRACTURE TOUGHNESS OF A PROPELLANT.

By performing tests at temperatures lower than 20°C, we can also notice an increase of the fracture toughness. This suggests that an equivalence between temperature and loading rate can be found. A way to describe this phenomenon is to use the concept of reduced time and write the following relationship :

$$\tilde{G}=\tilde{G}\left(t_{r,2},T_1\right)=\tilde{G}\left(t_{r,2}/a_T,T_2\right) \qquad /27/$$

a_T is the so-called shift factor. It can be given by an empirical relationship from William, Landel and Ferry.

$$\ln a_T = \frac{-C_1 \cdot \left(T - T_{ref}\right)}{C_2 + \left(T - T_{ref}\right)} \qquad /28/$$

where C_1 and C_2 are two constants. T_{ref} is a reference temperature which is generally considered 50°C higher than the glassy transition temperature. Ferry /4/ has proposed that the logarithm of the shift factor is function of the fraction of free volume f.

$$\ln a_T = \frac{B}{2.303 \cdot \left(\frac{1}{f} - \frac{1}{f_0}\right)} \qquad /29/$$

We suggest to consider that this volumic fraction has a linear dependence with the temperature.

$$f = f_0 + \alpha \cdot f \left(T - T_{ref}\right) \qquad /30/$$

(f_0 is the volumic fraction of free volume at reference temperature)

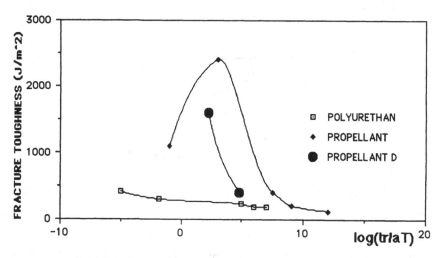

FIGURE 7 MASTER CURVE FOR THE FRACTURE
TOUGHNESS OF PROPELLANT AND POLYURETHAN

This leads to

$$\ln a_T = \frac{\dfrac{B.\left[T-T_{ref}\right]}{2.303.f_o}}{\dfrac{f_o}{\alpha.f}\left[T-T_{ref}\right]} \qquad /31/$$

We find the same expression as equation 23 with $C_1 = B/2.303\ f_o$ and $C_2 = f_o/\alpha f$. The experimental results are plotted in a master curve fracture toughness versus the logarithm of a_T with C_1 equal to 8.86 and C_2 to 101.6. It can be seen that for the propellant a transition appears for $\ln(tr/a_T) \sim 3$. This corresponds to a modified fracture process which is principally due to the fracture of the hard particles. One must notice that the results of the dynamic test do not fit the master curve very well. At the present time we have no good explanation for these results. First, we tried to verify the validity of our results using Split Hopkinson Bars not in Magnesium alloy but in PMMA. A special treatment of the signal recorded is required due to the fact that this material is viscoelastic.

It is also noticeable that we cannot find the glassy transition of polyurethan in the temperature and loading range explored.

In the present study, special attention was paid to geometrical correction in order to separate the material effect from the geometrical effect. In addition, the assumption to use the correction for the elastic stress intensity factor for non linear material was examined in the following way :

The variation of stored energy is proportional to the remote strain energy density.

$$-\Delta U = W_o^* . L\left(W_o^*, a/b, a/c, \ldots\right) . a^2/2 \qquad /32/$$

We assume that the $L(W^*O, a/b, a/c, \ldots)$ is the product of the two functions
$$L(W^*O, a/b, a/c, \ldots) = k(W^*O).F(a/b, a/c, \ldots) \qquad /33/$$

This is the assumption that material and geometrical effects are separated.

By analogy we consider that
$$a^2/2. F(a/b, a/c, \ldots) = \int a.f(a/b, a/c, \ldots).da \qquad /34/$$

Function f(a/b,a/c,....) plays for non linear material the same role as the elastic calibration function Y (a/b,a/c,....).

Function F(a/b,a/c,.....) can be obtained experimentally and compared to function G (a/b,a/c,.....).

$$G\left(a/b,a/c,...\right) = \int a.Y^2\left(a/b,a/c,...\right) da \qquad /35/$$

A comparison between the two functions shows (figure 7) that they are relatively close .

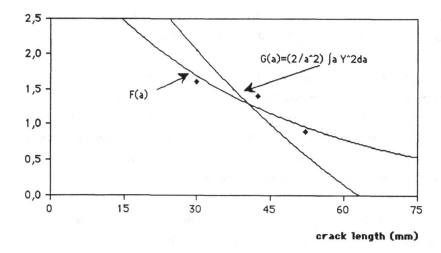

FIGURE 8 : COMPARAISON OF THE NON LINEAR
AND THE ELASTIC CALIBRATION FUNCTION

V) CONCLUSION :

The use of a global energetic criterion is an interesting way of determining the fracture toughness of such reologically complex materials as filled polymers. The viscoelastic behaviour of this material leads to using a master curve representation of the common influence of temperature and loading rate. It was found that the dynamic fracture toughness does not fit this master curve. Further investigations are necessary to determine if this is due to the change of the fracture process or to experimental difficulties in measuring the dynamic fracture toughness. A new generation of Split Hopkinson Pressure Bars using viscoelastic bars is under developement for this question.

REFERENCES

1) LIEBOWITZ H. and EFTIS J.
On non linear effects in fracture Mechanics .
Engineering fracture Mechanics p. 267 - 281 (1971)

2) ANDREWS E.H.
A generalized theory of fracture mechanics.
Journal of Material Sciences p. 887 - 894

3) NAIT ABDELAZIZ M , R NEVIERE, G PLUVINAGE,
Experimental investigation of fracture surface energy of a solid
propellant under different loading rates
Engineering Fracture Mechanics 1988, 31.6, 1009-1026

4) FERRY .J.D.
Viscoelastic properties of polymers,John Wileys and Sons Editors.
3rd Edition NewYork 1980.

DYNAMIC FAILURE MECHANISMS
IN GALLSTONES USING LITHOTRIPSY

S.J. Burns+, S. M. Gracewski++, N. Vakil * and A. R. Basu **
Rochester Center for Biomedical Ultrasound
University of Rochester
Rochester, NY 14627 USA

+   Material Science Program, Department of Mechanical Engineering.
++   Department of Mechanical Engineering.
*   Department of Medicine, Gastroenterology Unit.
**   Department of Geological Science.

ABSTRACT

Gallstones most frequently are twinned, cracked, very large grained crystals of cholesterol that have been grown in a saturated bile solution. In cross section, they are similar to spherulitic polymers although the composition of most gallstones is nearly pure cholesterol. The dynamic failure mechanism of "pulverization" during lithotripsy is at present only partially known. The literature suggest that cavitation next to the stone is the mechanism for stone destruction. Yet *in vivo* and *in vitro* observations show that many stones initially break approximately in half suggesting that internal flaws and dynamic loading, not cavitation with its surface removal processes, is the dominant mechanism. It is proposed that the stones break by crack propagation in a dominantly transient, reflected, stress field. Crack propagation and failure in gallstones is complicated by the large number of internal flaws, the low yield strength of the stones and the complex microstructure as it effects stress waves and dynamic fracture mechanisms. The physical properties of gallstones are presented from measurements of the elastic constants using literature wave speeds, hardness strengths and static fracture toughness. The static fracture properties have been measured on stones extracted from patients. These natural gallstones have very irregular shapes and a wide range of mechanical properties. While the surfaces of broken stones are very rough there is partial evidence of fatigue crack propagation. The cyclic crack propagation is probably due to stress intensity differences between compressive and tensile dynamic stresses on pre-existing cracks.

INTRODUCTION

The procedure of using a pulsed, compressive, focused, shock wave to pulverize a kidney stone or a gallstone is referred to as lithotripsy. Gallstone lithotripsy is successful in only about 20% of the patients treated using this procedure; kidney stone lithotripsy, on the other hand, is successful in a very large percentage of patients. Gallstones and kidney stones have very different physical properties: the former are quite soft, pliable yet brittle; the latter are hard and brittle. Gallstone patients who have invasive surgery, involving gall bladder removals i.e. cholecystectomies, require hospitalization and substantial time for recuperation. This procedure is estimated to cost 6 billion U.S. dollars per year, while lithotripsy may be done on an outpatient basis.

The fragmentation of human body stones by intense, acoustic, pressure waves is a dynamic fracture process. The stones are mechanically stressed with very rapid pressure pulses and break by dynamic failure as specified in this paper. Dornier first introduced extracorporeal shock wave lithotripsy, ESWL, and it was subsequently proposed that kidney stones fragmented when the pressure produced by the shock waves, focused at the stone exceeded the stone's "strength" [1]. Detailed analysis of the pressure pulse and the mechanism of failure were in this early work left unspecified. A body stone placed in pressurized fluid such as is found in a hydraulic ram that is slowly loaded would never break because in hydrostatic compression the stone's "strength" could never be exceeded. Failure is always associated with deviatoric stresses or tensile stresses. Thus, the dynamic aspect of lithotripsy is very important in fragmenting body stones.

Lithotripters

At present, there are several manufactures of lithotripter's. At the University of Rochester we have access to both an electrohydraulic Dornier HM-3 lithotripter and a piezoelectric Diasonic Therasonic lithotripter. The basic design of lithotripters is based on focusing stress waves. For example, a common design is a bowl in the shape of an ellipsoid of revolution with a pressure source located at one focus while the stone is ideally located at the second focus. The pressure source in the electrohydraulic lithotripter model is a spark plug (electrode) that has an electrical discharge of varying voltage and decay time applied across it. The electrical energy heats and expands gasses created at the electrodes of the spark plug. The expanding gasses are located in an aqueous liquid which in turn transmits a pressure pulse to part of the surface of the ellipse of revolution. See Figure 1 for a schematic drawing of this geometry. The pulse is reflected from the

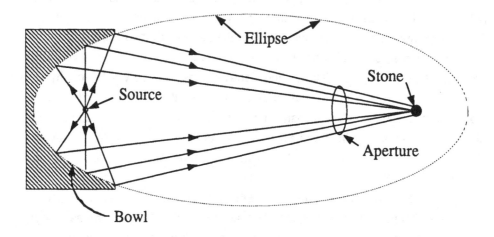

Figure 1. The pressure source is at the left focus of the ellipse; the patient's gallstone is placed at the right focus. The ray path to the stone is shown.

elliptical surface and then transmitted through a water bath to a flexible surface that is smeared with a jelly for coupling the transmission into the patient's body. The patient's stone is placed at the ellipse's second focus. The ray that strikes the stone travels, of course, the sum of the distances from the pressure source to any point on the surface of the ellipse plus the distance to the stone. Since all rays travel the same distance they arrive at the focus simultaneously creating a shock wave for non-invasive fragmentation of stones. The stone is placed ideally at the second focus of the ellipse by moving either the bowl or the patient. The aperture of the ellipse establishes the solid angle that the acoustic pulse subtends at the stone. If the aperture is small then a plane wave approximation is justified for the pressure pulse at the stone. The second unachievable extreme is a fully spherical wave, from the completely utilized ellipsoidal cavity, impinging on the stone. Figure 1 is a schematic representation of the pressure source, the elliptical surface and the stone. The stone is located by either x-ray radiography or by ultrasonics. These locating observations are used to position the lithotripter relative to the patient's stone so that the stone can be placed at the focus. Now, having briefly reviewed how an acoustic pulse impinges on a stone it is important to summarize the possible failure processes that could lead to stone fragmentation as cited in the literature.

FRAGMENTATION MECHANICS

Gallstones in general do not break easily in ESWL. The process or processes that can cause a stone to fragment are discussed in this section. Kidney stones break more readily than gallstones. It is not clear if the failure processes in both stones are the same. The literature has proposed several processes for stone failure. Several new or additional processes are also briefly noted below. However, it should be realized that the mechanism for stone destruction *in vivo* is not well documented nor fully understood.

i) Cavitation A small bubble in the fluid that surrounds the stone expands and violently collapses. If the bubble is near a solid surface, the bubble can collapse asymmetrically generating a high speed liquid jet that impinges on the solid [2-4]. The shock wave from the bubble collapse causes a pit in the stone. An accumulation of pits causes the stone to erode. The stone in this case is eroded from the surface towards the center.

ii) Spalling The compressive pressure peak is much larger than the tensile pressure peak for lithotripsy pulses. The back surface of the stone is a pressure release surface. The wave that is reflected back into the stone from this surface will be primarily tensile. Kolsky [5] in his book **Stress Waves in Solids** gives several examples in cones and cylinders of spalling fractures. The damage for lithotripsy will occur on the posterior side of the stone.

iii) Stress wave focusing An acoustic pulse incident on a sphere or disk is refracted and reflected inside the object. The surfaces where stress waves are focused are called caustics. If the stones are approximated as spheres then there will be caustic lines in the stones [6]. Fractures in polymethyl methacrylate (PMMA) spheres have been observed when placed near an explosive charge [7-11]. In this case there are several fracture surfaces that are closely related to the caustic wave surfaces in the sphere. This is an example of stress wave focusing *inside* the sphere. Caustics (focusing) will also occur inside gallstones.

iv) Fatigue crack propagation This assumes a pre-existing flaw or crack that is either interior to the stone or on the stone's surface. The repetitive stress pulses in both tension and compression cause a crack to propagate using fatigue crack propagation processes [12-14]. The change in the stress intensity value must exceed a threshold value. The rate of crack growth per cycle is expected to follow a power law in the stress intensity factor difference that is applied to the crack as it propagates. The stones would fracture by fatigue crack propagation.

v) <u>Tensile instability</u> The liquid surrounding the stone would again cause a localized failure in this case in the liquid bile. The tensile instability would give a subsequent void collapse. This process does not require a nucleus in the bile since the fluid has a free energy spinodal on the chemical, pressure state variable axis. Although, spinodals have been related to the "tensile strength" of water [15,16] there is no evidence for such processes in bile. These chemical instabilities need further verification, evidence and support. The stone would however erode from the outside inward for this process.

vi) <u>Critical pressure</u> The stone breaks when the pressure exceeds the "strength" of the stone. "Strength" in a hydrostatically pressurized stone does not exist. All failure processes in materials are always related to differential stresses for shearing the material or to tensile stresses that can cause fractures from existing flaws. The pulse shape and its interactions not just the amplitude are important for failure. Defects and the structure of the stone are closely connected to the failure mechanism.

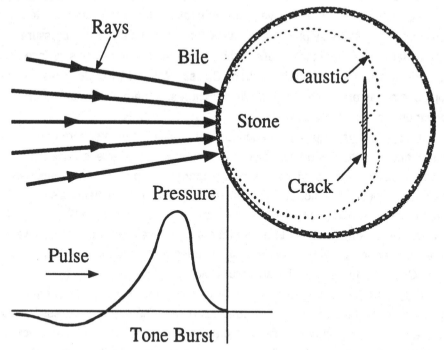

Figure 2. A tone burst travelling to the right generates refracted and reflected stress waves in the gallstone. The caustic curve is where the reflected (tensile) waves are focused. A crack near the caustic is shown.

119

The stresses in a solid spherical object that is pressurized by a harmonic infinite plane wave have been solved. A plane wave traveling in a fluid scatters from the spherical object and establishes the stress field in the solid [17]. For a lithotripter pulse the stress field is compressive until part of the wave is reflected from the surface of the sphere. The stress fields inside the sphere before reflections are compressive and only the reflected pulses are tensile [11]. Another way to recognize that the stresses are mainly compressive is to consider the Boussinesq point contact with a half space [18]. In this limit the stress field is completely compressive. The Hertz stress field between two contacting curved surfaces [18] is also mainly compressive. The free surfaces have small radial stresses, σ_{rr}, while the other stress components are either compressive, $\sigma_{\theta\theta}$, or zero. Directly beneath the contact all stresses are compressive. Thus, it is clear that the stress field in the stone during lithotripsy is mainly a compressive field and not the tensile stresses which are needed to aid in fracture.

The complex stress state in the stone being mostly compressive is one of the main contributors to the difficulty of fragmenting stones. Figure 2 shows the pressures applied to a stone. A dynamic compressive stress may reflect from a pre-existing crack and the reflected pulse will be tensile so that the stress field can cause some crack propagation. The caustic lines, on which the stress waves focus, and known fractures are also shown in Figure 2.

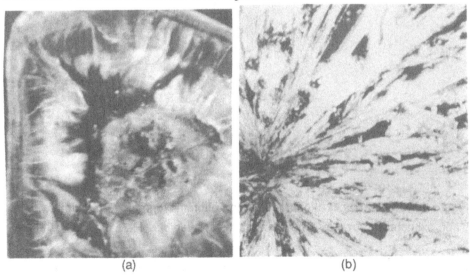

(a)   (b)

Figure 3. (a) Reflected light macrophotograph of a polished gallstone is shown.
(b) A thin section of a cholesterol gallstone with radial crystals shown in
transmission photomicroscopy. Both stones are 9mm in diameter.

Gallstone Microstructures

Gallstones are mainly composed of cholesterol crystals which mechanically are very soft, yet extremely brittle. Gallstones are quite different from kidney stones which are hard and brittle. The latter stones break readily in lithotripsy whereas the former are difficult to break. Figure 3 is a section of a gallstone. The stone has three major parts: first, is the core or nucleus near the center of the stone from which the stone started to grow; second are the long nearly radial crystals of cholesterol that contain chiral twins. This region is the center and main portion of the stone; the third part of the stone is the outer core or shell. The growth bands of the shell of the stone are like the growth rings of a tree. The gallstones are located in the biliary track in the body and are surrounded by bile which is saturated with cholesterol. The outer shell of the stone appears to be quite different from the long slender crystals that make up most of the stone as a shown in Figure 3. In Figure 4 a higher magnification photomicrograph shows that across the crystallites of cholesterol there are extensive flaws or cracks. These cracks are between the dominant (001) crystallographic planes in the material as can be seen by the color (shade) differences between chiral twins in each crystallite. The chiral twins are

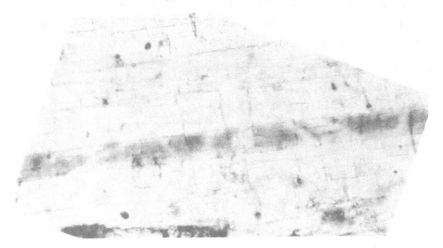

Figure 4. A high magnification transmission photomicrograph. The chiral twins are approximately horozintal. The short cracks are vertical. Field of view is 100 μm.

interfaces between right and left handed crystallographic axes. In our case, however, these interfaces are to be considered mechanically weak and the most likely sites for crack propagation in these stones. It should also be mentioned in passing that there are other types of gallstones (bilirubinate or pigment) but these are less likely to occur at least in the western hemisphere. Finally, the microstructure part of the stones is that they are non-homogeneous, contain large numbers of flaws, interfaces and cracks.

Mechanical Properties

The mechanical properties of these stones are very briefly listed below in Table 1. The compressional wave speed has been measured in gallstones. [19,20] These measurements plus the independently reported value of Poisson's ratio which is equal to 0.12, permits the evaluation of the elastic constants. The well known equations are (see for example [5])

$$\frac{E(1-v)}{(1+v)(1-2v)} = \rho V_p^2, \tag{1}$$

where v is Poisson's ratio, E is Young's modulus, ρ is the mass density and V_p is the compressional (longitudinal) wave speed. The shear modulus, G, is related to E for isotropic material by

$$G = \frac{E}{2(1+v)} \tag{2}$$

Finally, the shear wave speed, V_s, is found [5] from $V_s = \sqrt{G/\rho}$. These two wave speeds are related and their ratio depends only on v. Using equations (1) and (2) gives $V_p/V_s = \sqrt{2(1-v)/(1-2v)}$. Poisson's ratio as has been mentioned was reported [19] for gallstones. Using this static measurement of 0.12, the wave speed ratio is thus 1.5 or the compressional wave is 50% faster than the shear wave.

TABLE 1
Mechanical Properties of Cholesterol Gallstones
(Dry Condition)

Wave Speed, V_p. (km/s)	Hardness, H. (MPa)	Young's Modulus, E. (GPa)	Shear Modulus, G. (GPa)
2.356 [19]	58 [19]	5.8	2.6
2.082 [19]	19 [19]	4.5	2.0
1.767 [20]	--------	3.3	1.5
--------	39 [21]	--------	--------
--------	44 [22]	--------	--------

E from eqn. (1); G from eqn. (2); density, $\rho = 1.07 \times 10^3$ kgm/m^3; H is the hardness. $H \cong 3\sigma_0$ where σ_0 is the flow stress.

Table 1 is data taken from several separate sources and comments on the variability of the data are in order. The starting source of the material is, of course, the human body and there is no control over purity, structure, size and history after removal from the bladder (the stones dry out) all of these add to property variability. Indeed even within a single stone or a matched pair of stones from the same gall bladder there are large differences in the materials and mechanical properties.

Fracture Mechanics

The critical stress intensity factor, K_c, for crack propagation [12-14] has not previously been measured nor reported for gallstones. Ideally, a dynamic measurement would provide the critical stress intensity factor that must be applied to cracks in gallstones to initiate crack propagation. A dynamic high strain-rate value would differ from a static loading value. However, considering the intrinsic variability in this material initial static measurements will provide data for outlining the fracture response of the stones.

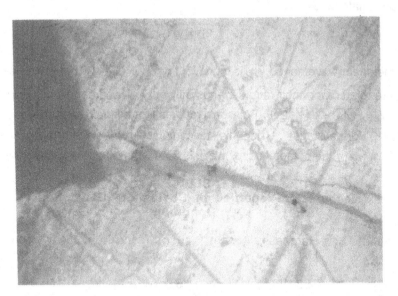

Figure 5 A Vickers impression with heavy loads gives microcracks at the corners of the impressions. These cracks are used to estimate the critical stress intensity factors for cracks propagation in gallstones.

The fracture toughness has been measured in our laboratory using microindentation techniques on polished surfaces of these small irregularly shaped gallstones. The samples were mounted in a cold curing epoxy, sectioned and polished. The flat sections were then indented with a 600 gram Vickers diamond to induce cracks around a microhardness impression. Figure 5 is a photomicrograph of a stone that has been indented and cracked. It can be seen that the cracks extend far from the hardness impression so this material is clearly very brittle. The quantitative fracture toughness from fracture hardness values is found using the following equation for K_c.

$$K_c = .025 \, E^{1/2} \, a \, \frac{P^{1/2}}{c^{3/2}} \tag{3}$$

In this case the cracking occurs on unloading due to residual stresses [23]. In the equation a is half the diagonal of the hardness impression, P is the load, E is still Young's modulus and c is the crack length. If the cracking occurs during loading [24] then ,

$$K_c = .04 \, E^{1/2} \, (c \, a)^{1/2} \frac{P^{1/2}}{c^{3/2}} \tag{4}$$

in both of these cases, K_c is very low. It of course means that the material is very brittle. Equations (3) and (4) give values of 100 and 200 kPa$\sqrt{m}$ respectively for the gallstones measured.

Fatigue Cracking

Gallstones that have been broken *in vitro* have been inspected for possible evidence of fatigue crack propagation. The observations that stones frequently fracture on a plane nearly normal to the incident acoustic rays and in the upper half of the stone i.e. on the part of the stone that is opposite the incident ray and near the caustic line. This suggests that for these stones fracture not cavitation is the major process for failure. Furthermore since most stones break after thousands of shots or acoustic tone bursts this suggests that fatigue or more specifically fatigue crack propagation is possible and likely. The evidence presented in this paper suggests that the plane strain, plastic zone size, r_p, at a crack tip in the gallstone would have a size of

$$r_p \cong \frac{1}{6\pi} \left(\frac{K_c}{\sigma_0}\right)^2 \tag{5}$$

Taking the critical stress intensity factor K_c as 150 kPa $\sqrt{m}$ and σ_0 as one third of the hardness, i.e. 15 MPa gives a plastic zone of about 5 μ m. Fatigue crack striations should be faint but observable with this size plastic zone. Striations were looked for in a scanning electron microscope on the fracture surfaces of stones that only broke into several parts. (Stones that are broken many times seem to be easier to repeatedly break and may have different fracture processes.) The evidence for fatigue striations is presented in Figure 6. These markings are not as clear as those that are seen on fatigued metal surfaces. In Figure 6, there are fatigue striation surface markings that suggest fatigue crack propagation is occurring. The spacing between markings suggests that hundreds or thousands of cycles are necessary for stones to completely break.

Figure 6. A scanning electron micrograph: gold coated, *in vivo* fractured, gallstone surface. Faint fatigue crack striation are diagonal lines. Field of view 50 μm.

CONCLUSIONS

Some preliminary information and observations have been made on cholesterol gallstones. The conclusion is that the stress waves in the gallstones are focused on caustic lines and furthermore that fracture is by fatigue crack propagation. Complete details and observations of dynamic stone failure *in vivo* remains as a challenging problem. Several measurements of physical properties have been used to deduce the mechanical properties that are necessary for describing possible dynamic failure mechanics in this important material. These preliminary mechanical values are reported in this paper.

ACKNOWLEDGEMENTS

Partial support from National Institutes of Health through USPHS Grant No. DK 39796 is acknowledged. Discussions with E. Carr Everback have been helpful.

REFERENCES

1. Chaussy, C., Brendel, W. and Schmiedt, E., Extracorporeally induced destruction of kidney stones by shock waves. Lancet, 1980, **20**, 1265-68.

2. Delius, M., Brendel, W. and Heine, G., A mechanism of gallstone destruction by extracorporeal shock waves. Naturwissenschaften, 1988, **75**, 200-1.

3. Coleman, A.J., Saunders, J.E., Crum, L.A. and Dyson, M., Acoustic Cavitation generated by an extracorporeal shock wave lithotripter. Ultrasound in Medicine and Biology, 1987, **13**, 69-76.

4. Barkun, A.N.G. and Ponchon, T., Extracorporeal biliary lithotripsy review of experimental studies and a clinical update. Annals of Internal Medicine, 1990, **112**, 126-37.

5. Kolsky, H., Stress Waves in Solids. Dover Publications, New York, 1963, pp. 186-196.

6. Rossmanith, H.P., International Seminar on Dynamic Failure of Materials. Technical University of Vienna, Vienna, January 1991. (personal communication).

7. Ting, T.C. T. and Lee, E.H., Wave-front analysis in composite materials. Journal of Applied Mechanics. 1969, **36**, 497-504.

8. Achenbach, J.D., Hemann, J.H. and Ziegler, F., Separation at the interface of a circular inclusion and the surrounding medium under an incident compressive wave. Journal of Applied Mechanics. 1970, **37**, 298-304.

9. Siva-Gomes, J.F., Al-Hassani, S.T.S. and Johnson, W., A note on times to fracture in solid perspex spheres due to point explosive loading. Int. J. mech. Sci.. 1976, **18**, 543-45.

10. Johnson, W. and Mamalis, A.G., Fracture development in solid perspex spheres with short cylindrical projections (bosses) due to point explosive loading. Int. J. mech. Sci., 1977, **19**, 309-14.

11. Lovell, E., Al-Hassani, S.T.S. and Johnson, W., Fracture in solid spheres and circular disks due to a "point" explosive impulse on the surface. Int. J. mech. Sci., 1974,**16**, 193-99

12. Caddell, R.M., Deformation and Fracture of Solids. Prentice-Hall, Inc., New Jersey, 1980, pp. 289-301.

13. Paris, P.C., Fatigue - an interdisciplinary approach. Proc. 10th Sagamore Conf. ,Syracuse University Press, Syracuse, NY 1964, pp. 107-32.

14. Hellan, K, Introduction to Fracture Mechanics, McGraw-Hill, Inc., New York, 1984, pp. 125-137.

15. Debenedetti, P.G. and D'Antonio, M.C., Stability and tensile strength of liquids exhibiting density maxima. A I Ch E Journal. 1988, **34**, 447-55.

16. Debenedetti, P.G. and D'Antonio, M.C., On the entropy changes and fluctuations occurring near a tensile instability. J. Chem. Phys., 1986, **85**, 4005-10.

17. Ying, C.F. and Truell, R., Scattering of a plane longitudinal wave by a spherical obstacle in an isotropically elastic solid. J. Appl. Phys., 1956, **27**, 1086-97.

18. Timoshenko, S. and Goodier, J. N., Theory of Elasticity, McGraw-Hill, New York, 1951, pp. 362-82.

19. Holtum, D. and Faust, U., Bestimmung physikalischer Kenngrößen menschlicher Gallensteine. Wechsel wirkungen: Aus Lehre und Forschung der Universität Stuttgart. Jahrbuch 1989, pp. 29-36.

20. Singh, V.R. and Agarwal, R., Ultrasonic studies of gall-bladder stones. Ultrasonics. 1989, **27**, 114-5.

21. Gracewski, S.M., Vakil, N., Everbach, E.C., Davis, M.E. and Burns, S.J., Microhardness properties of human gallstones and synthetic stones. Journal of Biomechanical Engineering. 1991, submitted.

22. Stramme, S.K., Cocks, F.H. and Gettliffe, R., Mechanical property studies of human gallstones. J. Biomed. Mat. Res., 1990, **24**, 1049-57.

23. Ostojic, P. and McPherson, R., A review of indentation fracture theory: its development, principles and limitations. International Journal of Fracture. 1987, **33**, 297-312.

24. Chia, K-Y., Indentation fracture and crack tip deformation in ionic and diamond cubic ceramics. Ph.D. Thesis, 1985, University of Rochester, Rochester, N Y, pp. 47-50.

STRESS WAVE FOCUSING INDUCED FRACTURE - A PHOTOELASTIC STUDY

H.P. ROSSMANITH and R.E. KNASMILLNER
Institute of Mechanics
University of Technology Vienna
Wiedner Hauptstraße 8-10/325, A-1040 Vienna, Austria

ABSTRACT

Reflection and superposition of stress waves from convex boundaries leads to stress wave focusing induced fracture in brittle materials. High speed recording techniques in conjunction with methods of photomechanics serve as a potential tool for visualization of highly complex interaction processes between stress waves and material inhomogeneities and discontinuities such as boundaries, interfaces, cracks etc.. Applications to non-surgical treatment of kidney stone crushing is discussed.

INTRODUCTION

The development of internal fractures in brittle solids due to wave focusing upon reflection of incident stress waves at convex surfaces is well known in the literature [1, 2, 3, 4]. Shock and explosive loadings give rise to sharply rising stress pulses which, when tensile and superimposed, may locally exceed the fracture strength of the brittle material and local fracture may occur.

In this contribution photoelastic experiments performed with Araldite B specimens and associated analytical wave constructions are described. High speed photography in conjunction with dynamic photoelasticity has been used for fringe pattern recording. The results of the high speed photographs are then interpreted in the light of relevant concepts of non-surgical treatment of kidney stones.

THEORY

When a homogeneous, isotropic, linear-elastic solid is subjected to localized explosive excitation, two body waves (the longitudinal or P-wave and the shear or S-wave) will spherically expand with constant velocities, (c_P and c_S), respectively, from the detonation site. In addition, Rayleigh surface waves (R-waves) will propagate with speed c_R along free boundaries of the specimen. These waves will propagate undisturbed unless interaction with inhomogeneities takes place. Upon reflection of the incident stress waves a highly complex wave pattern may emerge, which is composed of incident and reflected waves. An incident detonation-borne compressive elastic P-wave (S-wave) of finite pulse length will upon reflection at a free convex boundary give rise to the generation of reflected PP- and SP-waves (PS- and SS-waves), which due to their finite pulse length will gradually superimpose in time. The center of this transient region of wave superposition travels along a particular curve, the stress wave caustic. Shape and characteristics of the caustic depend on various parameters such as the geometry of the reflector and shape and profile of the stress wave pulse.

Employing high speed recording techniques as e.g. in dynamic photoelasticity instantaneous "snap-shots" of the stress wave patterns can be recorded, each of which contains points of the caustic, thus serves for path-finding to trace the caustic. Hence, the complete caustic can at no time photographically be captured in full but constructed only from sequences of e.g. isochromatic fringe patterns.

From arguments of stress wave superposition and from experimental evidence high stresses are expected along the caustic curve. If, as is the case with strict convex bodies, cusp formation takes place fracture may occur in the cusp region. Four different first order stress wave caustics will be formed in a convex body when a point disturbance is active at the surface. These caustics will be labelled PP, PS, SP and SS caustics. Higher order caustics due to multiple reflection of incident wavelets are possible but because of their low intensity they may be neglected.

EXAMPLES

Stress wave caustics in disk-shaped bodies

Consider the local stress concentration due to geometrical focusing in a disk-shaped specimen subjected to an explosive point excitation at the boundary. The caustics will be symmetric with respect to the diameter passing through the detonation site (Fig. 1).

The sequence of photoelastic recordings accompanied by their associated wave constructions based on geometrical ray theory shown in Fig. 1 illustrates the process of caustic formation in the disk specimen.

Frame #3, taken at $t = 88\mu s$ after detonation of the explosive ($150mgPbN_6$) shows the early phase where P- and S-waves propagate largely undisturbed through the body. Single and multiple reflection gives rise to the formation of reflected PP-, SP- and SPP-waves. The cusps of these curves propagate along their associated caustic curves. A symmetric pair of Rayleigh waves propagates along the free edges of the disk. The associated construction of wave maxima shows excellent agreement with the photoelastic fringe

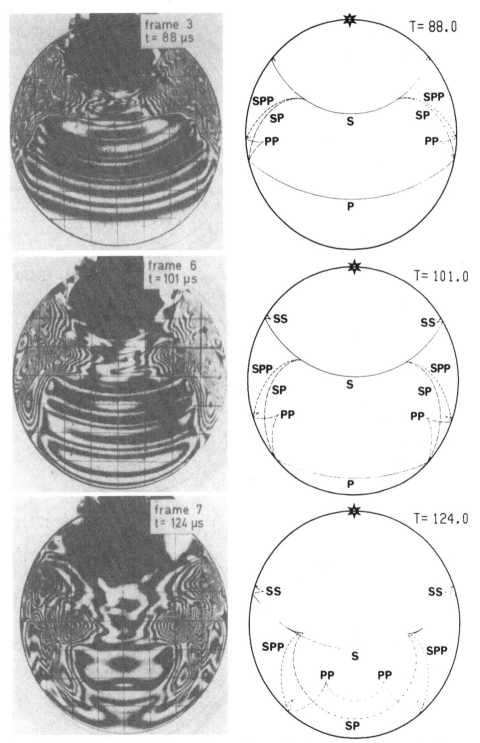

Figure 1: Isochromatic recordings and associated wave constructions for the formation of stress wave caustics in a circular disk (cont. next page).

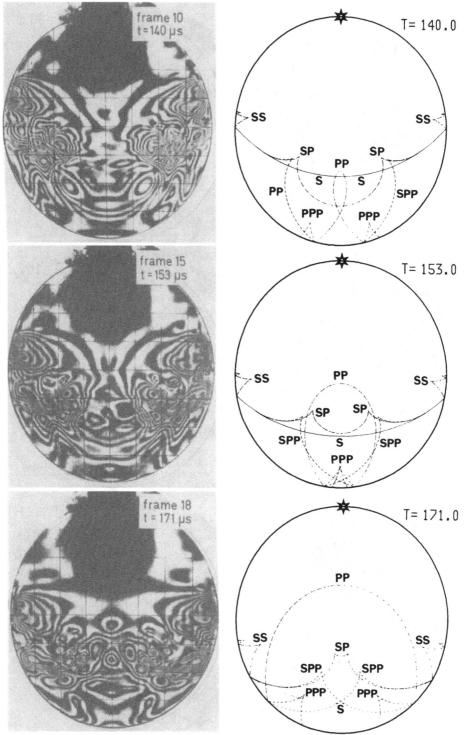

Figure 1(cont.): Isochromatic recordings and associated wave constructions for the formation of stress wave caustics in a circular disk.

pattern.

In frame #7, taken 36μs later, the P-wave has been reflected at the free surface of the disk and the two symmetric PP-cusps approach to merge at the site of the cusp of the PP caustic. In the isochromatic fringe patterns the PP-cusps are identified as two symmetric white regions enclosing a black area at about a distance of 2/3 of the disk diameter from the top. The SP-wave, generated by mode conversion of the incident P-wave also travels along the fictitious SP caustic with the object of forming another stress amplification at the site of the cusp of the caustic labelled SP.

The next frame, #10, taken at $t = 140\mu s$, shows several interesting features: formation of the cusp of the PP caustic, convergence of SP-maxima and the formation of a secondary (higher order) caustic by the multiple reflected PPP-waves. The latter event is fully demonstrated in the wave pattern of frame #15.

Finally, in frame #18 incipient formation of several stress wave caustics can clearly be recognized. First order caustics associated with SP- and SS-waves and second order caustics pertaining to PPP- and SPP-waves are about to cusp at their respective focusing sites. In addition, one may easily follow the spread and decay of the PP-wave as well as of the secondary PPP-wave. Fig. 2 shows the set of stress wave caustics along which the respective wave peaks travel to meet at the cusp points thus forming high stress concentrations by focusing. The figure holds for a disk-shaped body subjected to an explosive point disturbance acting at the boundary.

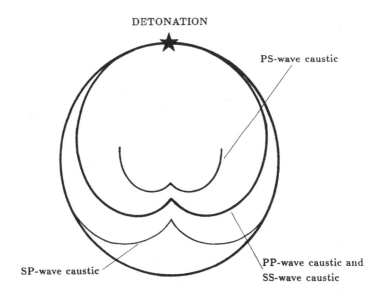

Figure 2: Set of stress wave caustics associated with an explosive point disturbance acting at the boundary.

It is interesting to note that during this wave interaction as well as caustic and cusp formation process the S-wave has not yet taken an active role with respect to the possibility of initiating fracture. Fracture may first occur just before frame #10 where the reflected symmetric tensile PP-waves superimpose and the amplitude of the highly localized stress intensification may reach the level of fracture strength. In terms of fracture mechanics, fracture initiation will occur when the stress-wave-induced applied stress intensity factor will be equal to the critical value, the fracture toughness of the material. Within the framework of fracture mechanics a small flaw or crack is assumed to be present in the cusp region. Fracture formation due to stress wave focusing has been reported in the literature, e.g. [1, 2, 3].

Influence of local non-convexity

If the reflecting boundary shows a localized non-convexity (imperfection) PP-cusp formation may be suppressed but there is still a chance for fracture initiation when the symmetric reflected tensile PP-waves superimpose. This is shown in the wave front constructions of Fig. 3, particularly in frame #3c associated with time $t = 90\mu s$ after detonation.

It is interesting to note that the "severity" of the imperfection does play a crucial role as to which reflected wave system will cusp. For the case shown in Fig. 2 the primary reflected waves PP, SP, SS and PS do not cusp whereas the secondary reflected wave systems, e.g. PPP, SPP, SSP, SSS, PSP, etc. do lead to cusp formations. This is clearly illustrated in the frames pertaining to later times.

The reader will agree that the resulting wave pattern ($t = 140\mu s$) bears considerable complexity. While representing an almost complete picture of stress wave focusing these patterns moreover fascinate by their intrinsic aesthetics.

Elliptical disk with internal excitation

Consider the special but important case where an elliptically shaped specimen is explosively excited at one focal point. Theoretically, this rotational symmetrical problem should give rise to one emerging P-wave only. This will be confirmed if the experiment is performed with very ductile model materials such as Lexan. In brittle behaving materials break-down of the borehole (needed for placing the explosive) also initiates a strong shear wave.

The sequence of wave front constructions based on geometrical ray theory is shown in Fig. 4. As one would expect from linear theory of acoustics the PP- and SS-waves will focus down on the dual focal point of the ellipse with no cusps being formed.

This is shown in frames #6 and #12, respectively, of Fig. 4. According to Snell's law, the SP- and PS-waves (and in fact all the other mode-converted reflected waves) will not focus on the focal point of the ellipse but converge to and cusp at their proper sites of stress intensification. For the SP-wave, focusing and cusping occurs over an interval of approximately $25\mu s$, beginning in frame #7 ($t = 80\mu s$) with two-stage formation of a double cusp proceeded by wave superposition which leads to strong stress intensifications in frame #8 ($t = 90\mu s$). Finally, two swallow-tail formations (PS) diverge as shown in frames #9 and #10. Focusing and cusping of the PS-wave has occurred approximately at time $t = 105\mu s$ as one can deduct from an investigation of the constructions shown in frames #10 ($t = 110\mu s$) and #11 ($t = 120\mu s$).

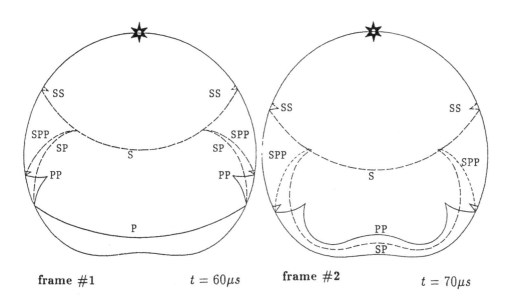

frame #1 $t = 60\mu s$ frame #2 $t = 70\mu s$

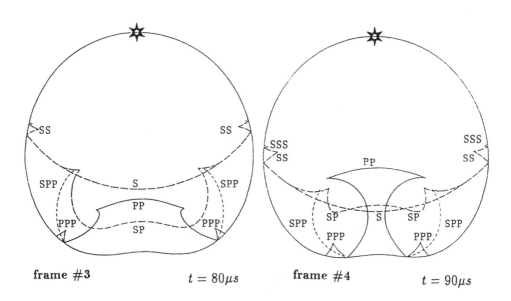

frame #3 $t = 80\mu s$ frame #4 $t = 90\mu s$

Figure 3: Stress wave focusing and amplification in an imperfect locally non-convex body due to an explosive point disturbance acting at the boundary (cont. next page).

134

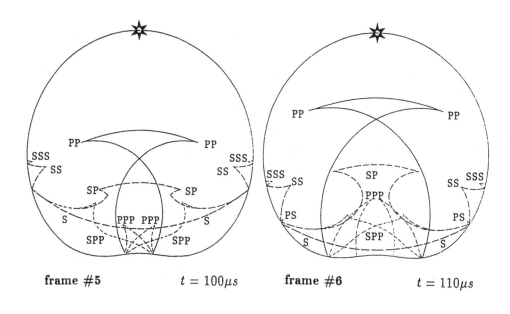

frame #5 $t = 100\mu s$ frame #6 $t = 110\mu s$

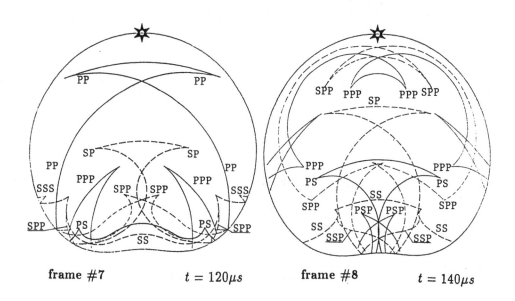

frame #7 $t = 120\mu s$ frame #8 $t = 140\mu s$

Figure 3(cont.): Stress wave focusing and amplification in an
imperfect locally non-convex body due to an explosive point
disturbance acting at the boundary.

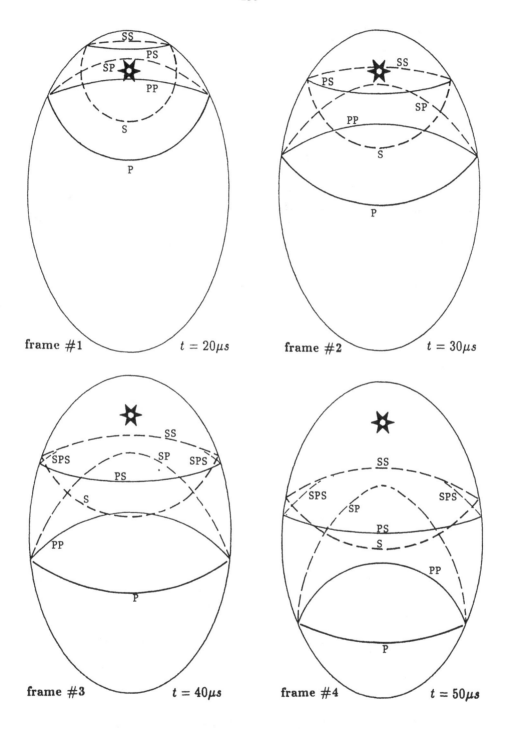

Figure 4: Sequence of wave front constructions associated with stress
wave focusing in an elliptical reflector/focuser due to an explosive point
disturbance located at one of the focal points (cont.next page).

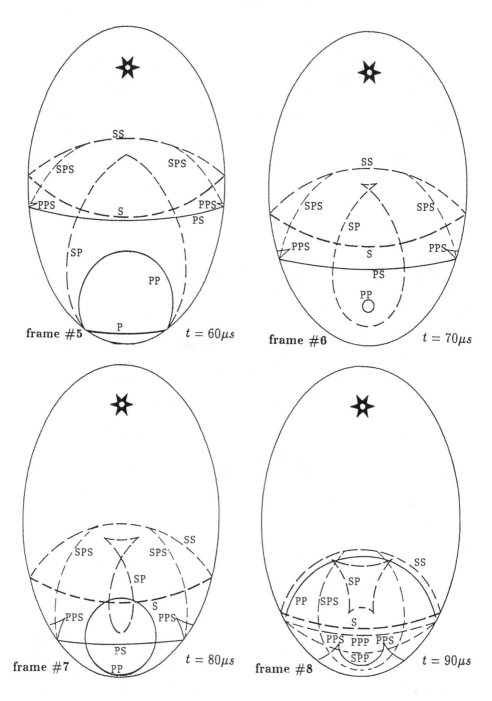

Figure 4(cont.): Sequence of wave front constructions associated with stress wave focusing in an elliptical reflector/focuser due to an explosive point disturbance located at one of the focal points (cont. next page).

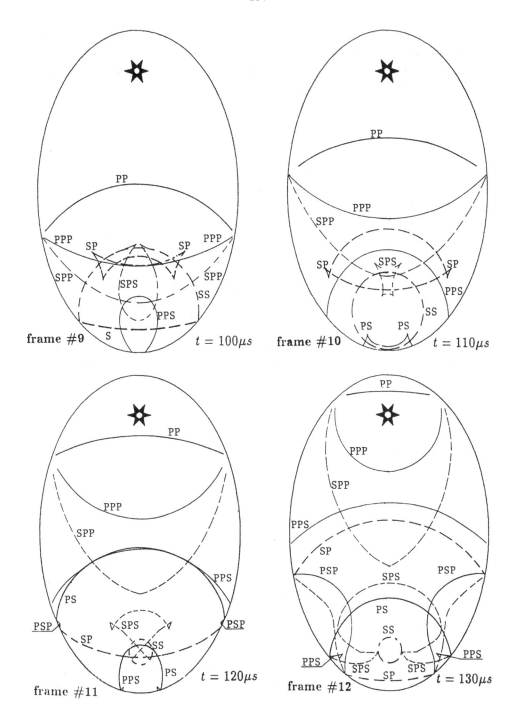

Figure 4(cont.): Sequence of wave front constructions associated with stress wave focusing in an elliptical reflector/focuser due to an explosive point disturbance located at one of the focal points.

It is interesting to note that the one focal point where the disturbance originated becomes the focusing center again for (pure mode) secondary waves such as PPP, SSS etc.. This becomes obvious when one compares frame #12 ($t = 130\mu s$) with frame #5 ($t = 60\mu s$). Mode-converted stress waves correspondingly develop self-interacting focusing patterns with associated amplifications.

DISCUSSION AND CONCLUSIONS

Dynamic photoelasticity is an extremely suitable tool for the investigation of stress wave focusing effects in convex and semi-convex brittle bodies susceptible to fracture initiation under high tensile stress amplifications. Both, spalling (with the fracture perpendicular to the axis of symmetry) due to reflected and superimposed tensile waves, as well as splitting (with the fracture plane along the line of symmetry) may be the result of a complex interplay of stress wave focusing and cusping and simple superposition.

The type of failure, i..e. spalling (normal crack) or splitting (parallel crack) will depend on the structure of the material as well as on the particular shape of the reflector/focuser. These results obtained from theoretical considerations and experimental recordings are in agreement with observations made with lithotripsy in connection with kidney stones and gallstones where normal and parallel cracks have been reported [5, 6].

Further detailed experimental and theoretical investigations are necessary for further elucidation of the physical processes involved in the failure of man-borne stones.

ACKNOWLEDGEMENT

The authors would like to acknowledge the financial support of this work (project # P 5814) granted by the *Austrian Science Foundation (FWF)*.

REFERENCES

[1] J.F. Silva Gomes and S.T.S. Al-Hassani. Internal fractures in spheres due to stress wave focussing. *Int.J. Solids Struct.*, 13:1007–1017, 1977.

[2] S.T.S. Al-Hassani and J.F. Silva Gomes. Internal fractures in solids of revolution due to stress wave focussing. In: M. Meyers et.al. (Editor), *Proc. Int. Conf. Shock Waves in Metals*, pages 169–18), Albuquerque, 1980.

[3] J.S. Rinehardt. *Stress Transients in Solids*. Hyperdynamics, Santa Fe, NM, USA, 1975.

[4] H.J. Ballmann, H.J. Raatschen, and M. Staat. High stress intensities in focussing zones of waves. In: P. Ladeveze (Editor), *Proc. of Local Effects in the Analysis of Structures*, pages 235–252. Elsevier, 1985.

[5] S.J. Burns, S.M. Gracewski, N. Vakil, and A.R. Basu. Dynamic failure mechanisms in gallstones using lithotripsy. This volume.

[6] K. Takayama. Private communication. Sendai, 1989.

NONLINEAR EFFECTS OF FRACTURE OF CERAMICS IN STRESS WAVES

VLADIMIR BELYAEV and OLEG NAIMARK
Institute of Mechanics of Solids,
Ural Dept. of the USSR Academy of Sciences,
1 Ak. Korolev Str., 614061 Perm, USSR

ABSTRACT

Relationships of ceramics fracture in shock waves are examined on a basis of a statistical-thermodynamic description of behaviour of defects. Constitutive equations reflect an essentialy non-linear character of kinetics of microcrack accumulation and take into account physical aspects of a fracture process. It is shown that the media with microcracks has characteristic features of non-linear systems with dissipation in which non-equilibrium (kinetic) transitions may occur. A process of damage localization is described. The kinetics of microcrack accumulation in a region of damage localization is characterized by explosive instability (a blow-up solution) and leads to formation of the new type of dissipative structures – blow-up structures. The dissipative structure formation is indentified with the generation of hotspots of a macroscopic fracture (macrocracks). The conditions of the formation of one or some structures is found. The modelling of the effects of the damage localization and of the kinetics of the dissipative structure development allows to describe the single and multiple spalling.

INTRODUCTION

The study of deformation behaviour and solid fracture under impact waves, including those initiated by intensive laser beams is related to nonlinear response analysis of systems beeing far from equilibrium. Such systems are characterized by a resonance dependence of the system response to external forces, which gives rise to self-sustaining dissipative structures with varying complexity [1,2]. Here, localization effects and blow-up regimes in media, containing microcracks are examined relative to the fracture analysis of ceramic plates subjected to impact wave loading and laser radiation flux. Theoretical treatment of fracture in solids at impact wave loads requires a thorough analysis of equations of state, their physical grounds and how completely they represent structural changes of material in a wide range of rates and intensities of loading. Since the direct observation of deformation behaviour and fractures under impact waves is not feasible in most cases of practical interest, the application of computational experiment methods to the analysis of these phenomena assumes ever greater importance.

CONSTITUTIVE EQUATIONS OF MEDIA WITH MICROCRACKS

One of the most interesting failure patterns is spallation produced by impact waves, including those initiated by intense heating due to absorption of laser radiation [3-6]. The spall is characterized by the small times ($\sim 10^{-7} - 10^{-6}$ s) and the large maximum tensile stresses, exceeding by several times the static ultimate strength. Despite this peculiarity, spallation is similar by its structural features to the process of quasi-static damage i.e. is folowed by multiple nucleation and growth of microcracks and localization of their clasters, making thereby an explicit evolutionary description of defective structures experienced deformation and fracture, most desirable. A statistical description of the behaviour of interacting microcrack ensemble in solids under load is given

elsewhere [7,8]. Here, the symmetrical tensor p_{ik} is assumed
as a structural parameter specifying the volume concentration
and preferred orientation. This tensor is defined by averaging
over the ensemble of microcracks, each of them beeing
characterized by the microscopic tensor $s_{ik} = s \, \nu_i \nu_k$:

$$P_{ik} = n \int s_{ik} W(s, \vec{\nu}) \; ds \; d^3\vec{\nu},$$

where n is a number of microcracks in the unit-volume, s is
the volume of microcrack; $\vec{\nu}$ is the normal vector to its
flanks; $W(s, \vec{\nu})$ is the function of the microcrack size and
orientation distribution.

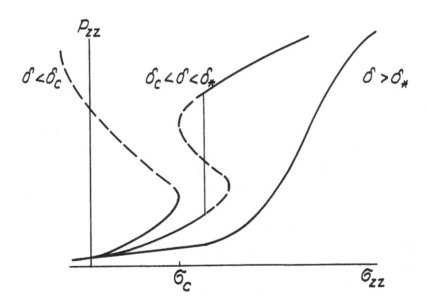

Figure 1. A response of solids to nucleation of microcracks.

Figure 1 gives a plot of the microcrack density tensor
component versus the stress σ_{zz} for specimens under uniaxial
tension at various values of the structural parameter δ. The
value of δ is determined by two characteristic scales: the
averaged measure of the material structure heterogencity and

correlation radius of the overstress fields produced by microcracks. Investigation, reported in [8], has readily shown the existance of three qualitatively different material responses to microcrack growth accoding to variation in δ: at $\delta > \delta_*$ this response is described by the monotonic dependence $p_{zz}(\sigma_{zz})$; in the range $\delta_c < \delta < \delta_*$ the dependence is characterized by the parameter p_{zz} jump, related to the orientation transition in a microcrack system; at $\delta < \delta_c$ the stress range is subdivided into the range of values $\sigma_{zz} < \sigma_c$, with p_{zz} beeing controlled by the thermodynamic branch and that of absolute instability $\sigma_{zz} > \sigma_c$ where the material is extremely sensitive to the microcrack growth.

The constitutive equations, interrelating kinetics of microcrack accumulation and relaxation properties of solids,

$$\sigma_{ik} = l_1 e^P_{ik} - l_2 dp_{ik}/dt,$$

$$\Pi_{ik} = l_2 e^P_{ik} - l_3 dp_{ik}/dt$$

have been obtained with methods of nonequilibrium thermodynamics, using the results of statistical description [8]. Here, $\Pi_{ik} = \partial F/\partial p_{ik}$ is the thermodynamical force, experienced by the system at the nonequilibrium value of p_{ik}, F is the free energy of microcracked medium; σ_{ik}, e^P_{ik} are the components of stress and rate of deformation tensors, l_α are the coefficients, which generally depend on p_{ik}; $d(\ldots)/dt$ is a tensor derivative with respect to time (Jaumann derivative). The developed model has been used to describe regularities of damage and fracture processes, developed in solids under creep conditions [7], of superplastic flow [9], low cycle fatigue damage [10] and fracture under stress wave loading [11-14].

LOCALIZED 'BLOW-UP' STRUCTURES IN SPALL DAMAGE PROCESS

The analysis has been made of spallation phenomenon in a ceramic plate, initiated by propagation of the plane one-dimensional stress wave in direction of z - axis.

In this case, the system of equation of state for microcracked medium, combined with the laws of mass and momentum coservation and equation of heat conduction has the form:

$$\sigma_{zz} = l_1 e^P_{zz} - l_2 \partial p/\partial t, \qquad \partial p/\partial t = (l_2 e^P_{zz} - \Pi_{zz})/l_3,$$

$$\partial \rho/\partial t = -\partial(\rho\, v_z)/\partial z,$$

(1)

$$\partial(\rho\, v_z)/\partial t = -\partial(\rho\, v_z^2 - \sigma_{zz}),$$

$$\partial T/\partial t + v_z \partial T/\partial z = a\, \partial^2 T/\partial z^2 + D(z,t)$$

Here ρ is the material density, v_z is the component of velocity vector, T is the temperature, a is the coefficient of thermal conduction, $D(z,t) = Q(t)\, \exp(-\varkappa z)$ is the absorption function, depending on the density of laser radiation flux $Q(t)$ and the absorption coefficient $\varkappa$. Tensor e^P_{zz} is calculated as difference $e^P_{zz} = e_{zz} - \partial u^\circ_{zz}/\partial t$, where $e_{zz} = \partial v_z/\partial z$ is the strain rate ; $u_{zz} = (K + \frac{4}{3}\mu)^{-1}[\sigma_{zz} + K\, \alpha\, (T - T_0)]$ is the elastic strain tensor, K and μ are the volumetric and shear moduli, α is the thermal expansion coefficient . The function Π_{zz} , represented in [10] in terms of statistical integrals, is approximated by the expression:

$$\Pi_{zz} = -\sigma_{zz} \exp(A_1 p^C_{zz}) + B\,(\delta - \lambda)(p_{zz} - p_0 \delta),$$

$$A_1 = A\, \text{sign}(\delta_c - \delta), \qquad C = \text{sign}(\delta_c - \delta).$$

The values of coefficients and approximation parameters are determined from quasistatic and dynamic testing data of ceramic (85% Al_2O_3, 15% SiO_2) rod [15].

The flux density of the single laser pulse is given by the law:

$$Q(t) = Q_0 \begin{cases} t/s_1, & 0 \le t \le s_1, \\ (\tau_p - t)/s_2, & s_1 < t \le \tau_p, \end{cases}$$

where Q_o is the amplitude of the radiation flux density, τ_p is the pulse duration, s and s_2 are times of pulse rise and decay, respectively. The equations (1) is supplemented by boundary conditions:

$$\sigma_{zz}(0,t) = \sigma_o(t), \qquad \sigma_{zz}(h,t) = 0, \qquad t \in [0,\infty)$$

and initial conditions:

$$v_z(z,0) = \sigma_{zz}(z,0) = p_{zz}(z,0) = 0, \qquad \rho(z,0) = \rho_o, \qquad z \in [0,h].$$

Numerical analysis has shown that the compressive stress wave produced by impact load is reflect from the free surface of the plate and due to its interaction with the target counterwave there develops a region of compressive stress, giving rise to intensive nucleation and growth of microcracks. This is followed by an asymptotic change $p_{zz}(p_{zz} \to \infty)$ while passing through the instability threshold $\delta = \delta_c$. In the examined situation the damage is characterized by nucleation of dissipative structures i.e. a space-localized region, in which microcracks grow at the essentially high rate than outside the structure, providing evidence for qualitative analogy to the phenomenon of heat localization under the blow-up conditions [16]. Thus, an organization of new type evidently appears in the medium, which couples a coherent space-time behaviour of the microcrack ensemble with dynamic and relaxation processes inside the system. It is reasonable to attribute the occurrence of such dissipative structure to the creation of conditions favourable for nucleation of qualitively new defect - an macroscopic crack.

As it follows from the quations (1) the damage evolution is defined, on the one hand, by the relaxation processes, and, on the other hand, by the intensity of the microcrack origin $\Pi_{ik} = \partial F/\partial p_{ik}$, which is estimated by the rate of the free energy decrease as the volume concentration of microcracks is increasing. The interaction of these competitional mechanisms governs the space-time evolution of the damage process.

The predicted distributions p_{zz} in figure 2 show a decrease in the half-width of the localized damage region, which allows to draw a qualitative analogy with the so-called LS-regimes of combustion in the non-linear medium [16]. It is characteristic that in both processes the size of localization region is determined not only by the properties of medium, but also by the energy of initial disturbance. A similar situation can be observed under conditions of impact waves, when the size of the stress pulse at the moderate amplitudes (4-5 GPa) defines, in principle, a coordinate and breaking moment. It should be noted, however, that the proper reason for the initiation of the macroscopic damage origin (dissipative structure, developing in the blow-up regime) is a microcrack distribution, caused by the stress, rather than the stresses by themselves. Under the spall conditions damage process usually covers a considerable sectional area of the material, while the progress in the blow-up regime is possible only inside the characteristic (fundamental) length with maximum value of τ_c - the time of blow-up [16]. The increased loading pulse gives rise to a number of structures localized at the neighbour fundamental lengths (figure 2b). In this case the appearance of more complicated dissipative structures is in accordance with the observed transition from a single spall to a multiple one [17].

In the analysis of damage process at intense loads the description of its stage character is of primary concern , since a transition to the avalanche growth of microcrack concentration implies the exhaustion of phisical strength in a solid. Investigation of localization phenomena and blow-up regimes reveals three stages in evolution of thermal disturbance . The analogous stages of microcrack accumulation may be distinguished in the present case of fracture. During the first, subcritical stage the growth p_{zz} is controlled by thermodynamical branches of the dependence $p_{zz}(\sigma_{zz})$; the second stage is associated with passing through the instability threshold and characterized by the time interval of coming to the self-similarity solution, consistent with the blow-up regime; the third stage is the blow-up regime proper,

146

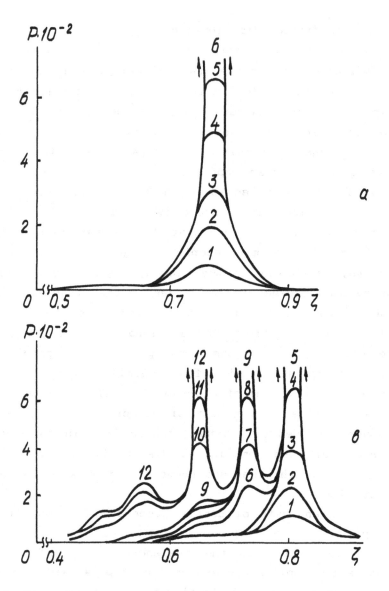

Figure 2. Distribution of microcrack density parameter $p = p_{zz}$ along the plate thikness $\zeta = z/h$ at variations times: a) $\sigma_o = 4 \times 10^9$Pa ; the curves 1-6 correspond to $t = 2.239$, 2.273, 2.289, 2.298, 2.306, 2.307 μs; b) $\sigma_o = 7 \times 10^9$Pa ; the curves 1-12 correspond $t = 2.1$, 2.2, 2.2036, 2.204, 2.2041, 2.35, 2.49, 2.5, 2.51, 2.58, 2.6, 2.61 μs.

which involves an avalanche localized growth of microcrack volume concentration, terminated by nucleation of a macroscopic crack and fracture of the plate.

SPALL DAMAGE UNDER RADIATION OF LASER PULSE

A rapid heating of a substence due to radiation may also result in the spall damage of material. In the analysis of spall phenomena in a layer of material subjected to radiation of a single laser pulse, a careful consideration is given to the solution of the equations (1) in the rangle $R \in [0 \le z \le h, 0 < t < \infty]$, which satisfies the boundary conditions:

$$\sigma_{zz}(0,t) = \sigma_{zz}(h,t) = 0, \qquad \frac{\partial T}{\partial z}(0,t) = \frac{\partial T}{\partial z}(h,t) = 0$$

and initial conditions:

$$T(z,0) = T_o, \ v_z(z,0) = \sigma_{zz}(z,0) = p_{zz}(z,0) = 0, \ \rho(z,0) = \rho_o.$$

Figure 3 demonstrates the predicted results for kinetics of microcrack accumulation and plate thickness distribution of stresses. A region of damage localization with abruptly increasing rate of microcrack accumulation is observed under conditions of compressive stress wave. Displacements of the local maximum of the microcrack concentration is due to propagation of the wave pulse front, though in the asymptotic range the maximum is stabilized and the process of microcrack accumulation in a local region exhibits characteristic features of the blow-up regime. As in the above mentioned cases, a creation of such origin procedes the nucleation of macroscopic fracture in the form of the frontal spall similar fracture situations may occur at a rather low concentration of energy in the absorbing layer when melting and evaporation of the substance is infinitesimal, while the absorption depth of energy radiation is large enough. The values of material physical constants, used in calculations are equivalent to the constants of coloured glasses (coefficient of absorption is ˜ $10 \ cm^{-1}$).

148

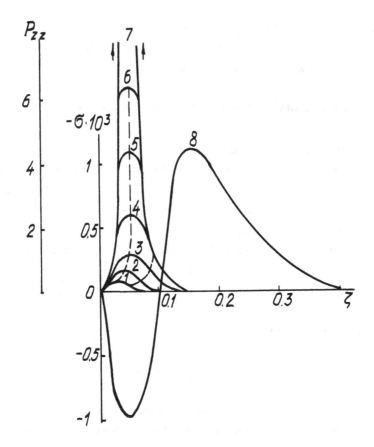

Figure 3. Kinetics of microcrack accumulation under condition
of tensile wave, initiated by the laser pulse. The
dependence p(t) (the curves 1-7) correspond to the
times   t = 0.239, 0.025, 0.0256, 0.0262, 0.02651,
0.02656, 0.02657 μs; 8 - is the stress distribution
σ = σ$_{zz}$/μ,   t = 0.025   μs.

The analysis of the space-time distribution of stresses
and volume density of microcracks shows that the applied
loading pulse, including thermal shock, initiates nucleation
of localized blow-up regimes in microcracked media.

REFERENCES

1. Nicolis, G. and Prigogine, I., Self-organization in Nonequilibrium Sistems. Wiley-Interscience, New-York, 1977, pp. 512.

2. Haken, H., Transition Phenomena in Nonlinear Sistems. In Stochastic Nonlinear Sistems in Phisics, Chemistry, and Biology. Proc. of the Workshop, Bielefeld, Fed. Rep. of Germany, October 5-11, 1980, eds. L. Arnold, R. Lefever, Springer-Verlag, Berlin, Heidelberg, New-York, 1981, 12-19.

3. Davison, L., Stevens, A.L., Continuum measures of spall damage. J. Appl. Phys., 1972, 43, N3, 41-59.

4. Curran, D.R., Shokey, D.A. and Seamon, L., Dynamic fracture criteria for a polycarbonate. J. Appl. Phys., 1973, 44, N9, 4025-37.

5. Seamon, L., Curran, D.R. and Shokey, D.A., Computational models for ductile and brittle fracture. J. Appl. Phys., 1976, 47, N 11, 4814-26.

6. Bakeev, A.P., Sobolev, A.P. and Yakovlev, V.I., Investigation of thermo-elastic stresses, appearing in a layer of material under action of laser pulse. Zh. Pricl. Mech. Tech. Phys., 1982, N 6, 92-98 (in Russian).

7. Naimark, O.B., Thermodynamics of deformation and fracture of solids with microcracks. In Preprint IMSS UNC AN SSSR., Sverdlovsk, 1982, N 22, 3-34. (in Russian).

8. Betekhtin, V.I., Naimark, O.B. and Silbershmidt, V.V., On fracture of solids with microcracks: experiments, statistical thermodynamics and constitutive equations. In Adv. Res. Fract. Proc. 7th Int. Conf. Fracture (ICF-7), Houston. Oxford, Pergamon Press, 1989, 6, 38-45.

9. Naimark, O.B., On void generation, constitutive equations and stability of superplastic deformation of metals. Zh. Prikl. Mech Tech. Phys., 1985, N 4, 144-150 (in Russian).

10. Naimark, O.B. and Davydova, M.M., On statistical thermodynamics of solids with microcracks and self-similarity of fatique failure. Probl. Prochn., 1986, N 1, 91-95 (in Russian).

11. Belyaev, V.V. and Naimark, O.B., Kinetic transitions in media with microcracks and failure of metals in stress waves. Zh. Prikl. Mech. Tech. Phys., 1987, N 1, 163-171 (in

Russian).

12. Naimark, O.B. and Belyaev, V.V., The kinetics of microcracks acculation and failure of solids in shock waves. Adv. Res. Fract. Proc. 7th Int. Conf. Fracture (ICF-7). Houston. Oxford, Pergamon Press, 1989, 6, 49-57.

13. Belyaev, V.V. and Naimark, O.B., Statistical thermodynamics of solids with microcracks and non-linear aspects of fracture in shock waves. In EUROMECH Colloquium 251. Application of Continuum Damage Mechanics. Abstr. Krakow, 1989, p. 12.

14. Belyaev, V.V., Modelling of non-linear phenomena in shock waves and fracture of solids with defects. Int. IMACS Conf. Abstr., Moscow-Vilnius, 1990, p. 232.

15. Bellendir, E.N., Belyaev, V.V. and Naimark, O.B., Kinetics of multicenter fracture under splitting-off conditions. Sov. Tech. Phys. Lett., 1989, 15, N 13, 90-93.

16. Samarskii, A.A., Numerical simulation and nonlinear processes in dissipative media. In Self-Organization. Autowaves and Structures Far from Equilibrium. Berlin, Springer Verlag, 1984, 119-129.

17. Breed, B.R., Mader, C.L. and Venable, D., Technique for the determination of dynamic - tensile - strength characteristic. J. Appl. Phys., 1967, 38, N 8, 394-414.

PHOTOMECHANICS OF SURFACE-CRACK – WAVE INTERACTION PHENOMENA

H.P. ROSSMANITH and R.E. KNASMILLNER
Institute of Mechanics
University of Technology Vienna
Wiedner Hauptstraße 8-10/325, A-1040 Vienna, Austria

ABSTRACT

High-speed recording techniques in conjunction with methods of photomechanics (e.g. photoelasticity and the method of caustics) is utilized for the visualization of complex interaction processes between Rayleigh surface waves and geometrical discontinuities. Results for the time history of the stress intensity factors $K_1(t)$ and $K_2(t)$ will be presented for normal and oblique surface breaking cracks.

INTRODUCTION

The interaction of elastic waves with inhomogeneities in structural components has become a topic of considerable interest in the past decade. With increasing optimization and rationalization in mining engineering, structural geomechanics and rock mechanics faster, safer and more efficient procedures for production of fractures and excavations are called for. Wave propagation and wave-induced fracture are phenomena of primary importance in mining engineering, in particular in conjunction with fragmentation blasting and percussion flaking. When the rock splitting process is done inefficiently in a quarry operation or in tunneling additional expenses in handling oversized fragments, repeated crushing, etc. is incurred to the mine operator or rock mechanics company. Knowledge of stress-wave interaction with pre-existing faults and cracks assists one to optimize rock breakage and consequently reduces cost in mining operations and avoids unwanted damage of near-by above ground and underground structures [1].

Irregular surface topographies or subsurface inhomogeneities such as canyons, cliffs and inclusions or cavities, respectively, can seriously affect the wave pattern during interaction. The subject has been treated in the literature as a problem of elastic wave diffraction. Application of the wave–inhomogeneity interaction problem may be found in a wide variety of engineering problems. The influence of local soil conditions on ground motion due to earthquakes has been noticed as well as concentration of earthquake damage to certain regions. Most often damage is due to the large amplification of seismic and/or surface waves associated with local topography and surface soil/rock properties. On the other hand, on the engineering side, deliberately produced barriers or inhomogeneities within the surface boundary layer, such as trenches, zones of rock scattered by explosives, etc. might reduce the intensity of dynamic stress concentration at cavities or other obstacles embedded and thus act as a vibration isolation [2, 3, 4, 5].

The dynamic analysis of wave propagation in materials with imperfections is extremely difficult because of the very complex wave patterns generated by the superposition and interaction of incident, reflected and diffracted body waves as well as surface waves [6].

Rayleigh waves play a dominant role in the transmission of seismic disturbances along the surface of the earth and therefore have been the subject of extensive research [7, 8]. On the other hand, Rayleigh waves may conveniently be employed in the field of nondestructive testing with acoustic surface waves. The scattering behavior of surface topographic features such as an array of grooves etc. with respect to the scattering of Rayleigh waves is fully utilized and exploited in the fabrication of microminiature acoustic surface wave devices such as filters, etc. [9, 10, 11].

Given appropriate scaling with respect to wave length and geometrical dimensions of the surface inhomogeneity, e.g. corrugation, grooves, canyons, etc., the interaction processes between elastic surface wave and surface inhomogeneity in the micro-world of miniature device engineering and in the macro-world of earthquake and tectonics engineering are alike. The phenomena observed in the interaction process crucially depend on the ratio of the wave length to the dimensions of the scatterer for harmonic waves. For a Rayleigh-pulse with a broad frequency spectrum the interaction process becomes more involved.

Significant progress in the treatment of the interaction of a surface wave with a single inhomogeneity has recently been made by several authors. In the theory of groove reflector arrays [12], a periodic weakly ondulative grating of shallow grooves in a semi-infinite elastically isotropic medium supports a traveling Rayleigh wave. For shallow grooves, the incident Rayleigh wave is weakly scattered with the wave energy partitioned into reflected and transmitted surface waves and scattered bulk waves that propagate into the elastic half space. Results of many investigations and studies indicate that topographic features such as shape, depth and for the case of an array the periodicity of the topographic features can have significant effects on the quantitative and qualitative partition of wave energy, distribution of wave patterns and spectral content. The interaction of Rayleigh waves with deep grooves or a canyon with steep cliffs with an aspect ratio of wave length to depth of canyon/groove much smaller than unity differs considerably from the shallow groove diffraction problem [13, 14].

Recently, enforced attention has been focussed on the interaction of elastic surface waves, Rayleigh waves, and edge cracks [9, 15, 16].

Surface-breaking cracks play a particularly important role in many fields of applications spanning the range from small surface flaws in surface technology and nondestructive testing in materials and structural integrity and quality assurance to large-scale surface cracks in jointed rock in geomechanics. Both, the effect of the interaction of earthquake induced or technically excited Rayleigh waves on these large-scale fractures as well as the interaction of tiny surface flaws in metallic specimens with ultrasonically generated

Rayleigh waves can appropriately be modelled in a photomechanics laboratory.

Dynamic photoelasticity and the method of caustics or shadowspot-technique are two suitable techniques for whole-field visualization of dynamic processes [17, 18, 19, 20]. Dynamic photoelastic experiments have provided whole field fringe patterns which permit a visual examination of the complex interaction which occurs when body waves and Rayleigh waves encounter a slit-type or crack-type discontinuity in a half-plane. The interaction process produces a multitude of reflected, diffracted, and transmitted response waves which are represented by the photoelastic fringe patterns. An early experimental dynamic-photoelastic investigation of the effect of Rayleigh wave interaction with various ensembles of surface-breaking cracks was conducted by J.W. Dally [21] and H.P. Rossmanith [22, 23]. The preceding decade has seen much theoretical and experimental activity in this research field.

The method of caustics or shadow patterns which is sensitive to stress gradients and changes in surface deformation has proved to be very powerful and applicable. Both effects, the influence of stresses and strains on the geometrical shape of the structural component (Poisson's effect) and the stress/strain-induced change of the optical properties of translucent solids and fluids by altering the refractive index tensor of the material, give rise to the formation of particular optical patterns. From these patterns, characteristic physical quantities can be derived [19, 20, 21, 24].

This study addresses the complex situation where a half-plane is weakened by a normal or inclined surface crack. Incidence, reflection, transmission and scattering of explosively generated body waves and Rayleigh-pulses as well as wave-induced initiation and fracture propagation will be studied. The importance of this problem stems from intimate applications in geophysics and tectonics [25, 26].

Rayleigh waves propagate along free surfaces such as crack faces and delaminations in layered rock and upon interaction with a crack a very strong mixed-mode time dependent stress intensity factor may result and lead to fracture initiation and,thus, promote delamination and cracking of interfaces and joints.

The time-variations of the stress intensity factors K_1 and K_2 due to the Rayleigh wave scattering about the tips of inclined surface cracks can be evaluated quantitatively and will be studied here.

BACKGROUND

Research in surface crack characterization has concentrated mainly on the measurement of relative amplitudes of scattered waves, evaluation of different arrival times of sequential waves at a receiving position and determination of spectral densities.

For many practical situations the time history of the stress intensity factors associated with wave diffraction about surface cracks is of utmost importance. Rayleigh waves in contrast to body waves carry their energy within a very shallow layer adjacent to the free surface. They are very sensitive to geometrical disturbances such as surface corrugations and surfaces-breaking or shallow embedded structural discontinuities [27, 28]. Surface cracks form a particularly interesting group of geometrical discontinuities with a strong potential for interaction. When a Rayleigh wave interacts with a surface crack several characteristic features of the interaction process can be distinguished:

- Reflection and diffraction with the entrant corner

- Downward propagation along crack face of the transmitted Rayleigh-pulse

- Swing of Rayleigh-pulse around the tip of the crack and climb-up along the shadow-side crack wall

- Reflection and diffraction with the shadow side entrant corner

- Interaction of the bulk part of the Rayleigh wave with the crack tip extending into the depth of the crack

- Post-interaction wave reflection processes

- Crack initiation and propagation

Each of these elementary phases constitutes in itself a complicated wave - crack interaction event and has previously been studied theoretically and in part also experimentally.

EXPERIMENTAL PROCEDURE

Dynamic photoelasticity was utilized with two-dimensional models. The photoelastic method provides visual information about the dynamic event over the entire region of interest and the photoelastic crack tip fringe patterns allow for qualitative interpretation and quantitative determination of the history of the stress intensity factors.

A large number of dynamic photoelastic investigations pertain to problems arising in geophysics and geomechanics. The modelling techniques employed in dynamic photoelasticity and caustics have become well known over the past 20 years and detailed descriptions may be found in the literature (see e.g. [17, 18, 19, 20]).

The models were machined from thin plates ($10mm$ thickness) of Araldite B, a birefringent polymer often used in photoelastic work. The three types of problems investigated are shown in Fig. 1: a) semi-infinite crack with model charge of $Q = 40mg\ PbN_6$ at distance $L = 185mm$ from the crack tip; b) surface crack of length $l_{cr} = 25mm$ normal ($\theta = 90°$) or inclined at various angles $\theta = 30°, 60°, 120°$ and $150°$ with model charged with $Q = 120mg\ PbN_6$ ($L = 150mm$) and c) sequence of surface breaking cracks. The framing rate of $\sim 200000 f/s$ guaranteed sufficient resolution in the time history of the stress intensity factors to capture the essentials of the dynamic interaction process. In all experiments the explosive was contained in a small charge holder which was rigidly fixed to the corresponding crack face Fig. 1a or the free outface Fig. 1b and c.

Dynamic loading of the model generates bulk waves and a Rayleigh-pulse which travel toward the crack. The model material becomes temporarily birefringent when subjected to a dynamic state of stress. Light field dynamic isochromatic fringe patterns were recorded using an improved version of a multiple-spark-gap Cranz-Schardin high-speed photographic camera. The extremely short illumination time of approx. $150ns$ allows recording of the movement of the isochromatic fringe pattern resulting in reasonably sharp photographs. Thus, a sequence of 24 photographs corresponding to distinct instances of time during the dynamic event is recorded which forms the basis for data reduction.

155

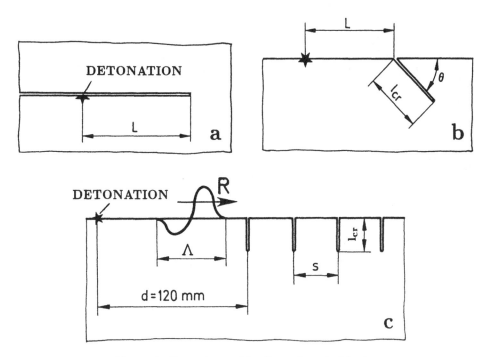

Figure 1: Geometry and loading of cracked body:

a) semi-infinite crack subjected to a one-sided crack face explosive loading,
b) inclined surface crack subjected to explosively generated Rayleigh wave
   excitation,
c) a sequence of normal surface breaking cracks.

RAYLEIGH WAVE (PULSE) CHARACTERIZATION

The incident system of elastic waves from which the Rayleigh wave will emerge is generated by the explosive excitation of the flawed boundary by detonating a small charge.

Upon detonation of the explosive two distinct circular-crested body waves, the longitudinal or P-wave and the shear or S-wave, extend with propagation speeds $c_P^2 = 2\mu/(\rho(1 - \nu))$ and $c_S^2 = \mu/\rho$ ($c_S < c_P$), respectively, into the bulk of the elastic solid. A second straight-crested shear wave, the von-Schmidt wave or head wave (V-wave) connects the P-wave with the S-wave in the dynamic isochromatic fringe pattern as shown in Fig. 2.

In addition, Rayleigh waves are generated which propagate along the free surface at a speed c_R ($c_R < c_S < c_P$). The speed c_R depends essentially on Poisson's ratio and a simple engineering approximation has been presented by Viktorov [29]. The pulse length of the explosively generated R-wave depends primarily on the properties of the explosive and on various experimental characteristics, such as charge containment, burning velocity, compactification etc. Excessive amount of explosive does not alter the R-wave markedly but generates unwanted additional fracturing at the explosive site. Hence, for a given

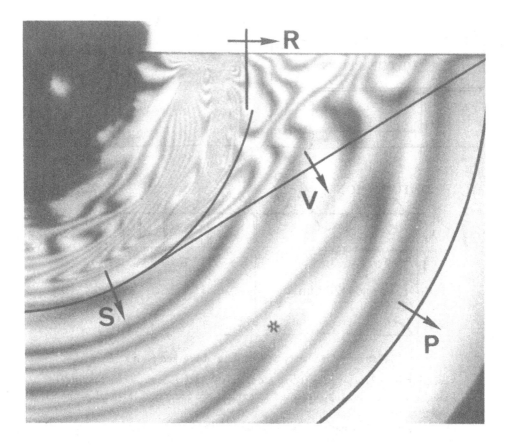

Figure 2: Dynamic isochromatic fringe pattern associated with elastic wave propagation from an explosive source at the free edge of a half-plane

explosive (PbN_6) the stress distribution of the pulse generated is predetermined and the dimensions of the model have to be scaled in order to comply with the requirements.

In the experiments to be described the R-pulse length is approximately $\Lambda_R = \Lambda'_R + \Lambda''_R = 12.5 + 12.5 = 25mm$, where Λ'_R and Λ''_R denote the length of the leading and trailing pulse halves, respectively. From an applications point of view the limit cases of a shallow crack ($d \ll \Lambda_R$) and a deep crack ($d \gg \Lambda_R$) are of importance. As the Rayleigh waves carry a significant amount of energy associated with its subsurface peak within a shallow layer just below the free surface, a strong interaction between the R-subsurface peak and the imperfection is to be expected when the pulse length Λ_R and the slit depth l_{cr} are comparable in dimensions. For the series of experiments the ratio d/Λ_R represents an important control variable. The superposition of reflected, refracted and diffracted longitudinal (P)-waves, shear (S)-waves and Rayleigh (R)-waves gives rise to highly complex wave patterns. In the case of a deep crack the R-wave interaction process may be viewed as a sequence of successive individual R-interaction problems, e.g. the reflection of an R-wave at the corner of a quarter-plane, followed by the interaction of a

R-wave with a crack tip. The dynamic interaction problem becomes very involved when the R-pulse length is comparable with the characteristic dimension of the model. This characteristic dimension is the depth l_{cr} of the edge slit or crack.

For a normal edge crack the P-wave front impinges nearly parallel to the crack faces. The trailing tensile pulse of the P-wave induces a stress intensity at crack tip A during the time interval $(L + \Lambda_P')/c_P < t < (L + \Lambda_P)/c_P$, where Λ_P' is the length of the compressive pulse of the P-wave. Λ_P is the overall length of the P-wave. For times $t > L/c_P$, the crack and the corner section play an important part in the interaction process. The wave front patterns may be constructed and the crack tip fringe patterns may be evaluated at each instant of time. Fracture initiation as a result of excessive high stress amplifications at the crack-tip during the wave diffraction process is observed. The region beyond the crack and the unloaded edge of the plate play no part in the response until the dilatational wave has reached the crack. Then, for a natural crack, the compressive leading pulse of the P-waves is transmitted across the closed crack and the PP-reflection will open the natural crack, whereas in the case of an edge slit with finite width the P-wave will be reflected at the slit wall and the corner region is reached by diffracted waves only.

DIFFRACTION OF R-WAVES FROM CRACK TIPS

The dynamic stress field generated in the vicinity of the tip of a semi-infinite crack by stress wave loading comprises bulk and surface waves. A sequence of theoretical wave front constructions and associated photoelastic fringe patterns pertaining to various stages of the wave diffraction process is shown in Fig. 3. The waves that are expected to be present during the phase where the R-pulse swings around the crack tip and returns along the upper crack face are sketched in Fig. 3c and shown in Fig. 3e.

By employing standard procedures for data reduction and stress intensity factor determination, one obtains the history of the stress intensity factors K_1 and K_2 as shown in Figs. 4. In frame #5 the leading tensile pulse of the R-waves swings around the crack tip and the turning process is almost finished in frame #12, i.e. after about $23\mu sec$. Hereby, the SIFs K_1 and K_2 undergo dramatic changes where the combination of (K_1, K_2) describes a complete loop as shown by the solid line in Fig. 5.

Theoretical solutions for the semi-infinite crack without reflective boundary have been worked out in Ref.[30], where the fundamental problem is Lamb's problem. Results have been reported to be extremely sensitive to the form of the input pulse, i.e. the time-pressure trace of the explosion. This relation is not readily known in an experiment and therefore direct comparison between the idealized theoretical solution and the experimental result is difficult.

Results of Fig. 5 show that for the Rayleigh-pulse diffraction about the tip of a crack is associated with a rather simple full cycle for K_1-K_2.

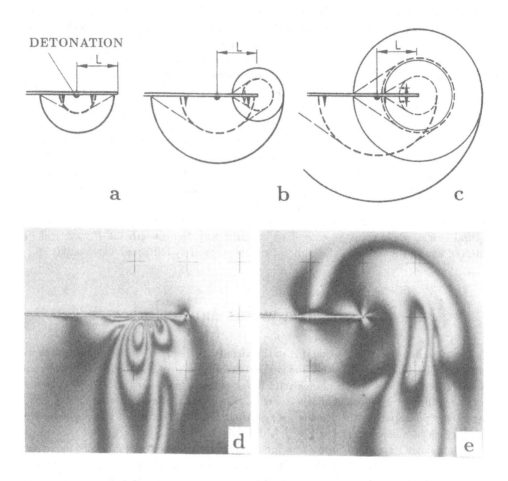

Figure 3: Theoretical wave front constructions (a,b,c) and experimentally recorded (isochromatic) wave patterns (d,e,f) associated with diffraction of explosively generated elastic waves about the tip of a semi-infinite crack:

    a)   wave expansion $s_P = c_P t < L$,

    b(d) P-wave diffraction $s_S = c_S t < L < s_P$ ($t = 14\mu s$, frame #3),

    c(e) S-wave diffraction $L < s_S < s_P$ and R-pulse reflection

        and diffraction ($t = 64\mu s$, frame #16).

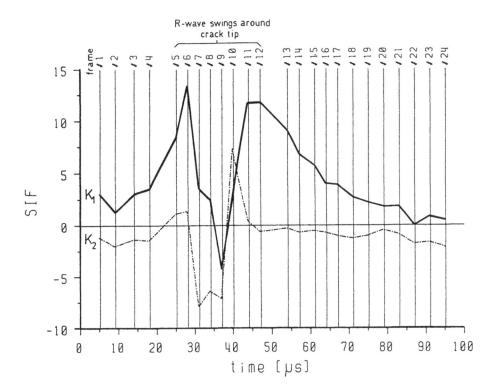

Figure 4: History of dynamic stress intensity factors K_1 and K_2 for Rayleigh wave diffraction about the tip of a semi-infinite crack due to point source excitation at one crack face only.

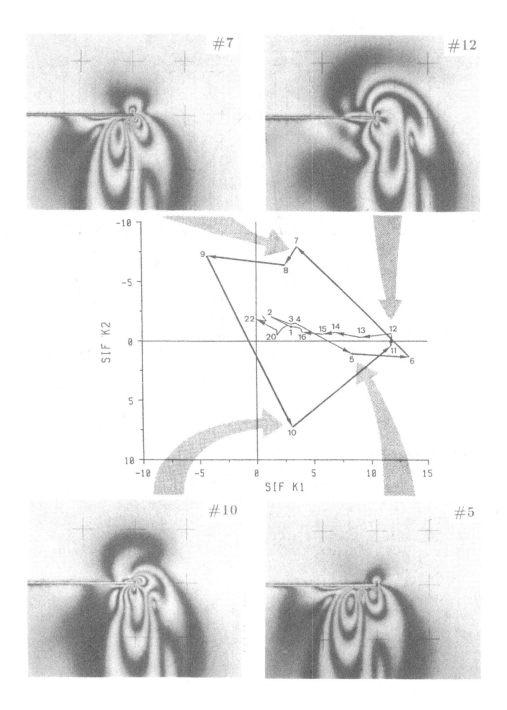

Figure 5: K_1–K_2 graph for Rayleigh wave diffraction about the tip of a semi-infinite crack due to point source excitation and associated experimental isochromatic recordings.

THE NORMAL SURFACE CRACK

Results of the preceding fundamental diffraction problems will be used for the study of the interaction of Rayleigh waves with normal and inclined surface cracks. Figure 6 shows the relevant theoretical wave patterns. The individual wave interaction processes such as interaction of P-wave, $vonSchmidt$-wave, leading edge of the R-wave, interaction of subsurface R-peak with crack tip, swing-around effect of the R-wave about crack tip and the interaction of the reflected waves along the crack line are represented by their characteristic features in the K_1, K_2-traces of Fig. 7. The most prominent feature is the turn of the R-wave about the tip of the crack expressed in the polygonal section frames #14 to #17 in Figs. 7 and 8. The interaction of the diffracted and reflected waves in the wake of the R-pulse with the crack tip occurs between frames #20 – #24. From earlier publications on the subject of R-wave interaction with a normal surface crack [31] experimental results pertaining to frequency spectral for the leading transmitted waves and the transmitted ultrasonic signal after interaction with slots of variable width are known. Experimentally obtained stress intensity factors have not been reported.

FRACTURE INITIATION

Upon incidence of the Rayleigh wave at the slit mouth (point C in Fig. 9) a reflected $R_r R$-wave travels back along the upper free surface of the quarter-plane and a transmitted $R_f R$-wave moves along the slit as shown in Figs. 9a and 9b. This kind of Rayleigh wave interaction problem has been investigated earlier [32, 33, 34]. The diffraction of the reflected P, S and V-waves at the bottom of the slit O induces fairly strong stress intensifications. Obviously, the local stress exceeded the fracture strength of the material and crack initiation in a direction toward the wave source took place. The crack OA propagates along a curved path because of transient wave scattering. This dynamic event is recorded in Figs. 9a to 9c, where a mixed-mode isochromatic fringe pattern as well as a mixed-mode pseudo-caustic centered at the moving crack tip A can be identified in Fig. 9a [34]. The smaller diameter of the fuzzy caustic at the (possibly arrested) crack tip in Fig. 9b implies a reduction of the stress intensity factor (possibly below the arrest value for the material utilized). Practically neither fuzzy caustic nor isochromatic "butterfly" fringes can be identified at crack tip A for the subsequent phases (Figs. 9c – 9f). While the crack OA is still moving, a nearly pure shear-mode stress intensification, to be identified by its characteristic shape ("∞") develops at the tip of the slit, O. Mode-2 crack initiation at an angle of approximately 90° to the plane of the slit is shown in Fig. 9b. As a result of the R-wave-slit interaction the $R_f R$-wave progresses along the face of the slit (Figs. 9c, frame #8). The leading pulse halves of the $R_r R$-wave and the $R_f R$-wave are in compression now. For the $R_f R$-wave this follows from the fact that the leading half of this wave is transmitted practically undisturbed across the contacting faces of the crack branch which extends from the slit tip O to point A (Figs. 9c and 9d). The compressive portion of the $R_f R$-wave turns around the bottom of the slit (frames #11 and #12 in Figs. 9d and 9e, respectively) and is guided by the extending crack OB (Fig. 9e). The trailing tensile half of the $R_f R$-wave causes crack opening along OA, hence,

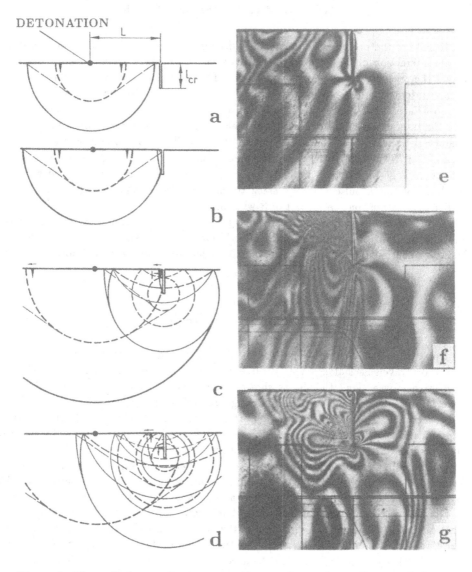

Figure 6: Theoretical wave front constructions and experimentally recorded (isochromatic) wave patterns associated with diffraction of explosively generated elastic waves about the tip of a normal surface crack:

    a)   wave expansion $s_P = c_P t < L$,

    b(e) P-wave reflection $s_P < \sqrt{L^2 + l_{cr}^2}$; $s_S < c_S t < L$, (frame #1),

    c(f) P-wave diffraction and S-wave reflection

        $s_P > \sqrt{L^2 + l_{cr}^2}$; $s_S < \sqrt{L^2 + l_{cr}^2}$, (frame #8),

    d(g) P-wave diffraction, S-wave diffraction and

        R-wave reflection and diffraction

        $s_S > \sqrt{L^2 + l_{cr}^2}$; $L < s_R = c_R t < L + l_{cr}$, (frame #13).

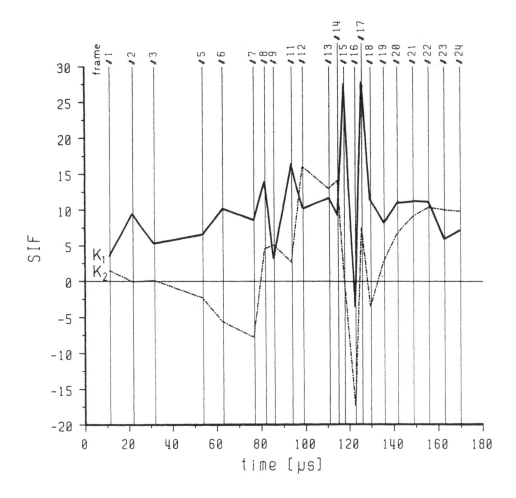

Figure 7: History of dynamic stress intensity factors K_1 and K_2 for Rayleigh wave diffraction about a crack tip of a normal surface crack ($\theta = 90^\circ$).

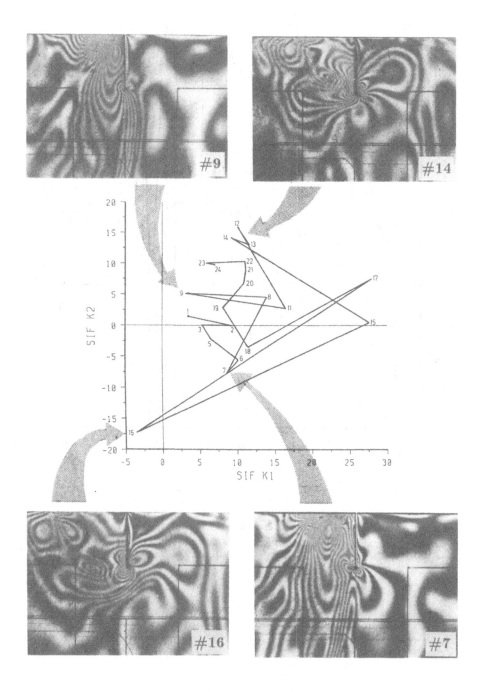

Figure 8: Trace of the dynamic stress intensity factors K_1 and K_2 for Rayleigh wave diffraction about a crack tip of a normal surface crack ($\theta = 90°$) and associated isochromatic recordings.

prevents further wave passage across the crack line. The last photograph (frame #16) of the sequence shown in Fig. 9 exhibits a remarkable detail: part of the kinetic energy of the transmitted compressive $R_d R$-wave is being converted into crack driving energy (giving also rise to the wavy crack path of crack OB in Fig. 9f) and the remainder of the wave disturbance turns around the (propagating?) crack tip to follow the free upper crack surface in the opposite direction. There is no mode-1 crack tip fringe pattern to be seen at crack tip B in Fig. 9f, and the small shear type pattern could well be caused by shear loading of the contacting crack surfaces. Although the propagation speed of a fracture is limited by the Rayleigh wave speed such high crack velocities have never been observed in photodynamic experiments. With an experimentally observed limiting crack speed of the order of c/c_r about 0.7, R-wave swing around a moving crack-tip becomes possible. From the changing crack tip positions (crack tip B) in Figs. 9e and 9f one may conclude that a dynamic R-wave turnover has taken place although the pattern shown in Fig. 9f, in all probability, pertains to a (at least) temporarily arrested crack.

Fig. 10 is associated with the phase of R-wave diffraction and reflection, and the fracture pattern development. A sequence of stress distributions (frames #9 to #16) of the trailing tensile pulse of the reflected $R_r R$-wave indicates that the $R_r R$-wave amplitude decays very slowly. On the left side of Fig. 10 one may note a slight increase of the tensile peak value in frame #10. This kind of amplification (about 15%) has been observed in other experiments with waves passing a nonwelded interface.

Complex transient strain fields may introduce a slight displacement discontinuity (e.g. a shift) along the crack faces OA^+ and OA^-. Then, part of the incident compressive pulse of the $R_f R$-wave will be reflected at the upper face $OA+$ of the crack before the small crack opening displacement has been compensated for by the wave-induced displacements. No stress distribution along the crack OB could be extracted from the fringe patterns.

In addition, crack tip positions as recorded in the individual frames (#7 to #14) have been indicated along the crack paths in Fig. 10. An average crack speed of $c = 736 m/s$ as compared to the Rayleigh wave speed $c_R = 1050 m/s$, $c/c_R = 0.7$, was obtained from an analysis of the position-time trace extracted from frames #7 up to #14.

In a second experiment the ratio l_{cr}/Λ_R was reduced to one half. Experimental results indicated a loss of interaction intensity with very little wave energy radiated from the slit upon V- and S-wave scattering. Results suggest that a critical ratio between R-wave length and slit depth for strong interaction to occur is of the order of unity. This effect has been reported earlier [17]. When the slit depth increases considerably ($\Lambda_R \ll l_{cr}$) the interaction problem investigated in Ref. [32] is obtained.

If the slit is replaced by a razor-blade cut natural crack, the leading compressive pulse of the P-wave is transmitted across the contacting crack surfaces and, therefore, no $P_r P^+$, $P_d P^+$ and $S_d P+$-waves will be generated in the case by the P^+-incidence. The + sign denotes the compressive leading P-pulse. The trailing P^--pulse is of very low amplitude and the $P_d P^-$- and $S_d P^-$-diffraction waves as well as the reflected $P_r P^-$-wave could not be identified. The R-wave interaction phase with the natural crack produced similar fringe patterns as have been recorded with the slit of finite width.

166

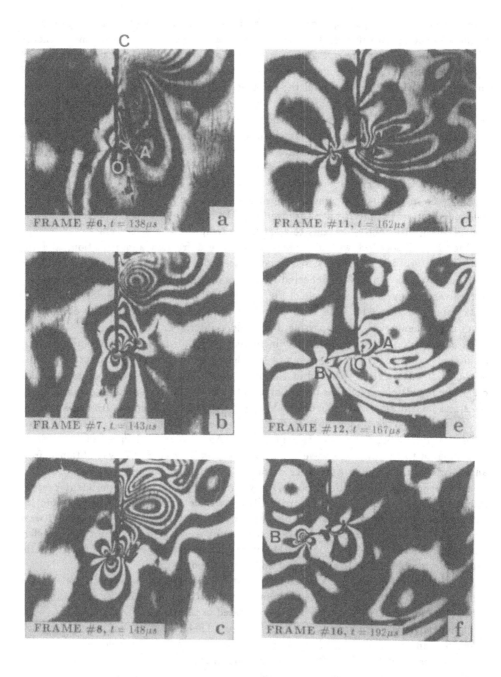

Figure 9: Sequence of dynamic isochromatic fringe patterns associated with the interaction of Rayleigh waves with an edge crack showing wave diffraction induced mode-2 crack initiation.

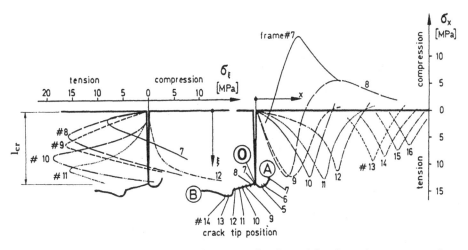

Figure 10: Distribution of stress tangent to the slit and free boundary associated with Rayleigh wave diffraction and reflection and fracture development.

ENSEMBLE OF SURFACE BREAKING CRACKS

Surface cracks have been of importance in many fields of engineering from thermal-induced micro-cracks during desiccation of cementitious materials to magneto-electrically induced surface damage in semi-conductor technology.

The single crack wave scattering problem has been the subject of a number of theoretical and experimental studies [31, 35].

Consider the model geometry given in Fig. 1c. The problem of Rayleigh wave scattering about a sequence of uniformly spaced edge cracks has been dealt experimentally by J.W. Dally [21] and H.P. Rossmanith and R.E. Knasmillner [22, 23]. The objective of this investigation was to develop a relationship between the geometry of the ensemble of cracks and the damping characteristics of Rayleigh waves. Fig. 11, taken from the experimental work of J.W. Dally [21], shows a sequence of four dynamic isochromatic fringe patterns associated with the scattering process of an explosively generated Rayleigh wave about the tips of a set of surface cracks. Frame #5, taken at 177μs after detonation, depicts the initial situation where the Rayleigh wave interacts with the surface corner of the first crack. The next photograph, frame #7 (212μs) shows a strong mode-2 singular stress pattern at the tip of the first crack. With the lower part of the wave propagating below the crack ensemble the state of stress at crack tip #1 changes from almost pure mode-2 to dominant mode-1, whereas crack tip #2 reveals a larger mode-2 contribution. Later in the event, photograph frame #16, taken at 408μs, shows residual stress intensifications at several crack tips due to global wave scattering effects.

A more recent study of this problem, by H.P. Rossmanith [13], by means of the method of caustics yields the sequence of caustics patterns shown in Fig. 12. Frame a (time 111μs illustrates the initial phase where the Rayleigh wave interacts with the first crack tip. The characteristic caustic pattern of the Rayleigh wave is identified with the black, quasi-semi-ellipse which approaches the leading crack of the ensemble from the left hand side.

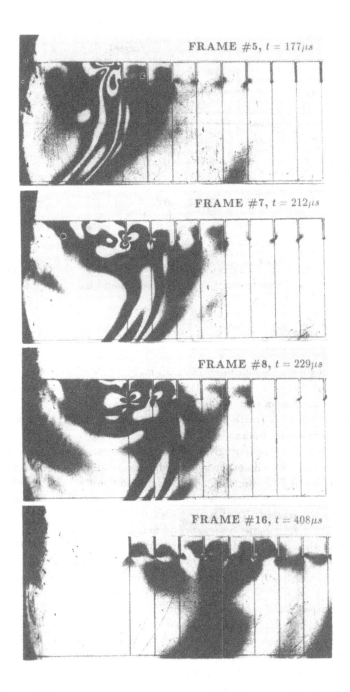

Figure 11: Sequence of four dynamic isochromatic fringe patterns associated with the scattering of an explosively generated Rayleigh surface wave with a regular array of surface-breaking cracks showing the variation of the singular crack tip stress pattern (taken from [21]).

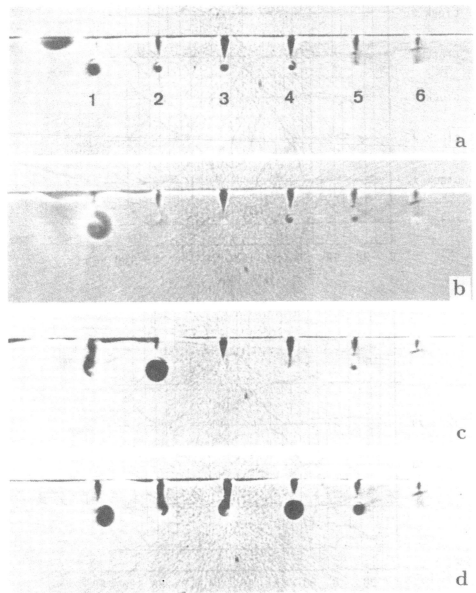

Figure 12: Dynamic high speed caustic recording of a Rayleigh wave diffraction about a sequence of surface-breaking cracks:
a) incident Rayleigh wave (time $t = 111\mu s$),
b) shear reversal and mixed-mode caustic at crack tip #1 (time $t = 132\mu s$),
c) body wave scattering from crack tip #1 and mixed mode excitation of crack tip #2 (time $t = 144\mu s$),
d) various types of excitation of surface cracks: crack tip #1 has initiated, shear reversal between pure mode-2 at crack tips #2 and #3, mode-1 stress field at crack tip #4 (time $t = 178\mu s$).

170

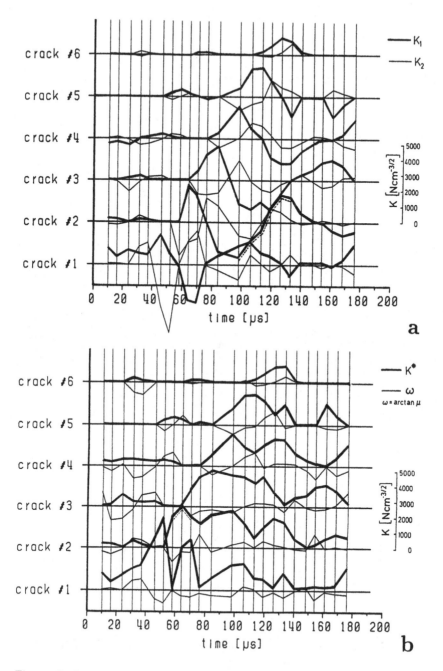

Figure 13: Dynamic caustic results obtained from experimental model studies showing the history of stress intensity factors K_1 and K_2 and their combinations for the diffraction of a Rayleigh wave about a sequence of edge cracks:
    a) distribution of stress intensity factors K_1 and K_2,
    b) distribution of complex stress intensity factor K^* and phase shift ω.

The stress field about the leading crack tip is one of mixed-mode with a dominant mode-1 contribution. When proceeding to the next frame, notice the strong shear stress reversal expressed by the different orientations of the caustics in frames Fig. 12a and Fig. 12b.

The body waves emanating from crack tip #1 in Fig. 12c compare well with the wave fringe pattern of Fig. 11, frame #8. Various types of mode excitations can be detected in frame Fig. 12d, where crack tip #1 has initiated and a fair mode-1 caustic is attached to the propagating branch crack. Crack initiation at a phase where the leading part of the bulk of the Rayleigh wave has already reached the fourth crack indicates a strong back effect in the scattering process. Also, observe the reversal of the shear stress at the adjacent crack tips #2 and #3.

Evaluation of the pure and mixed-mode crack tip caustics (Fig. 12) yields the distribution of the variation in time of the stress intensity factors K_1 and K_2 for the individual cracks of the comb-type structure as shown in Fig. 13a. These graphs reveal the damping capability of a series of edge cracks on the propagation of surface pulses. This is easily traced by comparing the peak values of K_1 and K_2 in Fig. 13a or the K^*-value in Fig. 13b for adjacent cracks consecutively labelled #1 to #6. In a series of experiments, the ratios between crack spacing, crack depth, and pulse length of the Rayleigh wave have been varied to learn that the dissipation of wave energy by diffraction and reflection of waves about the tips of the surface cracks increases with increasing crack depth and decreasing pulse length. In addition, crack initiation at crack #2 and at the last crack in the series was frequently observed.

THE INCLINED SURFACE CRACK

The diffraction characteristics for R-wave interaction with inclined surface cracks are very much different from the case of the normal surface crack and strongly depend on the type of entrant corner: obtuse $(0° < \theta < 90°)$ or acute $(90° < \theta < 180°)$. Theoretical wave patterns for $\theta = 60°$ and $120°$ are shown in Fig. 14.

For all surface wave/crack diffraction problems the ratio between wave or pulse length and the depth of the surface crack is a crucial quantity which controls the interaction pattern. The other characteristic quantity is the angle of inclination, θ.

In the case of obtuse diffractive corners $(\theta < 90°)$ (Figs. 14a,b,c) the bulk waves (longitudinal wave $=P$, shear wave $= S$, von Schmidt or head wave $= V$) will interact and be diffracted first at the corner before expansion has sufficiently progressed for the wave fronts to reach the crack tip. In this case the explosively generated initial R-wave which follows the bulk waves along the free surface and the crack face into the material starts to interact with the crack tip after diffraction of P- and V-waves is completed. The most prominent feature in the K_1–K_2 graph as well as in the sequences of isochromatic fringe patterns is associated with the run-off and swing around effect at the crack tip. The wave pattern associated with steep cracks forming obtuse entrant corner and extending more than one R-pulse length deep into the body show appreciable interaction between crack tip and subsurface peak of the R-wave (frame #11 – #13 in Fig. 15a and Fig. 15b) followed by the swing effect (frames #12 – #16 in Fig. 15a and Fig. 15b). Shallow inclined surface cracks forming very obtuse entrant corners yield a situation where the bulk of the R-wave energy contained and transported in a shallow subsurface layer strongly interacts

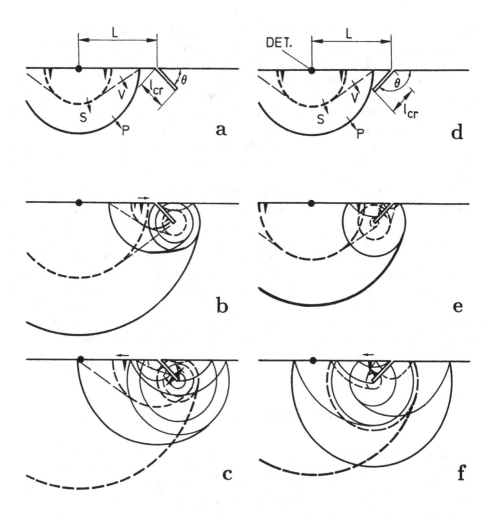

Figure 14: Theoretical wave front constructions for bulk and R-wave diffraction with an obtuse (a–c) and an acute (d–f) corner formed by inclined surface cracks:

a) and d) wave expansion $s_P = c_P t < s^*$,

b) and e) P-wave diffraction $s_S = c_S t < s^* < s_P$,

c) and f) S-wave diffraction $s^* < s_S$.

with the crack tip followed by fracture initiation (shown in frames #15 – #22 in Fig. 16a and Fig. 16b).

Surface wave diffraction about acute entrance corners ($\theta > 90°$) exhibits different characteristics. Here, the chop-off effect forms the main feature of the diffraction process. For deep cracks forming weak acute wedge-type entrant corners (R-pulse length/crack depth ≤ 1) the energetic part of the R-wave will be trapped in the wedge, the subsurface peak will interact with the crack tip and the bulk part of the R-wave will be split from the "head" of the R-wave. Associated wave front constructions are shown in Figs. 14d – 14f. New R-waves emerge at the crack tip as a consequence of P, S and V-wave diffraction with the crack tip. Interaction of the leading isotropic point (i.e. leading subsurface characteristic of R-pulse) yields a strong increase of K_1 and a sign-reversal in K_2 (shown in frames #4 – #8 in Figs. 17a and 17b). The subsurface peak hits the crack tip in frame #9 in Figs. 17a and 17b and the chop-off effect is completed in frame #12 in the same figures. The subsequent sequence of frames nicely demonstrates the outflow of waves trapped previously and diffracted in the acute entrant corner (shown in frames #13 – #24 in Figs. 17a and 17b). The situation is even more severe for the shallow inclined crack forming a very acute wedge trap ($\theta \sim 150°$). The interaction and diffraction processes between R-pulse and crack tip become more complex with pronounced interaction of the leading lobe (frames #4 – #5), leading isotropic point (frames #6 – #7) and trailing isotropic point (frames #8 – #10) where the R-pulse has completely passed in frame #10 of Figs. 18a and 18b.

A very strong and complex wave field is established after diffraction of the incident bulk waves and R-pulse at the apex of the wedge. Maximum wedge interaction (shown in frames #11 – #12) is followed by a wave "spill" effect (frames #13 – #17) which induces again a Rayleigh-pulse which propagates downward along the upper crack-edge. The interaction of this transmitted R-pulse with the crack tip is represented in Figs. 18a and 18b by the characteristic increase of K_1 and sign change in K_2 (frames #17 – #21(24)).

DATA ANALYSIS

The experimental recordings contain sets of sequentially recorded fringe patterns which represent two-dimensional "snap shots" of the dynamic process taken at a predetermined sequence of time instants. Each frame represents a different stage in the evolution of the scattering or interaction process and shows the total wave field at different degrees of complexity. The dynamic fringe patterns and caustics were analyzed by plotting the isochromatic fringe order distribution along particular sections of the boundary and by a new caustic data reduction procedure [19, 20, 36]. For a detailed discussion of the technique see [37] and the following paper by H.P. Rossmanith and R.E. Knasmillner on "Recent Advances in Data Processing from Dynamic Caustics" in this volume.

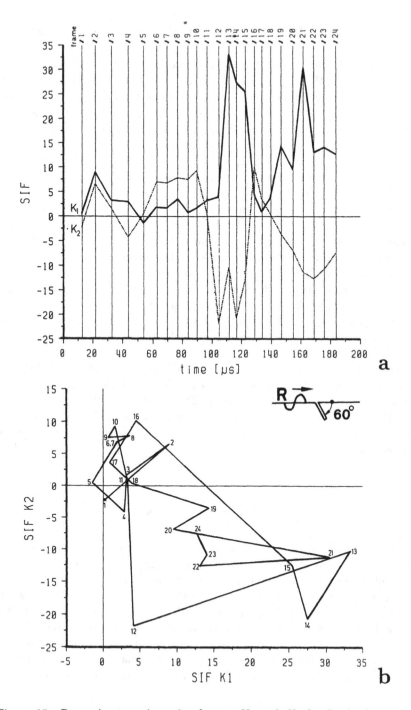

Figure 15: Dynamic stress intensity factors K_1 and K_2 for Rayleigh wave diffraction about crack tip of a steep inclined surface crack ($\theta = 60°$):

a) K_1 and K_2 vs. time,

b) K_1–K_2 trace.

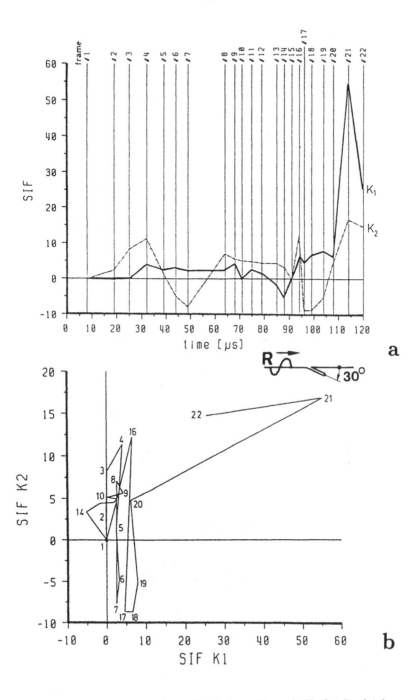

Figure 16: Dynamic stress intensity factors K_1 and K_2 for Rayleigh wave diffraction about crack tip of a shallow inclined surface crack ($\theta = 30°$):

a) K_1 and K_2 vs. time,

b) K_1–K_2 trace.

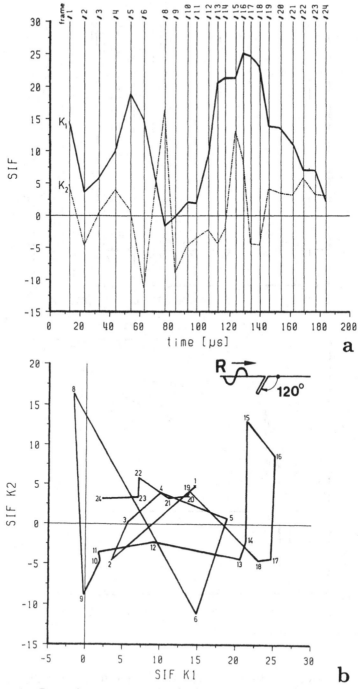

Figure 17: Dynamic stress intensity factors K_1 and K_2 for Rayleigh wave diffraction about crack tip of an inclined surface crack (deep crack, $\theta = 120°$):

a) K_1 and K_2 vs. time,

b) K_1–K_2 trace.

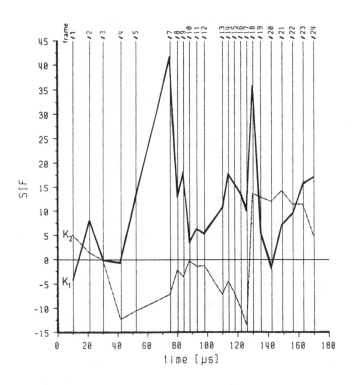

a

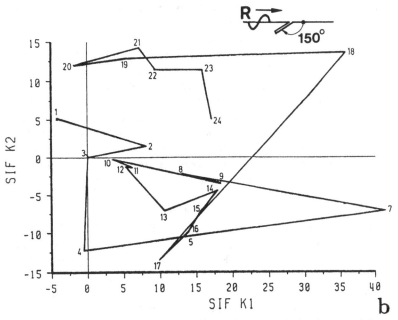

b

Figure 18: Dynamic stress intensity factors K_1 and K_2 for Rayleigh wave diffraction about crack tip of an inclined surface crack (shallow crack, $\theta = 150^\circ$):

a) K_1 and K_2 vs. time,

b) K_1–K_2 trace.

DISCUSSION AND CONCLUSION

The results illustrated in the foregoing investigations show that the interaction between surface waves and normal and inclined surface cracks is a very complex process which consists of a series of well-defined elementary wave interaction (reflection, transmission, diffraction) events, the sequence and dominance of which are controlled mainly by two parameters: the angle of inclination of the surface crack and the ratio between pulse length of the Rayleigh wave and the depth of the crack measured normal to the surface. Obtuse and acute entrant corner geometries induced by various crack inclinations yield different response wave patterns.

Dynamic photoelasticity and the method of caustics in conjunction with high-speed photography has been utilized here to generate sequences of dynamic isochromatic fringe patterns and caustic patterns illustrating "snap-shots" of the dynamic event. These fringe patterns have been digitized and the near field crack tip fringe patterns have been employed for determination of the stress intensity factors K_1 and K_2.

This is the key point of the entire analysis of experimental data. The Rayleigh wave, being a surface wave, exhibits its fringe pattern characteristics in a thin layer next to the free surface and therefore when interacting with a crack tip fundamentally interferes with the classical crack tip fringe pattern (Fig. 5) and most often impedes data extraction and reduction. A typical example is represented in Fig. 5(frame #7) where reliable data for the determination of K_1 and K_2 can only be extracted from data points located in the upper half of the interaction pattern. Overdeterministic systems of data points associated with a multi-parameter K-determination have been employed throughout this work [12, 37].

It should be noted that performing the experiment and problem analysis within the framework of the method of optical caustics does not eliminate the crucial fundamental problem of data extraction. Again, the caustic of the incident Rayleigh-wave [13, 14] will interfere with the crack tip caustic and evaluation of the "joint" caustic pattern will become extremely difficult if not impossible at present.

The interaction process produces a multitude of reflected, diffracted and transmitted response waves which are represented by the photoelastic fringe patterns. These diffracted waves result in a serious loss of energy available for fracture initiation and propagation. Experimental results suggest that the fractures are initiated and driven by the joint action of the incident head-wave and a multitude of reflected shear waves. The Rayleigh wave which exhibits most of its energy at or near the free surface is altered significantly by the flaw and its energy is partitioned among the reflected and transmitted $R_r R$-wave, respectively. It is observed that fractures are produced at the bottom of the slit by higher order shear waves that have been reflected along the vertical edge of the quarter-plane. Most of the fracture pattern has been established before transmitted $R_t R$-waves become operative in the sense that part of their energy is being converted in crack driving energy when the $R_t R$-wave has caught up with the advancing crack tip. This complicated transient energy conversion problem needs further study in the light of impact-induced fatigue crack propagation.

More experimental as well as theoretical work on Rayleigh wave interaction with cracks is required in order to fully understand the physical processes involved in wave diffraction and fracture initiation.

179

ACKNOWLEDGEMENT

The authors would like to acknowledge the financial support granted by the *Austrian Science Foundation* under project numbers # P 5814 and # P 6632.

REFERENCES

[1] H.P. Rossmanith (Editor). *Rock Fracture Mechanics, CISM Course #275.* Springer Verlag, 1989.

[2] T.W. May and B.A. Bolt. The effectiveness of trenches in reducing seismic motion. *Earthquake Engg. and Struct. Dyn.*, 10:195–210, 1982.

[3] A. Ohtsukim and H. Yamahara. Effect of topography and subsurface inhomogeneity on seismic Rayleigh waves. *Earthquake Engg. and Struct. Dyn.*, 12(1):37–58, 1984.

[4] F.J. Sanchez-Sesma et.al. Diffraccion de ondas P, SV y de Rayleigh en un semiespacio elastico. *J. of the Geotherm Engg. Div., ASCE*, 1982.

[5] H.L. Wong. Effect of surface topography on the diffraction of P, SV and Rayleigh waves. *Bull. Seism. Soc. Am.*, 72:1167–1183, 1982.

[6] J.D. Achenbach. *Wave Propagation in Elastic Solids.* Applied Mathematics and Mechanics. North-Holland Publishing Comp.-Amsterdam, 1973.

[7] E.A. Ash and E.G.S. Paige (Editors). *Rayleigh Wave Theory and Application*, volume 2, pages 1–356. Springer-Verlag Berlin, 1985.

[8] H. Lamb. On the propagation of tremors over the surface of an elastic solid. *Phil. Trans. Roy. Soc., London, UK*, Ser. A(203):1–42, 1904.

[9] J.D. Achenbach, A.K. Gautesen, and D.A. Mendelsohn. Ray analysis of surface wave interaction with an edge crack. *IEEE Transactions on Sonics and Ultrasonics*, SU-27(3):124–129, 1980.

[10] Y.C. Angel and J.D. Achenbach. Reflection and transmission of obliquely incident Rayleigh waves by a surface-breaking crack. *Acoustic Society Am.*, 75(2):313–319, 1984.

[11] S.R. Sehsadri. Effect of periodic surface corrugation on the propagation of Rayleigh waves. *Acoustic Society Am.*, 65(3):687–694, 1979.

[12] R.J. Sanford and J.W. Dally. A general method for determining mixed-mode stress intensity factors from isochromatic fringe patterns. *Engg. Fracture Mechanics*, 11:612–633, 1979.

[13] H.P. Rossmanith. The caustic approach to Rayleigh waves and their interaction with surface irregularities. *Special Volume of Int. J. Optics & Lasers in Engg.*, 1990 (in print).

[14] H.P. Rossmanith and R.E. Knasmillner. Diffraction of Rayleigh waves at surface irregularities. In A.W. Leissa (Editor), *Proc. of the 1ˢt Pan American Congress of Applied Mechanics (PACAM)*, pages 748–751, Rio de Janeiro, Brazil, January 1989.

[15] J.D. Achenbach, Y.H. Pao, and H.F. Tiersten (Publishers). Report of the workshop on "Application of elastic waves in electrical devices, nondestructive testing and seismology". Technical report, Northwestern University, May 1976. NSF-Report.

[16] D.A. Mendelsohn, J.D. Achenbach, and L.M. Keer. Scattering of elastic waves by a surface breaking crack. *Wave Motion*, 2:277–292, 1980.

[17] J.W. Dally. Introduction to dynamic photoelasticity. *Experimental Mechanics*, 20(12):409–416, 1980.

[18] A. Kuske and G. Robertson. *Photoelastic Stress Analysis*. John Wiley & Sons Ltd., Chichester, New York, Brisbane, Toronto, 1974.

[19] H.P. Rossmanith. Topics in photomechanics – cracks, waves and contacts. In Kawata K. Nisida M. (Editor), *Proc. Int. Symp. on Photoelasticity*, pages 135–144, Tokyo, Japan, 1986. Springer-Verlag, Berlin.

[20] H.P. Rossmanith. Dynamic photoelasticity and the method of caustics. In Wang W.C. (Editor), *Advanced Photomechanics*, pages 1–450. Dynamic Engineering Institute, National Tsing-Hua University, Hsinchu, Taiwan, 1988.

[21] J.W. Dally. Dynamic photoelastic studies of stress wave propagation. In *Proc. IUTAM-Conference on Elastic Wave Propagation*, Toronto, Ontario, Canada, 1977.

[22] H.P. Rossmanith and R.E. Knasmillner. Diffraction of Rayleigh waves at surface irregularities. *Applied Mechanics Review*, 42, Part 2(11):S223–S232, November 1989.

[23] H.P. Rossmanith and R.E. Knasmillner. Interaction of Rayleigh waves with surface breaking and embedded cracks. In Y. Kanto H. Homma (Editor), *Proc. of OJI Int. Seminar on Dynamic Fracture*, pages 236–262, Toyohashi University of Technology, Japan, August 1989. Chuo Technical Drawing Co. Ltd.

[24] J.F. Kalthoff. The shadow optical methods of caustics. In A.S. Kobayashi (Editor), *Handbook on Experimental Mechanics*, chapter 9, pages 430–500. Prentice Hall, Inc., 1987.

[25] J.A. Hudson and L. Knopoff. Transmission and reflection of surface waves at a corner. 2. Rayleigh waves (theoretical). *Geophysical Research*, 69(2):281–289, 1964.

[26] A.K. Mal and L. Knopoff. Transmission of Rayleigh waves at a corner. *Bull. Seism. Soc. Am.*, 56:455, 1966.

[27] H.P. Rossmanith and J.W. Dally. Rayleigh wave interaction with inhomogeneities – Part I: Near surface cavity and inclusion. *Strain*, 19:7–13, 1983.

[28] H.P. Rossmanith and J.W. Dally. Rayleigh wave interaction with inhomogeneities – Part II: The buried cavity and inclusion. *Strain*, 19:159–171, 1983.

[29] I.A. Viktorov. Rayleigh-type waves on a cylindrical surface. *Soviet Physics-Acoustics*, 4(2):131, 1958.

[30] L.M. Brock. Shear and normal impact loadings on one face of a narrow slit. *Solid Structures*, 18:467–477, 1982.

[31] C.P. Burger and A. Singh. Dynamic photoelasticity as an aid in developing new ultrasonic test methods. In *Proc. SESA-Meeting*, pages 87–98, Dearborn, MI, USA, May 1981.

[32] A.N. Henzi and J.W. Dally. A photoelastic study of stress wave propagation in a quarter-plane. *Geophysics*, 36(2):296–310, 1971.

[33] H.W. Reinhardt and J.W. Dally. Some characteristics of Rayleigh wave interaction with surface flaws. *Materials Evaluation*, 28:213–220, 1970.

[34] H.P. Rossmanith. Elastic wave interaction with a cracked quarter-plane. *Meccanica*, 20:127–135, 1985.

[35] L.M. Brock and H.P. Rossmanith. Analysis of the reflection of point force-induced crack surface waves by a crack edge. *J. Applied Mechanics*, 52:57–61, 1985.

[36] R.E. Knasmillner. Vielpunktmethode zur Bestimmung des Spannungsintensitätsfaktors mit Hilfe der Methode der Kaustik. *ÖIAZ*, 131:318–320, 1986.

[37] H.P. Rossmanith and R.E. Knasmillner. An interactive caustic-based method of stress intensity factor determination. In D. Firrao (Editor), *Proc. 8. European Conference on Fracture (ECF-8)*, volume 2, pages 1093–1098. EMAS, October 1990.

MEASUREMENT OF TRANSIENT CRACK TIP FIELDS USING THE COHERENT GRADIENT SENSOR

Ares J. Rosakis
Sridhar Krishnaswamy
Hareesh V. Tippur

Graduate Aeronautical Laboratories,
California Institute of Technology
Pasadena, CA 91125, USA

ABSTRACT

A shearing interferometric method (CGS) is used to study dynamic crack-tip fields in both transparent and opaque media. CGS is a full-field optical technique that provides real-time fringes representing crack-tip deformation or stress state. The results of the experiments are interpreted on the basis of a newly developed transient higher-order asymptotic analysis. The good agreement between the experimental results and the transient *higher-order* asymptotic analysis clearly indicates that the contribution of the non-singular terms to the total stress and deformation fields is significant.

INTRODUCTION

In this paper, we present a shearing interferometric technique –*coherent gradient sensing (CGS)* – which can be used to obtain *full-field* information of dynamic crack tip deformation in both opaque as well as optically isotropic transparent solids. Unlike the shearing interferometers that have been used in the past [1], the set-up discussed in this paper is particularly suited for real-time study of rapidly propagating cracks. Only the basic details of the CGS technique will be presented here. A more complete description of the method, an account of related optical techniques, as well as results of a companion study pertaining to static crack-tip deformation fields, may be found in [2,3]. The purpose of the present investigation is to map transient crack-tip deformation fields in materials such as PMMA, and to clarify certain issues regarding the interpretation of the experimental results on the basis of both existing asymptotic analyses as well as more complete transient higher-order descriptions of the crack-tip fields.

THE METHOD OF COHERENT GRADIENT SENSING

In Fig. 1 the schematic of the experimental set up used for transmission CGS is shown. A transparent, optically isotropic planar specimen is illuminated by a collimated beam of coherent laser light. The transmitted object wave is then incident on a pair of high density Ronchi gratings, G_1 and G_2, which have a pitch of p and are separated by a distance Δ. The diffracted wavefronts emerging from grating G_2 are collected through a filtering lens L_1. The frequency content (diffraction spectrum) of the field distribution on the G_2 plane is thus obtained on the back focal plane of the filtering lens. A filtering aperture is located on the back focal plane of lens L_1 so as to block all but the ± 1 diffraction order. The filtered beam is then further imaged by means of a system of lenses (denoted here by a single lens L_2 for simplicity) on to the image plane of a high-speed camera.

A simple modification of the above set up enables CGS to be used in reflection mode [3]. In this case, the specularly reflecting object surface is illuminated normally by a collimated beam of laser light using a beam splitter. The reflected beam, as in transmission mode, then gets processed through the optical arrangement which is identical to the one shown in Fig. 1.

A first order diffraction analysis is presented in [3] which, assuming that the specimen is under plane-stress conditions, demonstrates that the information obtained through the CGS method corresponds to in-plane stress gradients in transmission mode, and out-of-plane displacement gradients in reflection mode. Thus, in transmission mode, the CGS fringes are shown to be related to the gradients of stress quantities through,

$$ch\frac{\partial}{\partial x_1}(\hat{\sigma}_{11} + \hat{\sigma}_{22}) = \frac{mp}{\Delta}, \tag{1}$$

$$ch\frac{\partial}{\partial x_2}(\hat{\sigma}_{11} + \hat{\sigma}_{22}) = \frac{np}{\Delta}, \tag{2}$$

where c is a material dependent stress-optic coefficient, h is the nominal thickness of the specimen, $\hat{\sigma}_{11}$ and $\hat{\sigma}_{22}$ are the in-plane stresses, and n, m are the fringe orders.

In reflection mode, the CGS fringes are related to the gradients of the out-of-plane displacement field u_3 through:

$$\frac{\partial u_3}{\partial x_1} = -\frac{\nu h}{2E}\frac{\partial}{\partial x_1}(\hat{\sigma}_{11} + \hat{\sigma}_{22}) = \left(\frac{mp}{2\Delta}\right) \quad m = 0, \pm 1, \pm 2, \dots, \tag{3}$$

$$\frac{\partial u_3}{\partial x_2} = -\frac{\nu h}{2E}\frac{\partial}{\partial x_2}(\hat{\sigma}_{11} + \hat{\sigma}_{22}) = \left(\frac{np}{2\Delta}\right) \quad n = 0, \pm 1, \pm 2, \dots, \tag{4}$$

where, once again, we assume that the specimen is under plane stress conditions. Here, ν is Poisson's ratio and E is Young's modulus of the material.

EXPERIMENTS

Experiments were performed on PMMA specimens in both transmission and reflection modes. The specimen geometry used is shown in Fig. 2. The nominal specimen

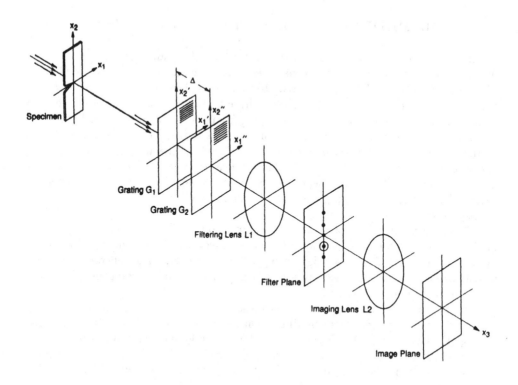

FIGURE 1: Transmission CGS set-up schematic.

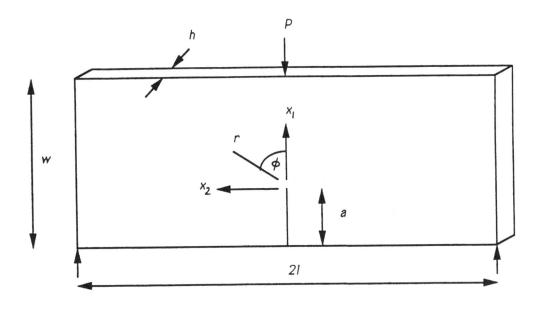

FIGURE 2: Specimen Geometry

dimensions were length $2l = 15.2$cm, width $w = 12.7$cm, and thickness $h = 1$cm. Specimens were made from a flat sheet of PMMA. A band saw, approximately 0.75 mm thick, was used to cut an initial notch of length $a = 25$mm in these specimens. In transmission mode, no further specimen preparation was needed. In reflection mode, an aluminum coating was applied to the PMMA specimen surface through a vacuum deposition technique in order to make it reflective.

The specimens were loaded in a three-point bending configuration. The loading device used to dynamically load the specimens was the Dynatup 8100A drop-weight tower. The experimental configuration is such that the crack is under Mode-I loading conditions.

A rotating-mirror type high-speed camera set-up was used to obtain a sequence of dynamic CGS interferograms. A Spectra Physics (model 166) argon-ion pulse laser (output power 2W at $\lambda = 514$nm in continuous wave mode) was used as the light source. The laser beam was expanded and collimated to obtain a beam 50mm in diameter which was centered on the initial notch tip of the specimen. The transmitted or reflected object wavefront was then processed through a pair of line gratings of density 40 lines/mm with a separation distance $\Delta = 30$ mm. The gratings were oriented with their principal direction parallel to the crack line in order to obtain the x_1-gradient information of the crack-tip fields. The resulting diffracted wavefronts were then collected, filtered and imaged onto a rotating-mirror high-speed camera through a series of lenses. The pulsing circuit of the laser was set to give 50ns exposure every 7 to 10 μs for a total of 1ms from the time of an input trigger which was synchronized to the moment of impact of the drop-weight with the specimen.

RESULTS

The time sequence of CGS interferograms mapping the transient crack tip deformation fields are analyzed in this section. In the first part, the analysis of fringe patterns is attempted under the assumption of K_I^d-dominance; that is, it is assumed that the stress and displacement fields all around the crack tip are well described by the asymptotic, square-root singular, dynamic stress-intensity factor (K_I^d) term. The inadequacies of this approach are then discussed, and a scheme for interpretation of CGS fringes using a more full-field description is described.

Fringe Interpretation Under K_I^d-Dominance Assumption

Freund and Clifton [4] have shown that the stress-field with reference to a Cartesian coordinate system moving with a propagating crack tip for all *plane* elastodynamic solutions for smoothly running cracks under mode-I conditions, can be asymptotically described by the square-root singular expression:

$$\hat{\sigma}_{\alpha\beta}(r,\phi) = \frac{K_I^d(t)}{\sqrt{2\pi r}} \Sigma_{\alpha\beta}^I(\phi,\dot{a}) + O(1), \quad \text{as } r \to 0. \tag{5}$$

Here (r,ϕ) is a polar coordinate system travelling with the crack-tip, $\Sigma_{\alpha\beta}^I$ are known universal functions of ϕ and crack velocity $\dot{a}$ (with respect to some fixed coordinate frame). Indeed, $K_I^d(t)$, the so-called dynamic stress-intensity factor in Mode-I, is the only factor which is dependent on the specific geometry and loading conditions. Any finite region around the crack tip where the above *asymptotic* field dominates (to within some acceptable error) will be referred to as a K_I^d-dominant region.

Transmission Mode: Figure 3 shows a representative sequence of CGS (transmission mode) interferograms for the case of a dynamically propagating crack in PMMA specimen (PD-11).

It was found that the crack propagated with an essentially constant velocity of about 0.25 the shear wave speed of the material. These fringe patterns correspond to the case when the diffraction gratings were oriented with their principal directions parallel to the x_1-axis. Thus, the fringes surrounding the crack-tip at each instant of time represent the x_1-gradient of $(\hat{\sigma}_{11} + \hat{\sigma}_{22})$.

The fringe patterns were digitized for analysis to get fringe order (m) and location (r,ϕ) with respect to the crack-tip at each instant in time. If we now assume that the crack-tip fields are K_I^d-dominant, we find, using equation (5) in the fringe relation (1) that:

$$ch\mathcal{D}(\dot{a})\frac{K_I^d}{\sqrt{2\pi}}r_l^{-3/2}\cos(3\phi_l/2) + o(r^{-3/2}) = \frac{mp}{\Delta}, \tag{6}$$

where a negative sign has been absorbed into the fringe orders. In the above:

$$\mathcal{D}(\dot{a}) = \frac{(1+\alpha_s^2)(\alpha_l^2 - \alpha_s^2)}{4\alpha_l\alpha_s - (1+\alpha_s^2)^2}, \tag{7}$$

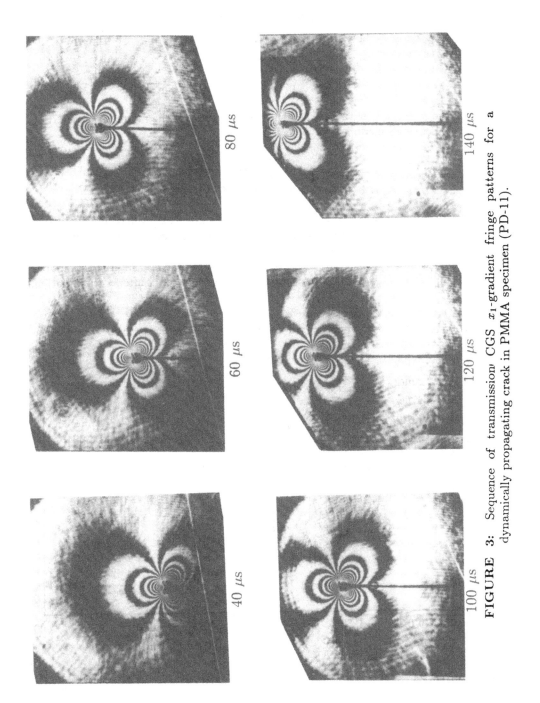

FIGURE 3: Sequence of transmission CGS x_1-gradient fringe patterns for a dynamically propagating crack in PMMA specimen (PD-11).

$$\phi_l = \tan^{-1}\{\alpha_l \tan \phi\}, \tag{8}$$

$$r_l = r \cos\phi\{1 + \alpha_l \tan^2 \phi\}, \tag{9}$$

$$\alpha_{l:s} = \left\{1 - \left(\frac{\dot{a}}{c_{l:s}}\right)^2\right\}^{1/2}, \tag{10}$$

and c_l and c_s are the longitudinal and shear wave speeds of the material, respectively. Now let us define a function $Y_1^d(r, \phi)$ as follows:

$$Y_1^d(r, \phi) = \left(\frac{mp}{\Delta}\right) \frac{\sqrt{2\pi}}{ch\mathcal{D}(\dot{a})} \frac{r_l^{3/2}}{\cos(3\phi_l/2)}, \tag{11}$$

It is apparent from the above and equation (6) that in regions where K_I^d-dominance holds, Y_1^d would be a constant equal to the instantaneous dynamic stress-intensity factor K_I^d to within experimental error.

Figure 4 shows a typical plot of Y_1^d against normalized radial distance (r/h) for one particular specimen (PD-11) for a time instant a few microseconds after crack initiation.

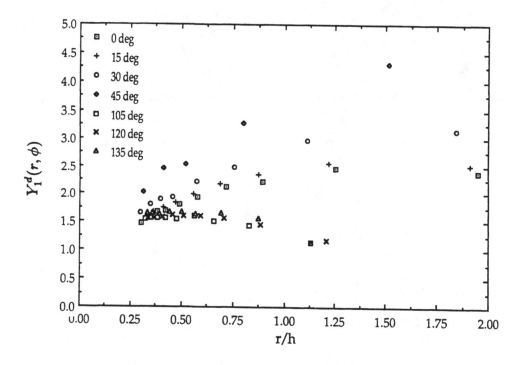

FIGURE 4: Radial variation of $Y_1^d(r, \phi)$ along various ϕ for a dynamically propagating crack in PMMA specimen (PD-11) corresponding to $20\mu s$ from crack-intiation when crack-length to plate width ratio $a/w = 0.23$.

As is apparent from the figure, there appears to be *no* region around the crack tip over which the function Y_1^d is constant. Indeed the spread in Y_1^d values from different locations is as much as 400%. Obviously, extraction of the dynamic stress intensity factor value cannot be based here on a simplistic assumption of near tip K_I^d-dominance. One other interesting point must be made. The instantaneous crack length a to plate width w ratio at the time shown in Fig. 4 was $a/w = 0.23$. The immediate question that arises is: would there be a region of *static* K_I-dominance around the crack-tip for the case of a *statically* loaded specimen of the *same* geometry? The answer is provided in Fig. 5 which shows a plot of the static counterparts of the quantities plotted in Fig. 4 for the case of a specimen (PS-9) with $a/w = 0.2$.

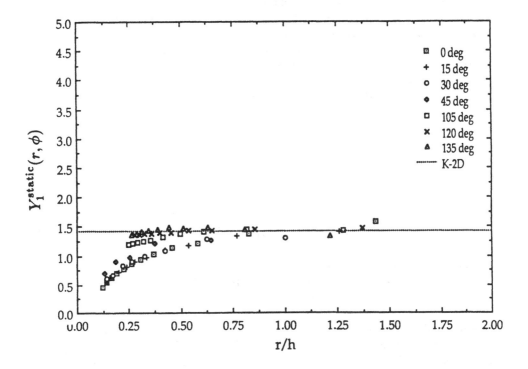

FIGURE 5: Radial variation of $Y_1^{static}(r, \phi)$ along various ϕ for a statically loaded crack in PMMA specimen (PS-9) with crack-length to plate width ratio $a/w = 0.2$.

In the region $(r/h) < 0.5$, Y_1^{static} does not seem to be constant and thus the field in this region does not appear to be K_I-dominant. This is consistent with other experimental investigations wherein such deviation has been attributed to near-tip three dimensionality [5,6]. Outside this three dimensional region, however, there appears to be a sizeable region of constant Y_1^{static} in the range $0.5 \leq (r/h) \leq 1.25$. Further, the constant value of Y_1^{static} in this range is in good agreement with the static stress intensity factor K_I^{2D} as obtained from applied boundary load measurements. This is a clear indication that *even though a specimen might exhibit a sizeable region of K_I-*

dominance under static *loading conditions, there might be no corresponding region of*
K_I^d-*dominance under dynamic conditions.*

The observed lack of K_I^d-dominance outside the near tip three dimensional region
could be due to many reasons. Fundamentally, it could be that the highly transient
nature of the deformation field in this finite geometry necessitates a more full-field
description of the crack tip fields. In addition, there is a possibility of rate-effects
affecting the properties of the material. If the *mechanical* properties of the material
are strongly rate-dependent, then one might have to use not an elastic analysis, but a
viscoelastic one. For PMMA, however, the viscoelastic response, especially at the high
loading rates and short time scales under consideration, is expected to be essentially
glassy. If, on the other hand, the rate-dependence of the *optical* properties is strong, then
the apparent lack of K_I^d-dominance could be an artifact of the choice of a *transmission*
optical technique. In order to rule out this latter possibility, dynamic crack tip fields in
the same material (PMMA) were mapped using CGS in *reflection* mode.

Reflection Mode: Figure 6 shows a sequence of the x_1-gradient fringe patterns for
PMMA specimen (PD-16) in reflection mode.

Interpretation of these fringes is once again done assuming K_I^d-dominance, under
which the reflection mode fringe relation (3) becomes:

$$\frac{\nu h}{2E}\mathcal{D}(\dot{a})\frac{K_I^d}{\sqrt{2\pi}}r_l^{-3/2}\cos(3\phi_l/2) + o(r^{-3/2}) = \left(\frac{mp}{2\Delta}\right), \tag{12}$$

Now, define the function:

$$Z_1^d(r,\phi) = \left(\frac{mp}{2\Delta}\right)\frac{1}{\mathcal{D}(\dot{a})}\frac{2E\sqrt{2\pi}}{\nu h}\frac{r_l^{3/2}}{\cos(3\phi_l/2)}, \tag{13}$$

Once again, in regions of K_I^d-dominance, Z_1^d would be a constant identically equal to
the value of the instantaneous dynamic stress intensity factor.

In Fig. 7, the function Z_1^d is plotted against (r/h) for one particular instant during
crack propagation in PMMA specimen (PD-16).

Once again, we note that there is no region around the crack tip where this function
is a constant, indicating that the crack tip fields were not K_I^d-dominant. Thus it is clear
that the observed lack of K_I^d-dominance in the PMMA *transmission* CGS experiments
cannot solely be attributed to rate dependence of the stress-optic coefficient. Indeed,
we will see in the next section that the CGS interferograms can be consistently interpreted on the basis of higher-order, *elastodynamic* fields, which would therefore seem
to preclude rate-effects (on mechanical or optical properties) as being the source of the
observed lack of K_I^d-dominance.

Fringe Interpretation Using Higher-Order Description of Fields

While dynamic caustic patterns have conventionally been analyzed under the often
unverified assumption of K_I^d-dominance, the use of higher-order terms has been the
recent practice in the method of photoelasticity [7]. However, all available higher-order
elastodynamic solutions thus far have been for the case of steadily propagating cracks,
and the applicability of such solutions to possibly highly transient problems has been
questioned. Indeed, the main criticism of this approach has been that such a procedure
may result in inappropriate time averaging of field quantities [8].

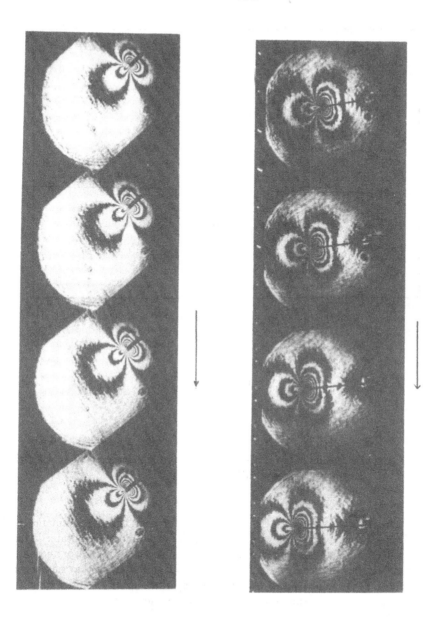

FIGURE 6: Sequence of reflection CGS x_1-gradient fringe patterns for a dynamically propagating crack in reflective PMMA specimen (PD-16).

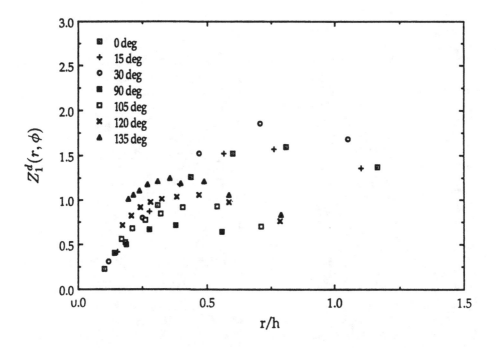

FIGURE 7: Radial variation of $Z_1^d(r, \phi)$ along various ϕ for a dynamically propagating crack in reflective PMMA specimen (PD-11).

In the following, we will relax the assumptions of near tip K_I^d-dominance and steady-state crack propagation. We will show that the experimentally obtained transmission and reflection mode CGS interferograms can be successfully interpreted on the basis of a transient, higher-order stress field for a propagating crack that has become available recently [9]. In addition, we will also identify the conditions under which dynamic crack propagation data can possibly be analyzed under higher-order steady-state approximation.

Transient Higher-Order Analysis: Freund and Rosakis [9] have extended the earlier interior asymptotic solution of Freund and Clifton [4] to provide a higher-order description of the transient stress-state at the vicinity of a dynamically propagating crack. Specializing their results to the case of constant crack velocity appropriate to our experiments, we find that the higher-order expression for $(\hat{\sigma}_{11} + \hat{\sigma}_{22})$ needed in the evaluation of fringe relations (1-4) is given by

$$
\frac{\hat{\sigma}_{11} + \hat{\sigma}_{22}}{2\rho(c_l^2 - c_s^2)} = \frac{3\dot{a}^2}{4c_l^2} A_0 \cos(\phi_l/2) r_l^{-1/2} + \frac{2\dot{a}^2}{c_l^2} A_1
$$

$$
+ \left[\left\{ \frac{15\dot{a}^2}{4c_l^2} A_2 + \left(1 - \frac{\dot{a}^2}{2c_l^2} \right) D^1(A_0) \right\} \cos(\phi_l/2) + \frac{\dot{a}^2}{8c_l^2} D^1(A_0) \cos(3\phi_l/2) \right] r_l^{1/2}
$$

$$
+ \left[\left\{ \frac{6\dot{a}^2}{c_l^2} A_3 + \left(1 - \frac{\dot{a}^2}{4c_l^2} \right) D^1(A_1) \right\} \cos(\phi_l) \right] r_l
$$

$$
+ \left[\left\{ \frac{35\dot{a}^2}{4c_l^2} A_4 + \left(1 - \frac{\dot{a}^2}{2c_l^2} \right) D^1(A_2) + \frac{1}{9} \left(1 - \frac{\dot{a}^2}{4c_l^2} \right) D^2(A_0) + \left(1 - \frac{\dot{a}^2}{2c_l^2} \right) \ddot{A}_0 \right\} \cos(3\phi_l/2) +
$$

$$
\left\{ \frac{3\dot{a}^2}{8c_l^2} D^1(A_2) + \frac{1}{6} \left(1 - \frac{\dot{a}^2}{4c_l^2} \right) D^2(A_0) + \frac{3\dot{a}}{8c_l^2} \ddot{A}_0 \right\} \cos(\phi_l/2) + \left\{ \frac{\dot{a}^2}{96c_l^2} D^2(A_0) \right\} \cos(5\phi_l/2) \right] r_l^{3/2}
$$

$$
+ \left[\left\{ \frac{12\dot{a}^2}{c_l^2} A_5 + \left(1 - \frac{\dot{a}^2}{2c_l^2} \right) D^1(A_3) + \frac{1}{16} D^2(A_1) + \left(1 - \frac{\dot{a}^2}{2c_l^2} \right) \ddot{A}_1 \right\} \cos(2\phi_l) \right.
$$

$$
\left. + \left\{ \frac{\dot{a}^2}{2c_l^2} D^1(A_3) + \frac{1}{8} \left(1 - \frac{\dot{a}^2}{4c_l^2} \right) D^2(A_1) + \frac{\dot{a}}{2c_l^2} \ddot{A}_1 \right\} \right] r_l^2 + o(r_l^2),
$$

$$\tag{14}$$

where

$$
D^1(A_k) = -\frac{(k+3)\dot{a}}{c_l^2 \alpha_l^2} \frac{d}{dt}(A_k), \quad k = 0, 1, 2, \ldots,
$$

$$
D^2(A_k) = D^1(D^1(A_k)), \tag{15}
$$

$$
\ddot{A}_k = \frac{1}{c_l^2 \alpha_l^2} \left(\frac{d^2}{dt^2} A_k \right),
$$

and ρ is density of the material. Although the above expression is for the special case of constant crack velocity, the coefficients A_k in the above transient field are allowed to be time-varying as opposed to the steady-state approximation where the corresponding coefficients must be constant. Note also that the *spatial variation* of the terms associated with coefficients A_k are identical to the ones that would be obtained from a higher order *steady state* analysis with A_0 being related to the dynamic stress-intensity factor and so on. Furthermore, it is seen that the higher-order transient expression contains additional terms whose coefficients depend on time derivatives of lower-order coefficients. For example, $D^1(A_0)$ is proportional to the first time derivative of the stress-instensity factor history and so on.

Transmission Mode: Using the transient field given in equation (14) in the x_1-gradient transmission mode fringe relation (1), and re-arranging terms, we get an expression of the form:

$$\left(\frac{mp}{\Delta}\right)\frac{1}{\mathcal{D}(\dot{a})}\frac{\sqrt{2\pi}}{ch}\frac{r_l^{3/2}}{\cos(3\phi_l/2)} = K_I^d + \left\{\beta_2\frac{\cos(\phi_l/2)}{\cos(3\phi_l/2)} + \beta_3\frac{\cos(5\phi_l/2)}{\cos(3\phi_l/2)}\right\}r_l+$$

$$\left\{\beta_4\frac{1}{\cos(3\phi_l/2)}\right\}r_l^{3/2}+$$

$$\left\{\beta_5\frac{\cos(\phi_l/2)}{\cos(3\phi_l/2)} + \beta_6 + \beta_7\frac{\cos(7\phi_l/2)}{\cos(3\phi_l/2)}\right\}r_l^2+$$

$$\left\{\beta_8\frac{\cos(\phi_l)}{\cos(3\phi_l/2)}\right\}r_l^{5/2} + o(r^3),$$

16)

where K_I^d, the dynamic stress-intensity factor, as well as $\beta_2...\ \beta_8$ are constants to be determined. The left hand side of the above equation is nothing but the function Y_1^d defined in equation (11). Under K_I^d-dominance, Y_1^d would have been a constant equal to the instantaneous stress intensity factor K_I^d. If significant higher-order transient terms exist, then the variation of Y_1^d would be given by the right hand side of (16), which for simplicity will be denoted by $G_1^d(r, \phi; K_I^d, \beta_2..., \beta_8)$.

A least-squares procedure analogous to the one described in [3] was used to fit the above function G_1^d to the experimental data Y_1^d obtained from the CGS interferograms. Since equation (14) is obtained from a two dimensional analysis, data from the near tip three dimensional region were excluded. This was done using the results from a three dimensional elastodynamic finite element analysis of this specimen configuration as described in [10].

Figure 8 shows the results for one particular time instant during crack propagation in specimen (PD-11). The agreement between the fitted function G_1^d (based on the transient expansion (14) to $O(r^2)$) and the experimental data in Fig. 8 is indeed seen to be remarkably good. Figure 9 conveys a more visual picture of the agreement between the transient analysis and the data. Here, a reconstructed fringe pattern from the fitted function G_1^d is shown superposed (as broken lines) on one side of the corresponding CGS interferogram. Also shown on the other side as solid lines is the reconstructed fringe pattern obtained from a least-squares analysis based on the assumption of K_I^d-dominance (equivalent to using only one term in equation (14)). The crosses on this side represent the experimental data obtained from the CGS interferogram. It is seen that the transient analysis agrees very well with the experimental data in the range ($0.5 <$ $r/h < 2.0, -\pi < \phi < \pi$), whereas the K_I^d-dominance assumption is clearly inadequate. This was seen to be the case for other time instants as well. This conclusively shows that *the observed lack of K_I^d-dominance in the two-dimensional region outside the near-tip three-dimensional zone is due to the important contribution of higher-order terms to the total stress and deformation fields around the crack-tip.*

The errors associated with interpretation of experimental data on the basis of K_I^d-dominance can now be evaluated. Figure 10 shows the time variation of the dynamic stress-intensity factor during crack propagation in specimen (PD-11) as obtained from the time sequence of x_1-gradient CGS fringe patterns. The solid line represents the value for the dynamic stress-intensity factor when the fringes were interpreted on the basis of the higher-order transient analysis. The dotted line corresponds to the value for the dynamic stress-intensity factor obtained when the fringes were analyzed under the assumption of K_I^d-dominance. It is clear that interpretation of experimental data

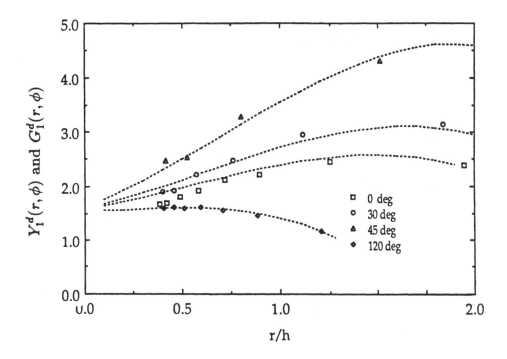

FIGURE 8: Comparison of the radial variation of the experimental data $Y_1^d(r,\phi)$ and the analytical fit $G_1^d(r,\phi)$ for various ϕ for transmission mode CGS pattern for specimen (PD-11).

on the basis of K_I^d-dominance leads to substantial errors in the measured values for the dynamic stress-intensity factor.

Reflection Mode: A higher order analysis of reflection mode CGS interferograms can be done along essentially the same lines as described above. Using the higher order expression for stresses from equation (14) in the reflection mode fringe relation (3), leads after rearrangement to:

$$\left(\frac{mp}{2\Delta}\right)\frac{1}{\mathcal{D}(\dot{a})}\frac{2E\sqrt{2\pi}}{\nu h}\frac{r_l^{3/2}}{\cos(3\phi_l/2)} = G_1^d(r,\phi;K_I^d,\beta_2...\beta_8). \tag{17}$$

Here G_1^d is the same expression as the right hand side of equation (14) and the left-hand side of the above expression is the quantity Z_1^d previously defined in equation (13). Once again, the experimental data Z_1^d can be fitted to the function G_1^d through the least-squares procedure discussed earlier.

Figure 11 shows the reflection mode CGS results for one particular time instant during crack propagation in PMMA specimen (PD-16). As in the transmission mode case, the agreement between the fitted function G_1^d and the data Z_1^d is good outside the near-tip three-dimensional region. This is more graphically brought out in Fig. 12 where

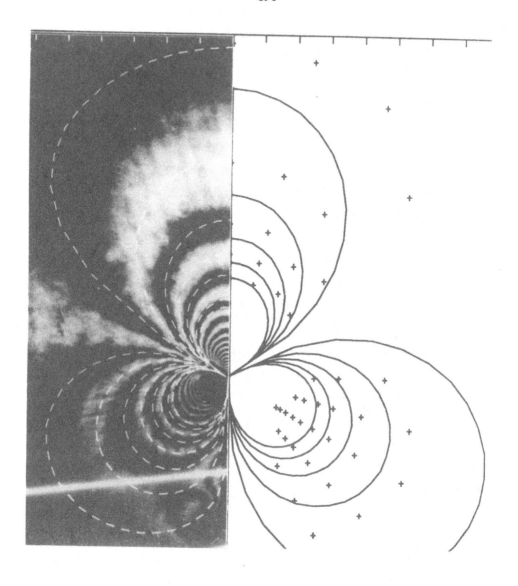

FIGURE 9: Synthetic fringe patterns reconstructed from one-term (crosses) and higher-order (broken lines) transient analysis superposed on the corresponding transmission mode x_1-gradient CGS interferogram for specimen (PD-11).

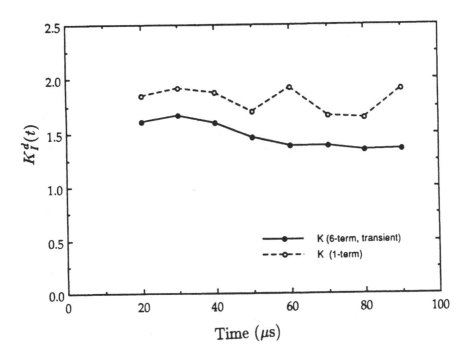

FIGURE 10: Time variation of the dynamic stress-intensity factor as obtained from one-term (broken lines) and higher-order (solid lines) transient analysis of CGS sequence for specimen (PD-11).

the reconstructed fringe pattern from the higher-order analysis is shown superposed (as broken lines) on the CGS interferogram on one half of the figure.

Also shown on the other side as solid lines is the reconstructed fringe pattern obtained from a least-squares analysis based on the assumption of K_I^d-dominance. The crosses on this side represent the experimental data obtained from the CGS interferogram. Again, it is seen that K_I^d-dominance assumption is clearly inadequate. We thus note that *a consistent interpretation of both transmission and reflection mode CGS fringe patterns is possible through inclusion of transient, higher-order effects.*

Steady-State Higher-Order Analysis: We next look at the possible applicability of higher-order, steady-state elastodynamic solutions [7] in the interpretation of CGS fringe patterns. A detailed examination of the transient, higher-order stress-field (equation (14)) indicates that the use of a steady-state, higher-order expansion would be appropriate only when (i) the time derivatives of the coefficients corresponding to the higher-order terms are all negligibly small, and (ii) the crack-velocity is essentially constant. Intuitively, these conditions imply that crack-tip state is relatively quiescent. If the dynamic crack propagation event is deemed quiescent, we may analyze the CGS interferograms on the basis of the *steady state,* higher order expansion for the stress field, the results of which should be accepted only if the use of such a procedure can be justified *post*

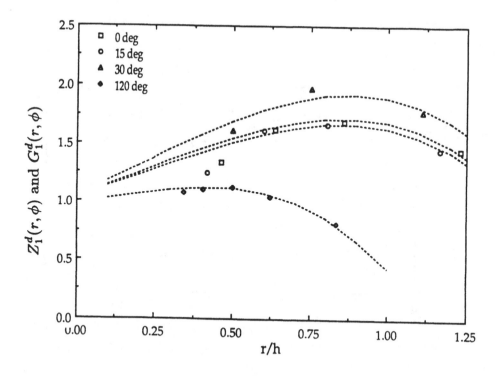

FIGURE 11: Comparison of the radial variation of the experimental data $Z_1^d(r,\phi)$ and the analytical fit $G_1^d(r,\phi)$ for various ϕ for reflection mode CGS pattern for specimen (PD-16).

facto.

Using the results of Dally, Fourney and Irwin [7] for the stress-field in the fringe relation (1) we find, on rearrangement of terms:

$$\left(\frac{mp}{\Delta}\right)\frac{1}{\mathcal{D}(\dot{a})}\frac{\sqrt{2\pi}}{ch}\frac{r_l^{3/2}}{\cos(3\phi_l/2)} = K_I^d + \sum_{N=3}^{\infty} A_N r_l^{\left(\frac{N-1}{2}\right)}\frac{\cos\left(\frac{N}{2}-2\right)\phi_l}{\cos(3\phi_l/2)}, \qquad (18)$$

where the stress intensity factor K_I^d, and the coefficients of the higher order terms $A_3...A_N$ are constants to be determined. Once again, the left hand side of equation (18) is the function Y_1^d, and the right hand sides will be denoted by $F_1^d(r,\phi,K_I^d,A_3...A_N)$. As before, the experimental data Y_1^d can be fitted to the function F_1^d by means of a least-squares procedure to extract the various coefficients including the stress intensity factor. Obviously, the infinite series defining F_1^d needs to be truncated at some point. Unfortunately, since the individual terms of the series are not mutually orthogonal, the values obtained for the corresponding coefficients could be affected by the choice of number of terms used. A consistent scheme for truncation of the series is, therefore, needed. The choice of the optimal number of higher order terms to use is made here on the basis of a criterion described in detail in [3]. Note that this is not a trivial issue,

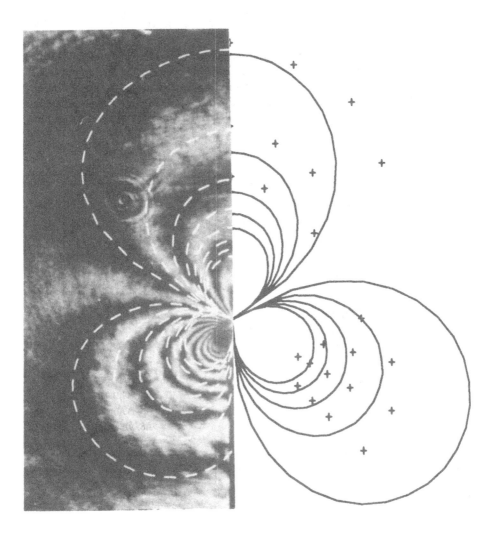

FIGURE 12: Synthetic fringe patterns reconstructed from one-term (crosses) and higher-order (broken lines) transient analysis superposed on the corresponding reflection mode x_1-gradient CGS interferogram for specimen (PD-16).

since it appears that erroneous results could be obtained if substantially more than the optimal number of terms is used.

The sequence of CGS interferograms from specimen (PD-11) were analyzed using the above procedure. Typically a five term expansion (in equation 18) was found to be necessary to adequately describe the crack tip fields. Figure 13 shows the *time* variation of the dynamic stress intensity factor and that of the other four higher-order coefficients as measured from x_1-gradient CGS interferograms through the least squares procedure described earlier.

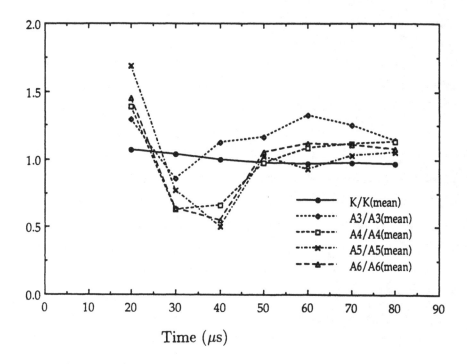

FIGURE 13: Time variation of the dynamic stress intensity factor and coefficients of higher-order terms obtained through a *steady-state* interpretation of CGS sequence for specimen (PD-11). Note that the quantities are normalized by their time-average values for convenience.

For convenience of presentation, the time varying coefficients have all been normalized by their time-average values. Note that for times sufficiently close to the crack initiation event, the field does not appear to be very quiescent as indicated by the large variation in the measured coefficients. The use of a steady-state analysis and the consequent results are mutually inconsistent at these early times, and thus the results of the steady-state approximation are not acceptable. However, at later times, the extracted coefficients do not appear to be very strongly time varying. Also, during the time shown, the crack propagated with essentially a constant velocity of $\dot{a}/c_l \approx 0.13$.

Thus, the use of the higher-order *steady-state* expansion and the consequent results are at least mutually consistent and may not be entirely unjustified at these later times. Of course, for propagation times close to initiation, or whenever the field does not appear to be very quiescent, use of the transient higher order expansion (14) would be more appropriate.

DISCUSSION

The results of this investigation point out the pitfalls associated with the extraction of dynamic fracture data under the assumption of K_I^d-dominance. That violation of assumed K_I^d-dominance could be the source of certain incompatible experimental results in the literature [6,11-14] has been the contention of many recent investigators [7,10,15,16].

In this paper, we have shown that it is possible to meaningfully interpret dynamic CGS fringe data even in situations of no K_I^d-dominance. To do this, the data outside of the near tip three dimensional region must be appropriately interpreted taking into account the contribution of transient, higher-order terms to the total stress and deformation fields around the crack-tip.

Contrary to conventional wisdom, K_I^d-dominance in the vicinity of a dynamically propagating crack appears to be the exception rather than the rule. It is thus critically important that *interpretation of experimental data under assumed K_I^d-dominance, steady-state or two-dimensional conditions be carefully justified prior to attributing physical credence to the observed phenomena.*

CONCLUSION

The salient points discussed in this paper can be summarized as follows:

1. A shearing interferometric technique (CGS) is proposed as a viable alternative to current methods in the experimental study of dynamic crack propagation. CGS is a full-field technique that can be used in reflection mode for opaque materials, and can also be used in transmission mode for optically *isotropic* transparent materials.

2. Results from dynamic crack propagation in PMMA specimens using CGS in reflection and transmission mode show that interpretation of experimental data on the basis of near tip K_I^d-dominance could lead to substantial errors due both to near-tip three-dimensionality as well as higher-order effects in the two-dimensional region.

3. It is further shown that a consistent interpretation of experimental data obtained from the two-dimensional region is possible using higher-order analytical description of the transient crack-tip fields.

REFERENCES

1. **Balas J., and Drzik, M.,** (1983) Staveb, Cas., 31, c 6/7, VEDA, Bratislava.

2. **Tippur, H.V., Krishnaswamy, S., and Rosakis A.J.,** (1989a), "A Coherent Gradient Sensor for Crack-Tip Deformation Measurement: Analysis and Experimental Results," GALCIT SM-Report 89-1, California Institute of Technology, (to appear in the *International Journal of Fracture*).

3. **Tippur, H.V., Krishnaswamy, S., and Rosakis A.J.,** (1989b), "Optical Mapping of Crack Tip Deformation Using the Methods of Transmission and Reflection Coherent Gradient Sensing," GALCIT SM-Report 89-11, California Institute of Technology, (submitted to the *International Journal of Fracture*).

4. **Freund L.B., and Clifton R.J.,** (1974), "On the Uniqueness of Plane Elastodynamic Solutions for Running Cracks," *Journal of Elasticity*, vol. 4, No. 4, pp 293-299.

5. **Rosakis A.J., and Ravi-Chandar K.,** (1986), "On Crack-Tip Stress State: An Experimental Evaluation of Three-Dimensional Effects," *International Journal of Solids and Structures*, vol. 22, No. 2, pp 121-134.

6. **Nigam H., and Shukla A.,** (1988), "Comparison of the Techniques of Transmitted Caustics and Photoelasticity as Applied to Fracture," *Experimental Mechanics*, vol. 28, No. 2, pp. 123-133.

7. **Dally J.W., Fourney W.L., and Irwin G.R.,** (1985), "On the Uniqueness of $K_{ID} - \dot{a}$ Relation," *International Journal of Fracture*, vol. 27, pp 159-168.

8. **Knauss W.G., and Ravi-Chandar K.,** (1985), "Some Basic Problems in Stress Wave Dominated Fracture," *International Journal of Fracture*, vol. 27, pp 127-143.

9. **Freund L.B., and Rosakis A.J.,** (1990), "The Influence of Transient Effects on the Asymptotic Crack-Tip Field During Dynamic Crack Growth," Eleventh National Congress of Applied Mechanics, Tucson, Arizona, May 1990.

10. **Krishnaswamy, S., Rosakis, A.J., and Ravichandran G.,** (1988), "On the Extent of Dominance of Asymptotic Elastodynamic Crack-Tip Fields; Part II: Numerical Investigation of Three-Dimensional and Transient Effects," GALCIT SM-Report 89-22, California Institute of Technology, to appear in the *Journal of Applied Mechanics*.

11 **Kalthoff J.F.,** (1983), "On Some Current Problems in Experimental Fracture Mechanics," in 'Workshop on Dynamic Fracture,' edited by Knauss W.G. et al., California Institute of Technology, pp 11-35.

12. **Ravi-Chandar K., and Knauss W.G.,** (1984), "An Experimental Investigation of Dynamic Fracture-I: Crack Initiation and Arrest," *International Journal of Fracture*, vol. 25, pp 247-262.

13. **Rosakis A.J., Duffy J., and Freund L.B.,** (1984), "The Determination of Dynamic Fracture Toughness of AISI 4340 Steel by the Shadow Spot Method," *Journal of the Mechanics of Physics of Solids*, vol. 32, pp 443-460.

14. **Zehnder A.T., and Rosakis A.J.,** (1989),, "Dynamic Fracture Initiation and Propagation in 4340 Steel Under Impact Loading," *International Journal of Fracture*, (to appear).

15. **Ravi-Chandar K., and Knauss W.G., (1987)**, "On the Characterization of the Transient Stress Field Near the Tip of a Crack," *Journal of Applied Mechanics*, vol. 54, pp 72-78.

16. **Krishnaswamy, S., and Rosakis, A.J., (1988)**, "On the Extent of Dominance of Asymptotic Elastodynamic Crack-Tip Fields; Part I: An Experimental Study Using Bifocal Caustics," GALCIT SM-Report 89-22, California Institute of Technology, to appear in the *Journal of Applied Mechanics*.

FE-INVESTIGATION ON CAUSTICS FOR MEASURING IMPACT FRACTURE TOUGHNESS

SHIGERU AOKI and TADASHI KIMURA

Department of Mechanical Engineering Science,
Tokyo Institute of Technology
O-okayama, Meguro-ku, Tokyo 152, JAPAN

ABSTRACT

A finite element simulation of an impact fracture toughness test under high loading rates is performed. An elastic cracked plate subject to dynamic loads is considered and caustics in reflection are generated by making assumption of plane stress condition. The crack is assumed to begin to propagate when the theoretically exact value of $K_I(t)$ reaches a critical value K_{crit}. The time history of the "measured" dynamic stress intensity factor $K_I(t)$ is determined from the diameter of the generated caustics. The impact fracture toughness K_{Id} is determined from the peak value of the time history of the "measured" $K_I(t)$. It is found that the impact fracture toughness thus determined increases with the decrease in time-to-fracture, and depends on the incident wave form when the loading rate is high. These results are discussed based on the time dependent stress field near the crack tip.

INTRODUCTION

Due to the dynamic linear fracture mechanics, a crack under dynamic loading begins to propagate when the dynamic stress intensity $K_I(t)$ reaches the impact fracture toughness K_{Id}, i.e.,

$$K_I(t) = K_{Id} \qquad (1)$$

and the K_{Id} values have been measured for many materials

[1],[2]. However, it has been reported recently that in the short time-to-fracture, the K_{Id} values indicate an enormously sharp increase or depend on the incident stress wave form [3]–[5]. These peculiar behaviors of K_{Id} raise a serious question about the extent to which the basic criterion, equation(1), can be validly used [6].

A modified criterion for fracture initiation has been proposed by Kaltohoff et al. [3]–[5] which requires $K_I(t)$ to exceed a critical value K_{crit} for a certain minimum time (i, e, a fracture incubation time). This criterion coincides approximately with equation(1) for a relatively long time-to-fracture, and yet can explain the above-mentioned peculiar experimental data for a short time-to-fracture. The existence of the incubation time is, however, a hypothetical assumption and physical background is not yet explained.

In the present study, a numerical simulation of an impact fracture test using caustics is performed in an attempt to explain the peculiar behaviors of K_{Id} and to discuss about the physical meaning of the incubation time. The dynamic stress and strain fields near the crack tip are analyzed by the finite element method, and the time history of caustics is simulated.

NUMERICAL PROCEDURES

Finite element analysis

Let us consider an elastic 60mm × 60mm rectangular steel plate with a central crack (30mm in length), and apply a time-dependent uniform pressure $\sigma_o(t)$ on the crack surfaces, as shown in Fig.1. The pressure $\sigma_o(t)$ is assumed to vary like a ramp or step function. In order to avoid the effect of the edge of the plate on the stress and strain fields near the crack tip, we restrict our attention to the period of time in which the stress waves due to $\sigma_o(t)$ travel through the plate, reflect at the edge and come back again near the crack tip.

The finite element mesh of a quarter of the plate is shown in Fig.2, where the 4-node isoparametric elements are used. Analyses are carried out with the assumption of the plane stress condition by using the finite element program "ABAQUS". The material constants are shown in Table 1. The crack node release technique [7],[8] is employed for modeling crack propagation. Consider a crack passing the location of node A and proceeding to the location of node B as shown in Fig.3. The nodal force is relaxed from the initial value f_o according to the following equation:

$$f(t) = f_0 \left(1 - \frac{b(t)}{l}\right) - \sigma_0(t) b(t) \qquad (2)$$

where l is the distance from node A to node B and b(t) is the distance from the crack tip and node A at time t. The effect of the pressure on newly created surfaces is represented by the second term in the right hand side of equation(2).

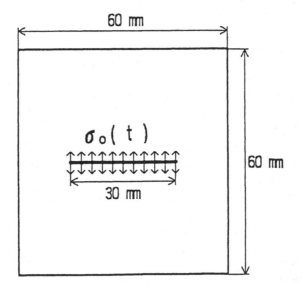

Figure 1. A center cracked plate subjected to a dynamic uniform pressure $\sigma_0(t)$ on crack surfaces

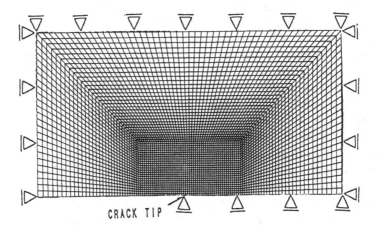

Figure 2. Finite element mesh

Simulation of caustics

Let us consider a flat plate with a through-crack. One surface of the plate, which is assumed to be reflective, occupies a region in x_1, x_2 plane at $x_3=0$ in an undeformed state. When loads are applied to the plate, due to the resulting change in thickness a small depression near the crack tip is made on the surface of the plate, as shown in Fig.4(a).

A family of incident light rays parallel to the x_3 axis is reflected on this deformed surface. The virtual extensions of the reflected rays form a bright curve (caustic curve) on the virtual screen as shown in Fig.4(a). The reflection process can be considered as a mapping of point (x_1, x_2) of the surface of the specimen in undeformed state onto point (X_1, X_2) of the virtual screen $(x_3=-Z_0)$. The mapping equations are given by [3],[9].

TABLE 1
Material constants

Young's Modulus	210 GPa
Density	7.86 g/cm^3
Poisson's Ratio	0.29

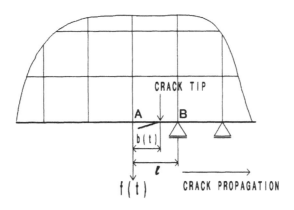

Figure 3. Crack node release technique to simulate crack propagation

$$X_i = x_i - 2 Z_o \frac{\partial w}{\partial x_i} \qquad (i=1, 2) \tag{3}$$

where w is the x_3-component of the displacement at point (x_1, x_2). Some points on the specimen surface map onto the caustic curve. The locus of these points is called the initial curve.

Representing w by using $K_I(t)$ under the assumption of plane stress condition and employing equation(3) lead to the relationship between the maximum transverse diameter D(t) of the caustic and $K_I(t)$ of a stationary crack [3],[9]:

$$K_I(t) = \frac{1}{10.7} \frac{E}{Z_o \nu B} \left\{ D(t) \right\}^{5/2} \tag{4}$$

where E is the Young's modulus, ν the Poisson's ratio, B the thickness of the plate.

In usual experiments the caustics are taken by a high speed camera, then the $K_I(t)$ is obtained by using equaton(4). In the present numerical simulation caustics are constructed by

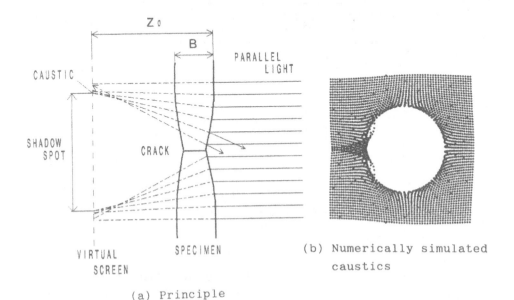

(a) Principle

(b) Numerically simulated
caustics

Figure 4.   Caustics method

using equation(3) with the values of w, which are obtained by the finite element plane stress analysis, and $K_I(t)$ is determined from $D(t)$ by using equation(4). It is necessary to make a little correction in equation(4) for a rapidly propagating crack [10]. However, let us use equation(4) without correction even for a rapidly propagating crack, because we will focus on the K_{Id} (i, e., the value of $K_I(t)$ at the onset of crack propagation) in the present investigation.

In order to check the accuracy of the present simulation, let us apply the static displacements represented by K_{app} on the boundary of the plate (Fig.2, B = 0.005m). The caustic obtained is shown in Fig.4(b). The ratio of static stress intensity factor K determined by equation(4) to K_{app} is shown in Table 2, where C_1 is the longitudinal wave speed, K_{app}=68.5 MPa$\sqrt{m}$ and t_o=7μs ($Z_o/(C_1 t_o)$=5, 12.5 and 25 correspond to Z_o=0.2m, 0.5m and 1m, respectively). It is found that K agrees well with K_{app} without depending on Z_o.

RESULTS

Time history of caustics

A linearly increasing uniform pressure (ramp load) is applied on the crack surfaces from time t=0. The crack is stationary for $t < t_o$ (t_o= 7μs) and it begins to propagate at t = t_o. The amplitude of the ramp load is chosen in such a way that the theoretical value of $K_I(t)$ reaches K_{crit} at time t = t_o, i.e., $K_I(t_o)$ = K_{crit} (K_{crit}=68.5 MPa$\sqrt{m}$). The crack velocity v is assumed to be 0.1 C_1.

Let S_h and S_r denote the maximum and minimum values of X_1-coordinate of caustics as shown in the inlet in Fig.5. The time variations of S_h, S_r and $(S_h+S_r)/2$ are shown in Fig.5, where $Z_o/(C_1 t_o)$ = 12.5. It is found that the apparent change of caustics due to crack initiation is observed at $t/t_o \fallingdotseq$ 1.16.

TABLE 2
Static stress intensity factor K obtained by numerically
simulated caustics

$Z_o/(C_1 t_o)$	5	12.5	25
K/K_{app}	0.996	0.999	1.001

This means that the crack initiation is detected a little later
than the actual one in case where the loading rate is high and
the time-to-fracture is short (less than $10\mu s$).

The time variations of $K_I(t)$ obtained from D(t) by using
equation(4) (which will be referred to as $K_I^m(t)$) and the
theoretical value of $K_I(t)$ (which will be referred to as
$K_I^{th}(t)$) are shown in Fig.6. The results for $v = 0.5C_1$ are also
plotted in the same figure for comparison. The $K_I^{th}(t)$ has a
peak at $t/t_o = 1.0$, while $K_I^m(t)$ has a peak at $t/t_o \doteq 1.16$.

Since the time $(t/t_o \doteq 1.16)$ at which $K_I^m(t)$ has a peak
coincides with the time at which the crack initiation is
detected (Fig.5), the value of $K_I^m(t)$ at the peak gives the
measured K_{Id}. It follows from this that the measured K_{Id} is
greater than K_{crit} for the case where the time-to-fracture is
short. It is also found in Fig.6 that the measured K_{Id} is
independent of the crack speed v and that $K_I^m(t) < K_I^{th}(t)$ for
$t/t_o < 1.0$.

Effect of test conditions on measured K_{Id}
The time variations of $K_I^m(t)$ for various Z_o's are shown in
Fig.7. The test conditions except for Z_o are the same as those
for Fig.5. It is found that the deviation of $K_I^m(t)$ from
$K_I^{th}(t)$ and the overestimation of K_{Id} become more significant

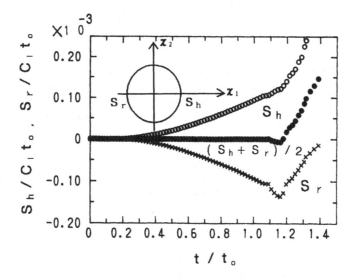

Figure 5.   Time variation of location of caustics

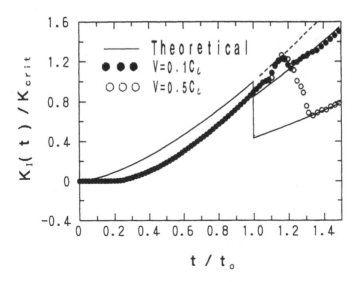

Figure 6. Effect of crack velocity on "measured" K_{Id}

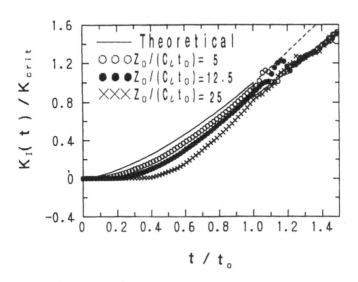

Figure 7. Effect of location of virtual screen on "measured" K_{Id}

with increase in Z_o, while in a static case K agrees well with K_{app} without depending on Z_o as shown in Table 2.

Figure 8 compares the results for a ramp load and a step load. The result for a ramp load is the same as in Fig.6 for v = $0.1C_1$ and in Fig.7 for $Z_o(C_1 t_o)$ = 12.5. The amplitude of the step load is chosen in such a way that $K_I^{th}(t_o)=K_{crit}$ (t_o= 7 μ s). Other test conditions for the step load are identical with those for the ramp load. It is observed that there exists a difference between the K_{Id} value obtained under a ramp load and that under a step load. This means that the K_{Id} values measured by caustics depend on incident stress wave form. Note that the delay time of detection of crack initiation coincides with each other.

The effect of loading rate on K_{Id} is shown in Fig.9. The solid circles indicate the same result in Fig.8 for a ramp load, while the hollow circles are the result for a more rapid ramp load, under which $K_I^{th}(t)$ reaches K_{crit} at t = 4 μ s. Other test conditions are identical with each other. It is found that K_{Id} is overestimated more under higher loading rate, and also that there is not a significant difference in delay time of detection of crack initiation. It is suggested from these results that the correct K_{Id} value (i. e. K_{crit}) would be obtained in case where the time-to-fracture is large (at least more than 15 μ s) under low loading rate.

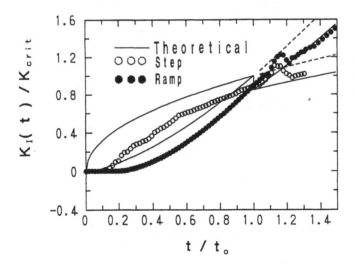

Figure 8.   Effect of stress wave form on "measured" K_{Id}

DISCUSSIONS

$K_I^m(t)$ before crack initiation

Figures 10(a) and (b) compare the exact [11], numerical and approximate (by the singular term) solutions of the stress σ_{22} ahead of the crack tip ($x_2=0$, $x_1>0$) for a step load. The amplitude of the step load σ_a is identical with that in Fig.8. Figure 10(a) is represented in an ordinary scale, while Fig.10(b) in a logarithmic scale. The numerical solution is found to agree well with the exact solution. It is found that a flat stress field spreads widely behind the wave front, i, e., $x_1/(C_1 t) = 1.0$, and also that the radius R of the $K_I^{th}(t)$-field may be represented by $R \doteqdot 0.02 C_1 t$.

The time variations of the radii of the initial curve for $Z_0/(C_1 t) = 12.5$ and $K_I^{th}(t)$-field R are shown in Fig.11. Although R approaches to r_0 with increase in time, R is less than r_0 even at time $t=t_0$ ($t_0 = 7\mu s$). It follows that $K_I^{th}(t)$-field is not developed enough to cover the initial curve even at the crack initiation time. This explains the result in Fig.8 that $K_I^m(t) < K_I^{th}(t)$. Similar explanation may be applicable to the results that $K_I^m(t) < K_I^{th}(t)$ for the ramp load (Figs.6, 7 and 9).

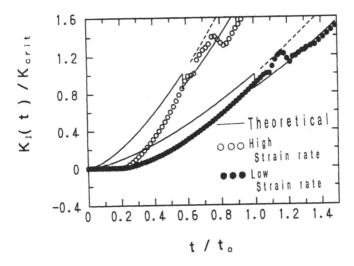

Figure 9. Effect of loading rate on "measured" K_{Id}

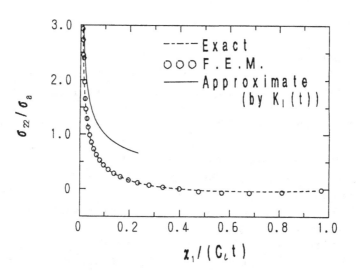

(a)Representation in ordinary scale

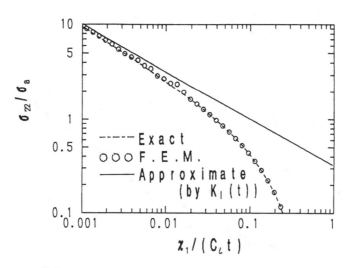

(b)Representation in logarithmic scale

Figure 10.   Stress distribution ahead of the tip of a crack
subjected to a step load $\sigma_0(t)$ on its surfaces

The radius of the initial curve r_o increases with increase in Z_o, while R is independent of Z_o. This explains the result in Fig.7 that the discrepancy of $K_I^m(t)$ and $K_I^{th}(t)$ increases with increase in Z_o. Note that the stress in the region just behind the wave front is compressive as shown in Fig.10(a) for a step load. The change of thickness of specimen in this region is, therefore, positive. Because this region remains near the crack tip for $t/t_o << 1.0$, the caustics are not formed ($D(t) = 0$) and hence r_o can not be defined. This is reason why there is no plot for $t/t_o << 1.0$ in Fig.11. The situation may be similar to that for a ramp load. This explains the results in Fig.6~9 that $K^m(t)=0$ for $t/t_o << 1.0$.

$K_I^m(t)$ after crack initiation

The stress field after crack initiation (Fig.12(a)) can be described as the sum of those in Fig.12(b) and (c), where p(t) denotes the pressure which is equal to the negative value of tensile stress acting on the plane ahead of the crack tip ($x_2 = 0$, $x_1 > 0$) in Fig.12(b). The stress in Fig.12(c) is equal to the change of stress due to crack extension. Although the front of the stress wave generated by crack extension from $t = t_o$ extends at time t to the distance $C_1(t - t_o)$ from the original crack tip, the flat stress field is thought to spread widely just behind the wave front for $v = 0.1~0.5$ C_1.

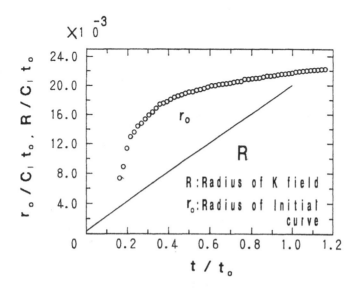

Figure 11.   Time variation of radii of initial curve for $Z_o/(C_1t_o) = 12.5$ and $K_I^{th}(t)$-field R

Because the stress in this field is small, it does not have an apparent effect on the caustics. The stress which has an apparent effect on caustics spreads so slowly ($0.2 \sim 0.3$ $c_1 t$) that it takes a considerable time for this stress to reach

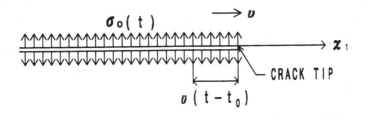

(a) Stress after crack extension

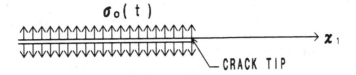

(b) Stress for a stationary crack subjected to pressure
$\sigma_0(t)$ on its surfaces

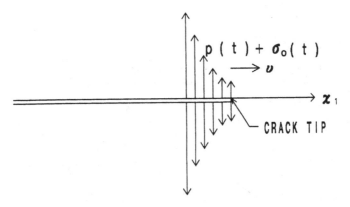

(c) Stress for a propagating crack subjected to pressure
$\{p(t) + \sigma_0(t)\}$ on the newly created surfaces

Figure 12. Superposition of stress field

217

the initial curve which has already developed large at t $\doteqdot$ t_o as shown in Fig.11. Until this stress reaches the initial curve, the caustics coincide with those for Fig.12(b) and continues to develop. This explains the delay of detection of crack initiation shown in Fig.5 and the overestimation of K_{crit} shown in Figs.6~9.

Because r_o increases with increase in Z_o, the delay time increases and hence the K_{crit} value is more overestimated, the results shown in Fig.7 can be similarly explained. Since the increment in $K_I^m(t)$ during the delay of detection of crack initiation is equal to that in Fig.12(b), the measured K_{Id} depends on the incident stress wave and loading rate. This explains the results in Fig.8 and 9.

CONCLUDING REMARKS

A finite element simulation of measuring K_{Id} by the caustics method in the range of short time-to-fracture is performed in this study. It is found that there exists a delay time to detect crack initiation because of the slowness of the development of the stress high enough to have an apparent effect on caustics. The peculiar experimental data that K_{Id} depends on incident wave form and loading rate are explained by taking the delay time of detection of crack extension by the caustics method into consideration. The delay time is independent of incident wave form and loading rate, and may explain to some extent the meaning of the incubation time. More study would be necessary to explain fully the physical meaning of the incubation time which might exist in the microscopic fracture behaviors.

For long time-to-fracture (at least longer than 15μs for steel), the effect of delay time to detect crack initiation is not significant, and hence the basic criterion equation(1) is useful. In case where a crack initiates near the mid-thickness of the specimen, it takes a considerable time for the stress wave due to crack initiation to reach the specimen surface. The delay time to detect crack initiation is, therefore, more significant than that obtained in the two-dimensional simulation.

218

REFERENCES

1. Rosakis, A.J., Analysis of the optical method of caustics for dynamic crack propagation. Eng. Fract. Mech., 1980, 13-2, pp. 331-343.
2. Kobayashi, A.S. and Lee, O.S., Elastic field surrounding a rapidly tearing crack. In Cracks And Fracture, ed. Hoagland, R.G. and Gehlen, P.C., STP 803, ASTM, 1983, pp. 1-21.
3. Kalthoff, J.F., Fracture behavior under high rates of loading. Eng. Fract. Mech., 1986, 23, pp. 289-297.
4. Shockey, D.A., Kalthoff, J.F. and Erlich, D.C., Evaluation of dynamic crack instalibity criteria. Int. J. Fract., 1983, 22, pp. 217-231.
5. Homma, H., Shockey, D.A. and Murayama, Y., Response of crack in structural materials to short pulse loads. J. Mech. Phys. Solids , 1983, 31, pp. 261-279.
6. Aoki, S., On the mechanics of dynamic fracture. JSME Int. J. Ser.I, 1988, 31-3, pp. 487-499.
7. Malluck, J.F. and King, W.W., Fast fracture simulated by conventional finite elements: A comparison of two energy-release algolithms. In Crack Arrest Methodology And Applications, ed. Hahn, G.T. and Kanninen, M.F., STP 711, ASTM, 1980, pp. 38-53.
8. Kanninen, M.F., A dynamic analysis of unstable crack propagation in the DCB specimen. Int. J. Fract., 1974, 10-3, pp. 415-431.
9. Rosakis, A.J., Zehnder, A.T. and Narasimhan, R., Caustics by reflection and their application to elastic-plastic and dynamic fracture mechanics. Optical Eng., 1988, 27-8, pp. 596-610.
10. Nishioka, T. and Kittaka, H., A theory of caustics for mixed-mode fast running cracks. Trans. JSME Ser.A. (in Japanese), 1987, 54-501, pp. 1061-1068.
11. Ma, C.C. and Freund, L.B., The extent of the stress intensity factor field during crack growth under dynamic loading conditions. J. Appl. Mech., 1986, 108, pp. 303-310.

SOME EXPERIMENTS ON MEASUREMENT OF
DYNAMIC STRESS INTENSITY FACTOR OF FAST PROPAGATING CRACKS

SHINICHI SUZUKI and SHINGO FUKUCHI
Department of Energy Engineering
Toyohashi University of Technology
Hibarigaoka 1-1, Tempaku-cho, Toyohashi, 441 JAPAN

ABSTRACT

The extent of three-dimensional (3-d) stress field in the vicinity of a fast propagating crack is measured in the direction of 72 degree from the propagation direction. Also measured is the distance from the crack tip to the boundary between the singular stress field and the region of higher order terms, in the same direction. They are measured with the method of pulsed holographic recording and reconstruction of a caustic light beam of a fast propagating crack. The crack is of the opening mode and, propagates at a speed of several hundred m/s. At the moment of the measurement, the crack is not interacting with the stress waves emitted from the crack itself. The extent of the 3-d stress field is about half the specimen thickness, which result is same as that of stationary cracks. The distance from the crack tip to the region of higher order terms was about 4mm, in the present study.

INTRODUCTION

When brittle material is broken by external tensile force, a fast propagating crack of the opening mode often appears, whose speed is at several hundred meters per second. The dynamic stress intensity factor K_I of the cracks has been measured with some optical methods. In them, the caustic method has been developed as the most useful method for K_I measurement, because of relatively simple optical system compared with the other method [1-10]. The caustic method is based on the assumption that the initial curve of a caustic is in the singular field which satisfies the plane stress condition.

We consider that the stress field shown in Fig.1 exists around a crack propagating through a plate specimen. There exists the singular field which satisfies the plane stress condition, around the crack tip. The three-dimensional

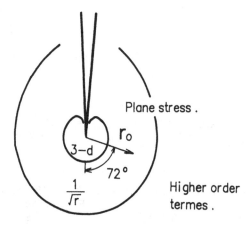

Figure 1. Stress Fields around a fast propagating crack.

(3-d) stress field, however, appears in the vicinity of the crack tip [5-11]. And, outside of the singular field, there is the region where effect of the higher order terms appears. If the initial curve of a caustic is in the singular field of plane stress, the caustic method gives the true K_I value even though the initial curve radius varies. But, the caustic method doesn't give the true K_I value if the initial curve is in the 3-d stress field or in the region of higher order terms. It is therefore important to know that how far the 3-d field spreads from the crack tip and how close the region of higher order terms is to the tip.

In the present study, we measure the extent of the 3-d stress field of a crack propagating through a PMMA plate specimen, in the direction of 72 degrees from the propagation direction of a crack. And also measured is the distance from the crack tip to the boundary between the singular field and the region of higher order terms, in the same direction. For the measurement, we utilize the method of pulsed holographic recording and reconstruction of a caustic light beam, and, obtain the following results. The extent of the 3-d stress field is about half the specimen thickness. The distance from the crack tip to the region of higher order terms is about 4mm, in the present study.

EXPERIMENTAL METHODS

As mentioned in the previous section, the caustic method gives the real K_I value whenever the initial curve of a caustic is in the singular field of

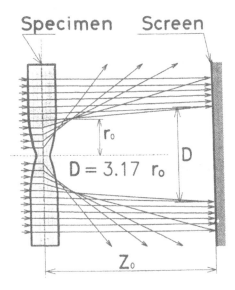

Figure 2.   The principle of the caustic method.

plane stress.   But, the K_I value given by the caustic method deviates from the real K_I value if the initial curve is in the 3-d stress field or in the region of higher order terms.   Utilizing the fact, we can know the area where the singular field is developed.   The experimental procedure is described in this section.   The real K_I value of a crack is obtained through the measurement of the crack opening displacement, COD, along the crack.

Caustic Method

Figure 2 shows the principle of the caustic method.   A crack propagates through the PMMA specimen perpendicularly to the paper plane.   The parallel light beam is incident upon the specimen surface perpendicularly.   The rays which pass near the crack tip deflect outward from the crack tip, because of decrease of the thickness and the refractive index of the PMMA specimen, which decrease is due to stress concentration around the crack tip.   On the screen behind the specimen, there is hence a shadow spot where no ray falls on.   And a bright curved line appears just outside the shadow spot, which bright line is called a caustic.   Following backward the rays which make the caustic, we can draw a circle on the specimen.   The circle is called the initial curve.   The theory of the caustic method gives the following two equations,

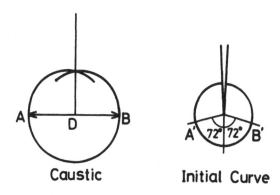

Figure 3.   Caustic curve and initial curve.

$$K_I^{Cau} = A(v)\ D^{5/2}\ /\ Z_o\ , \qquad (1)$$

$$D = 3.17\ r_o\ , \qquad (2)$$

where K_I^{Cau} is the dynamic stress intensity factor given by the caustic method, v is the crack speed, A(v) is a coefficient including dynamic effect, D is the transverse diameter of the caustic, Z_o is the distance between the specimen and the screen, and r_o is the initial curve radius.   We can know the dynamic stress intensity factor K_I^{Cau} and the initial curve radius r_o from the measured caustic diameter D through the above equations. The K_I^{Cau} is in agreement with the real K_I value if the initial curve is in the singular field of plane stress.   But, if the initial curve is in the 3-d stress field or in the region of higher order terms, then the K_I^{Cau} has different value from the real K_I value.   Utilizing the fact, we can know the region where the singular field is developed.   In the caustic method, measured is the transverse diameter of the caustic, that is between A and B in Fig.3.   The ray which reaches the point A on the caustic is the one which passes the point A' that is the intersection of the initial curve and the radius which makes an angle of 72 degree with the crack propagation direction, on the specimen.   It is similar for the ray that arrives at the point B on the caustic.   The extent of the singular field is, in the present study, measured in the direction of 72 degree from the propagation direction of a crack.

Specimens

Figure 4 shows the specimen used in the present study.   It is 190mm long, 250mm wide, and 3, 5 and 10mm in thickness.   There is a notch in the

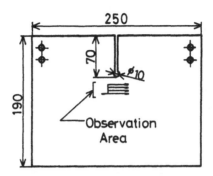

Figure 4.   PMMA plate specimen.

specimen and a fast propagating crack arises at the notch tip. Holographic recording of a caustic light beam is carried out when the crack is propagating in the observation area. When a crack appears suddenly or, propagates rapidly, stress waves are emitted from the cracks [12]. The stress waves are reflected by the side boundaries of the specimen and, go back to the crack it self. The specimen is so large that the holographic recording of a caustic light beam is carried out before the stress waves emitted at the crack initiation reach the crack. Hence the crack doesn't know the existence of side boundaries of the specimen. In the sense of the wards, we measure the caustic diameter of the crack propagating through an infinitely large specimen.

Holographic Recording of Crack and of Caustic Light Beams
Figure 5 shows the optical system for holographic recording of crack opening and a caustic light beam. A crack propagates through the PMMA specimen SP, perpendicularly to the paper plane. When the crack is propagating in the observation area, the Q-switched ruby laser PL oscillates once and emits a pulsed light beam. The light beam is half reflected by the beam splitter BS1 and, become a parallel light beam. The parallel light beam is half transmitted through the beam splitter BS3 and falls onto the specimen surface perpendicularly. The light beam is partly reflected by the specimen surface, and the reminder is transmitted through the specimen. The light beam reflected by the specimen surface is half reflected by the beam splitter BS3, passes through the lens L5 and, falls onto the holographic plate HP2. this is the object beam for holographic recording of the crack opening. The object beam makes the real image RI of the crack behind the holographic plate. The light beam transmitted through the beam splitter BS1 is diverged and collimated through the lenses L2-4, is half reflected by the beam splitter

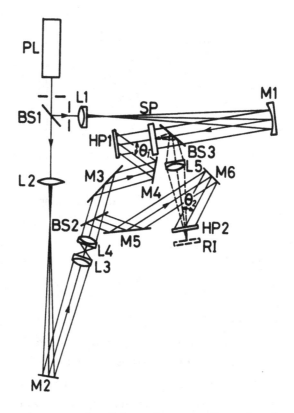

Figure 5.   The optical system for holographic recording of a fast propagating crack and a caustic light beam.

BS2, and, impinges on the holographic plate HP2 obliquely.   This is the reference beam for holographic recording of the crack opening.   In this manner the crack is recorded as a hologram at an instant during the propagation [10].

The light beam transmitted through the specimen becomes a caustic light beam and falls onto the holographic plate HP1 perpendicularly.   This is the object beam for holographic recording of the caustic light beam.   The light beam transmitted through the beam splitter BS2 impinges on the holographic plate HP1 obliquely.   This is the reference beam for holographic recording of the caustic light beam.   In this manner, the caustic light beam is recorded as a hologram simultaneously.

Reconstruction of Cracks

Figure 6 shows the method for reconstruction and microscopic photographing of the crack.   After development, illuminated with the He-Ne laser beam,

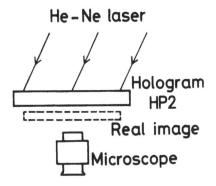

Figure 6. Reconstruction and microscopic photographing of a crack.

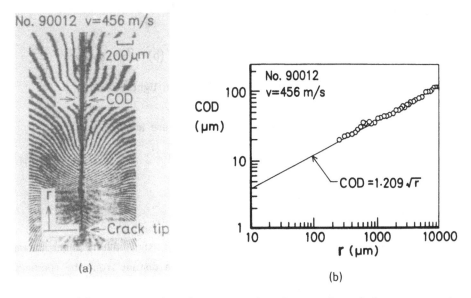

Figure 7.   (a) An Example of microscopic photographs of fast propagating cracks.   (b) Opening displacement of the crack shown in (a).

the hologram HP2 reconstructs the real image RI of the crack [10].   The real image is magnified and photographed through a conventional microscope. An example of the microscopic photographs are shown in Fig.7(a).   The dark region is the opened crack.   We can measure the crack opening displacement, COD, along the crack.   The result of the COD measurement is shown in the Fig.7(b).   The measured COD is proportional to the square root of the distance r from the crack tip.   Accordingly we can obtain the real value K_I^{Real} of the dynamic stress intensity factor from the COD measurement

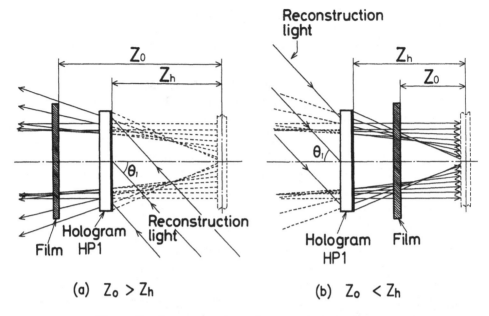

Figure 8. Reconstruction of a caustic light beam.

through the following formula of linear elastic fracture mechanics,

$$COD = K_I^{Real} \, B(v) \, r^{1/2} , \qquad (3)$$

where $B(v)$ is a coefficient including dynamic effect.

Reconstruction of Caustic Light Beams

Figure 8 shows the method for reconstruction of the caustic light beam. At first, we reconstruct the caustic light beam more distant from the specimen than the hologram, Fig.8(a). After development, the hologram HP1 is illuminated with the reconstruction beam which is identical to the reference beam. Then the hologram reconstructs the caustic light beam. Putting a film at the position Z_o, we can take the photograph of the caustic at Z_o. Caustics are photographed at all Z_o larger than Z_h, with varying the position of a film.

Next, the hologram is illuminated with the reconstruction beam which is incident on the hologram at the same angle but in the opposite direction, Fig.8(b). Then the hologram reconstructs the caustic light beam between the hologram and the specimen. Putting films in the caustic light beam, we can obtain the caustics at various positions between the specimen and

No. 90012

Crack speed 456 m/s

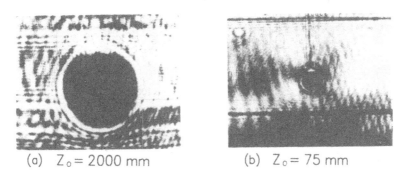

(a) $Z_0 = 2000$ mm   (b) $Z_0 = 75$ mm

Figure 9.   Examples of caustics reconstructed from a hologram.

the hologram.

Figure 9 shows two examples of the caustics reconstructed from a hologram through the method described above. The caustic in (a) is an example in the case Z_0 is larger than Z_h. On the other hand, the caustic in (b) is that in the case of Z_0 smaller than Z_h. We can, in this manner, obtain the photographs of caustics at all Z_0. Caustic diameters are measured from the photographs, and the dynamic stress intensity factor K_I^{Cau} and the initial curve radius r_0 are calculated through the equations, (1) and (2).

RESULTS AND DISCUSSIONS

Figure 10 shows the result of K_I^{Cau} versus the initial curve radius r_0, in the case specimens are 3mm in thickness. The horizontal scale indicates the initial curve radius r_0 normalized with the specimen thickness t. The vertical scale indicates the K value obtained through the caustic method, K_I^{Cau}, normalized with the real K value, K_I^{Real}. The value of K_I^{Cau} is in agreement with that of K_I^{Real} in the region r_0/t is between 0.5 and 1.3. It means that the singular field of plane stress is developed in the region, $0.5 < r_0/t < 1.3$. But K_I^{Cau} deviates below from K_I^{Real} in the region r_0/t is smaller than 0.5. It is due to the existence of the 3-d stress field in the vicinity of a crack tip. The extent of the 3-d field may be proportional to the specimen thickness, accordingly, we can say that the extent of 3-d

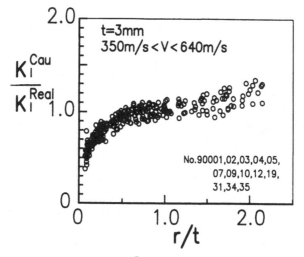

Figure 10. $K_I{}^{Cau}$ versus r_o, t=3mm.

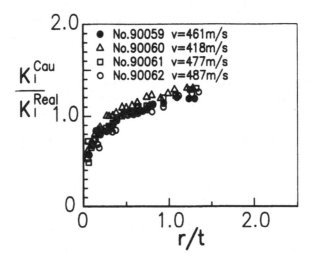

Figure 11. $K_I{}^{Cau}$ versus r_o, t=5mm.

stress field is about half the specimen thickness. The result is same as that of stationary cracks. On the other hand, the $K_I{}^{Cau}$ deviates above from the $K_I{}^{Real}$ in the region r_o/t is greater than about 1.3. It is the effect of higher order terms. The distance R_h from a crack tip to the boundary between the singular field and the region of higher order terms is independent of the specimen thickness. It is, hence, said that the distance R_h is about 4mm in the present study.

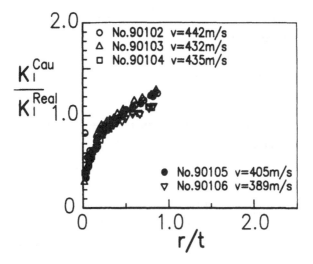

Figure 12. K_I^{Cau} versus r_o, t=10mm.

Figure 11 shows the result of cracks in the specimens 5mm in thickness. There is the 3-d stress field in the region r_o/t is smaller than 0.5. And the effect of higher order terms appears in the region r_o is greater than 4mm, that is, r_o/t is greater than 0.8. The region of singular field of plane stress becomes smaller than that of the specimens 3mm in thickness.

Figure 12 shows the result of cracks in the specimens 10mm in thickness. The 3-d stress field appears within the region r_o/t is smaller than about 0.5, like the two above cases. On the other hand, the effect of higher order terms appears in the region r_o is greater than 4mm, that is, r_o/t is greater than 0.4. The facts mean that the singular field of plane stress has vanished. It must thus be said that applying the caustic method to measure the dynamic stress intensity factor is impossible in the case of the specimens with 10mm thick, used in the present study.

Generally speaking, the distance R_h from a crack tip to the region of higher order terms depends on specimen shape and on crack length. As previously mentioned, the cracks in the present study were not interacting with the stress waves which were emitted from the cracks themselves and, reflected by the specimen boundaries. The cracks didn't, thus, know the existence of side boundaries of specimens. In the sense of the words, the cracks can be regarded as those in infinitely large specimens. Specimen shape may not, therefore, affect the result. On the other hand, there remains the possibility that crack length affects the result that R_h is about 4mm.

The radius R_h might increase as the crack length increases with time. The problem is, however, remained in the future.

CONCLUSIONS

The pulsed holographic method described in the present study makes it possible to measure the extent of the three dimensional (3-d) stress field in the vicinity of a fast propagating. The method can also measure the distance R_h from the crack tip to the boundary between singular field of plane stress and the region of higher order terms. The measurement is carried out in the direction of 72 degree from the propagation direction of the crack. The 3-d stress field spreads as far as the distance half the specimen thickness, which fact agrees with that of stationary cracks. The distance R_h was about 4mm in the present study. The distance R_h might increase as the crack length increases with time.

ACKNOWLEDGEMENT

The study was performed during the tenure of Grant-in-Aid for General Scientific Research (01550071) of the Ministry of Education, Science and Culture. One of the authors, Suzuki,S., expresses his gratitude for the grant. The authors also wishes to acknowledge Nakane,K., who is a student of our university and worked together in the study.

REFERENCES

1. Manogg,P., Schattenoptische messung der spezifischen bruchenergie wahrend des bruchvorgangs bei plexiglas, In Proc. Int. Con. Physics Non-Crystalline Solids, Delft, The Netherlands, 1964, 481-90.

2. Beinert,J.F. et al., Neuere ergebnisse zur anwendung des schattenfleckver-fahrens auf stehende und schnell-laufende bruche, In Proc. 6th Int. Con. Exp. Stress Anal., VDI-Report No.313, germany, 1978, 791-8.

3. Theocaris,P.S. and Kasamanis,F., Response of cracks to impact by caustics, Eng. Fracture Mech., 1978, 10, 197-210.

4. Kalthoff,J.F., On the measurement of dynamic fracture toughnesses - A review of recent work, Int. J. Fracture, 1985, 27, 277-98.

5. Yang,W. and Freund,L.B.. Transverse shear effects for through-cracks in an elastic plate, Int. J. solids Structures, 1985, 21, 977-94.

6. Rosakis,A.J. and Ravi-Chandar,K., On crack-tip stress state: An experimental evaluation of three-dimensional effects, Int. J. Solids Structures, 1986, **22**, 121-34.

7. Shimada,H. and Sasaki,T., Determination of the stress intensity factor by the method of surface reflected caustics (in Japanese), Trans. Japan Soc. Mech. Eng. (A), 1983, **49**, 579-88

8. Shimizu,K. et al., Some propositions on caustics and an application to the biaxial-fracture problem, Exp. Mech., 1985, **25**, 154-60

9. Suzuki,S. and Hosoda,Y., Holographic recording of caustic light beams of fast propagating cracks (in Japanese), Trans. Japan Soc. Mech. Eng., 1989, **55**, 1974-78.

10. Suzuki,S. et al., Pulsed holographic microscopy as a measurement method of dynamic fracture toughness for fast propagating cracks, J. Mech. Phys. Solids, 1988, **36**, 631-53.

11. Suzuki,S., Three-dimensional measurement of opening displacement of rapidly propagating cracks in PMMA, In the Proc. Appl. Stress Anal., ed. Hyde,T.H. and Ollerton,E., Elsevier Applied Science Publishers, London, 1990, 26-35.

12. Suzuki,S. and Nakajima,T., Development of laser inducing technique for fast propagating cracks in PMMA, In PVP-Vol.160 Dynam. Fracture Mech., ed. Homma,H., Shockey,D.A. and Yagawa,G., ASME,1989, 79-84.

EXPERIMENTAL AND NUMERICAL STUDIES ON FAST CURVING FRACTURE

T.NISHIOKA, R.MURAKAMI, T.MURAKAMI AND K.SAKAKURA

Department of Ocean Mechanical Engineering
Kobe University of Mercantile Marine
5-1-1 Fukae Minamimachi, Higashinada-ku, Kobe, 658, JAPAN

ABSTRACT

This paper summarizes recent experimental and numerical studies on fast curving fracture. The following topics are included; (1) Experimental techniques for high-speed photography of caustic pattern during fast crack curving, (2) a higher-order theory of caustics for mixed-mode fast running cracks, (3) a component separation method for determining mixed-mode stress intensity factors by using the path independent dynamic J integral, and (4) finite element simulation techniques for fast curving crack propagation. Pertinent experimental and numerical results are also presented.

INTRODUCTION

First, experiments of dynamic crack curving were carried out[1] using the double cantilever beam (DCB) specimens which have slightly unsymmetrical geometries with respect to the crack axis. High-speed photographs of caustics patterns in the fast curving fracture tests were taken by a laser caustic method which can be quickly synchronized to the initiation of a brittle fracture event. The laser caustic method consists of a high-speed camera, a He-Ne laser, and an appropriate trigger system with an acoustic optical modulator. The trigger system is activated when a conductive paint line surrounding the initial notch-tip is broken due to the initiation of crack propagation. By using the high-speed photographs, the instantaneous coordinates of crack-tip and characteristic dimensions of the caustic patterns were evaluated.

In order to extract the stress intensity factors K_I and K_{II} from the slightly unsymmetrical caustic patterns for dynamically curving cracks, the present authors[2] have developed a theory of caustics for mixed-mode fast running cracks. In this theory, only the singular stress field for a dynamically propagating crack was used to derive the analytical solution for the caustic and initial curves. However, an unsymmetrical caustic pattern with respect to the crack axis may be caused not only by the mode II singular stress field, but also by the mode II higher-order stress fields.

From this point of view, a higher-order theory of caustics for mixed-mode fast running cracks has been developed by the present authors[3]. This will be explained briefly in this paper.

The state of the art of computational methods in dynamic fracture mechanics has been greatly advanced, as reviewed in [4], by the development of finite element procedures to model propagating singularities and of the path independent dynamic J integral[5] which characterizes the strength of a propagating crack-tip field. However, the applications of these techniques have been limited to the cases of straight crack propagation, due to various difficulties in finite element modelling of fast curving crack propagation.

To overcome these difficulties, the authors have developed a moving finite element method together with a concept of element-controlling plane based on Lagrangean-element mapping, and a component separation method[6] for determining the mixed-mode stress intensity factors using the components of the dynamic J integral, J_k' .

Generation phase simulation for the fast curving fracture test is carried out using the above techniques. In the finite element simulation, the formation process of caustic pattern is also simulated for the dynamically curving crack. Shapes of simulated caustic patterns agree well with the caustic patterns in the high-speed photographs, and with those based on the higher-order theory of caustics.

FAST CURVING FRACTURE TESTS

In the present experiments[1], an optically isotropic material, polymethyl methacrylate (PMMA) was used. The geometry of a double cantilever beam (DCB) specimen is shown in Fig.1. The pin holes were made such that one of their centers is located at a distance of f from the center line of the specimen (the center line of the initial notch). Thus, the distance (f-25)mm indicates the amount of deviation from the axis of symmetry. Because the stress distribution in the specimen becomes slightly unsymmetrical due to the deviation, it may be easier for fast curving fracture to occur. Three specimen geometries were prepared as indicated in Table 1, while the radii of the initial notch-roots were kept constant (1.0 mm).

The optical set-up for the transmitted caustics is shown in Fig.2. A He-Ne laser (50 mW) was used as the light source of the caustics. A conductive paint line for a trigger was placed surrounding the initial notch-tip.

A displacement-controlled load was applied to the specimen through a wedge between the pin-holes. When the conductive paint line is broken due to the initiation of crack propagation, the trigger signal is sent to the acoustic optical modulator. The direction of the laser beam is then changed in an extremely short time. The laser beam reaches the high-speed camera (Cordin Dynafax 350, 35000 frames/sec), after passing through the specimen. The transmitted light is focused on the lens of the high-speed camera. In the present experiments, high-speed photographs of the caustic patterns were taken at time intervals of about 30 μsec.

EXPERIMENTAL RESULTS

The fracture paths appearing in the fast curving fracture tests are shown in Fig.3. As seen in Fig.3(a), in the case in which the wedge was loaded symmetrically, the crack was apt to propagate straight again after

234

34

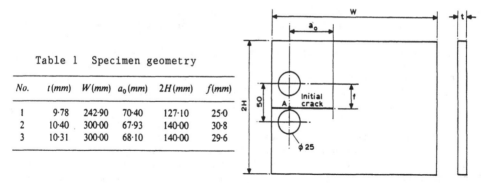

Table 1   Specimen geometry

No.	t(mm)	W(mm)	a_0(mm)	$2H$(mm)	f(mm)
1	9·78	242·90	70·40	127·10	25·0
2	10·40	300·00	67·93	140·00	30·8
3	10·31	300·00	68·10	140·00	29·6

Fig.1   Double cantilever beam specimen

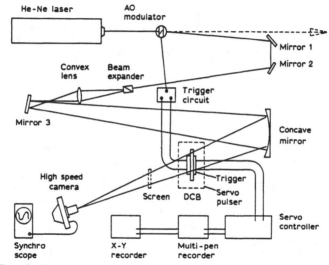

Fig.2   Experimental set-up for a laser caustic method

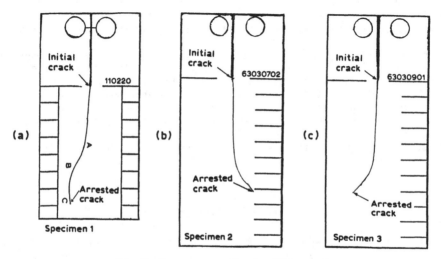

Fig.3   Fracture paths in DCB specimens

the first curving. This experimental result was unexpectedly obtained in a series of experiments for mode I dynamic crack propagation and arrest[7].

Contrary to this case, when the wedge was loaded unsymmetrically with respect to the initial crack line, the crack further departs from the initial crack line as the crack propagates after the first curving takes place (see, Figs.3(b) and (c)). However, comparing Figs.3(b) and (c), it is seen that the direction of the deviation does not have much influence on the initial propagation direction. Rather, the initial direction seems to be determined by the defects or microcracks around the notch root, which are caused by making the notch.

A series of high-speed photographs of the caustic patterns obtained for specimen 3 are shown in Fig.4. The photographs for the period of t=0 ∿ 391 μsec show the caustic patterns under the mode I condition. The caustic patterns for the period of $t \geq 421$ μsec were formed during clockwise curving of the crack. In the cases of the curving cracks, the caustic curves become open shapes. The open shapes depend on the rotation of the crack direction[1].

The time variations of the crack-tip position, a_x and a_y, measured from the point A (see Fig.1) in the tangential and perpendicular directions to the initial crack line, are shown in Fig.5 for all the cases.

Each a_x-t as well as a_y-t curve is least-squares fitted by a polynomial of the ninth order. Differentiating the polynomials for the a_x-t and a_y-t curves with respect time t, the components of the crack velocity in the x and y directions, C_x and C_y, are obtained. The crack velocity in the instantaneous crack propagation direction C_n is then determined by the following relation:

$$C_n = C_x\cos\theta_c + C_y\sin\theta_c \qquad (1)$$

where θ_c is the angle between the instantaneous crack propagation direction and the x-axis. Figures 6(a),(b) and (c) show the time variations of C_x, C_y and C_n in specimens 1,2 and 3, respectively.

The crack velocity C_n in specimen 1 decreases immediately after the crack extension, as shown in Fig.6(a), while the peak of C_n is seen after a certain amount of crack extension in specimen 2 and specimen 3. The maximum crack speeds observed in these specimens are about 300 m/sec.

DYNAMIC J INTEGRAL AND COMPONENT SEPARATION METHOD

Nishioka and Atluri[5] have derived the dynamic J integral which possesses the following salient features : (i) it has the physical meaning of the energy release rate, (ii) it has the property of the so-called path independent integral, which gives a unique value for an arbitrary integral path surrounding the crack-tip, and (iii) it is invariant with respect to the shape of an infinitesimal process zone (the near-field path)[8]. Therefore, it always has the unique relation to the stress intensity factors. These features are very important for the theory and the numerical analyses in dynamic fracture mechanics[4].

For a crack propagating at an angle θ_0 measured from the global X_1 axis (see Fig.7), the energy release rate can be expressed by

$$G = J'_1\cos\theta_0 + J'_2\sin\theta_0 \qquad (2)$$

and

$$J'_k = \int_{\Gamma+\Gamma_c}[(W + K)n_k - t_iu_{i,k}]dS + \int_{V_\Gamma}[\rho\ddot{u}_iu_{i,k}-\rho\dot{u}_i\dot{u}_{i,k}]dV \qquad (3)$$

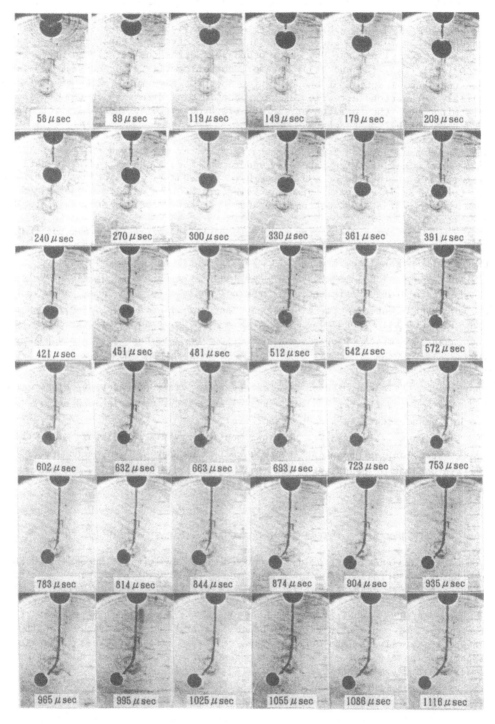

Fig.4 High-speed photographs of the caustic pattern for dynamic crack curving

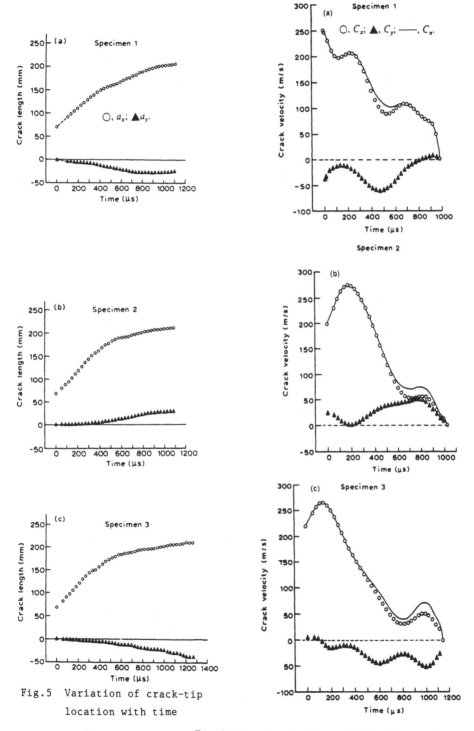

Fig.5   Variation of crack-tip
         location with time

Fig.6   Variation of crack velocity with time

where $J'_k(k=1,2)$ has the meaning of the energy release rate due to a unit crack extension along the global coordinate X_k $(k=1,2)$; W and K are the strain and kinetic energy densities, respectively; t_i are the traction forces; u_i, $\dot{u}_i$ and $\ddot{u}_i$ are the displacement, velocity and acceleration of the material point; ρ is the mass density; and the other nomenclatures are shown in Fig.7.

The crack-axis components of the J' integral can be evaluated by the following coordinate transformation :

$$J_k^{'0} = \alpha_{k\ell}(\theta_0) \, J_\ell^{'} \tag{4}$$

where, for an in-plane problem, $\alpha_{11}=\alpha_{22}=\cos\theta_0$, $\alpha_{12}=-\alpha_{21}=\sin\theta_0$. The $J_k^{'0}$ integrals are related to the instantaneous stress intensity factors as [5]:

$$J_1^{'0} = \frac{1}{2\mu} \{A_I(C)K_I^2 + A_{II}(C)K_{II}^2 \} \tag{5.a}$$

$$J_2^{'0} = - \frac{A_{IV}(C)}{\mu} K_I K_{II} \tag{5.b}$$

where C is the instantaneous crack velocity ; and the detailed expressions for the crack-velocity functions A_I, A_{II}, A_{IV} are given in [5].

Nishioka and Atluri[9] have developed the following direct method. K_I and K_{II} are solved directly from Eqs.(5.a and b) as

$$K_I = \pm \left(\frac{\mu}{A_I(C)} \{(J_1^{'0}) \pm [(J_1^{'0})^2 - (A_I A_{II}/A_{IV}^2)(J_2^{'0})^2]^{1/2}\} \right)^{1/2} \tag{6.a}$$

$$K_{II} = \pm \left(\frac{\mu}{A_{II}(C)} \{(J_1^{'0}) \mp [(J_1^{'0})^2 - (A_I A_{II}/A_{IV}^2)(J_2^{'0})^2]^{1/2}\} \right)^{1/2} \tag{6.b}$$

The signs of K_I, K_{II} can be determined by monitoring the signs and magnitudes of the crack opening displacements δ_I and δ_{II}. The crack opening displacements are expressed by

$$\delta_I = (u_2^0)^+ - (u_2^0)^- , \quad \delta_{II} = (u_1^0)^+ - (u_1^0)^- \tag{7}$$

The signs of K_I ,K_{II} correspond to the signs of δ_I, δ_{II}, respectively ($K_I > 0$, if $\delta_I > 0$, and so on). The term in braces { } in Eq.(6.a) is determined as follows: if $|\delta_I/\beta_1| \geq |\delta_{II}/\beta_2|$, take + [ ]; and otherwise, take -[ ], with similar choices for K_{II}.

However, it was discovered in [9] that, in order to employ the direct method, it is necessary to model accurately the crack-tip region with elements that possess the $r^{-1/2}$ stress singularity known to exist at a sharp crack-tip. Consequently, the use of regular elements without the singularity results in somewhat inaccurate values for the stress intensity factors. This can be attributed to the numerical inaccuracy of the energy densities (W+K) which are necessary in the evaluation of the $J_2^{'0}$ integral or $J_k^{'}$ $(k=1,2)$ integrals.

Contrary to this, since the values of $(n^0{}_1)^\pm$ on Γ_C vanish in the vicinity of the crack-tip $(r \to 0)$, the $J_1^{'0}$ integral is not affected by the

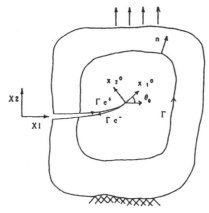

Fig.7 Elastic body containing a curved crack

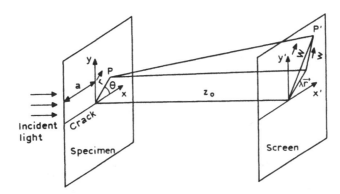

Fig.8 Geometrical relation for transmitted caustic method

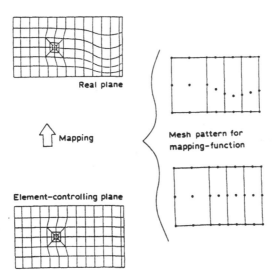

Fig.9 Mapping between the real plane and the element-controlling plane

modelling of the crack-tip region, and can be evaluated accurately, even by the global components of the J' integral $(J_1'^0 = J_1' \cos\theta_0 + J_2' \sin\theta_0)$. Therefore, the $J_2'^0$ integral can be evaluated accurately, if the component ratio $\gamma = J_2'^0/J_1'^0$ is accurately determined.

Using Eqs.(5.a and b), the component ratio can be expressed by

$$\gamma = -2A_{IV}/(A_I/\alpha + \alpha A_{II}) \tag{8}$$

where $\alpha = K_{II}/K_I = (\delta_{II}\beta_1)/(\delta_I\beta_2)$ and $\beta_1 = \sqrt{1-(C/C_d)^2}$, $\beta_2 = \sqrt{1-(C/C_s)^2}$; C_d and C_s are the dilatational and shear wave velocities, respectively.

Using the above relations in the formulae of the direct method given by Eqs.(5.a and b), we obtain

$$K_I = \delta_I \{ \frac{2\mu J_1'^0 \beta_2}{A_I(\delta_I^2 \beta_2 + \delta_{II}^2 \beta_1)} \}^{1/2} ; \quad K_{II} = \delta_{II} \{ \frac{2\mu J_1'^0 \beta_1}{A_{II}(\delta_I^2 \beta_2 + \delta_{II}^2 \beta_1)} \}^{1/2} \tag{9.a,b}$$

It is noted that the signs of the stress intensity factors are automatically determined by the signs of the crack opening displacements.

A HIGHER-ORDER THEORY OF MIXED-MODE CAUSTICS FOR DYNAMICALLY PROPAGATING CRACKS

The stress concentration around the crack tip causes a reduction in the thickness of the specimen and a change in the refractive index of the material. As a consequence, the light ray transmitted through the point P on the specimen is deflected outward and reaches the point P' on the screen, as shown in Fig.8.

The position vector $\vec{W}$ of the image point P' is given by

$$\vec{W} = \lambda\vec{r} + \vec{w} \tag{10}$$

where λ is the magnification of the optical system. The vector $\vec{w}$, which indicates the deviation of the light ray on the screen, is given as

$$\vec{w} = -z_0 \text{grad}\Delta s \tag{11}$$

where z_0 is the distance between the specimen and the screen. Δs is the change in the optical path length caused by the object. Using elasto-optical relations, the path length change Δs can be related to the principal stresses in the specimen. If we consider a transmitted caustic of an optically isotropic material or a reflected caustic of an opaque material, Δs is a function only of the sum of principal stresses, $\sigma_1 + \sigma_2$. The following equation expresses Δs in a unified fashion, including the method of reflected caustics:

$$\Delta s = C_0 t (\sigma_1 + \sigma_2) \tag{12}$$

where C_0 is the elasto-optical constant, and the value of which is given by (Beinert and Kalthoff[12]) for various materials; t is the thickness of the specimen.

The outward deviation of the incident rays around the crack-tip creates a dark spot on the screen. This dark spot is surrounded by a bright light concentration, which is called the " caustic curve". The caustic curve consists of a series of light rays passing through a particular curve on the specimen. This curve is called the "initial curve" of the caustic curve. The initial curve for a static crack problem is usually represented by a circle around the crack-tip. The initial curve can be determined by the condition that the mapping is not invertible between $\vec{r}$ on the specimen and $\vec{W}$ on the screen. Mathematically, this condition can be expressed by the vanishing point of the Jacobian:

$$J = \frac{\partial(W_x, W_y)}{\partial(x,y)} = \frac{\partial W_x}{\partial x}\frac{\partial W_y}{\partial y} - \frac{\partial W_x}{\partial y}\frac{\partial W_y}{\partial x} = 0 \tag{13}$$

where W_x and W_y are the components of the vector $\vec{W}$ with respect to x' and y', respectively. Substitution of the determined initial curve into Eq.(10) then leads to the expression for the caustic curve.

As mentioned previously, the caustic pattern can be evaluated by the stress field in the vicinity of the crack tip. The general solutions for the stress and displacement fields near the elastodynamically propagating crack tip, were obtained by Nishioka and Atluri[5] in a unified fashion for all the three fracture modes. The inplane components can be expressed by

$$\sigma_x = \sum_n \frac{K^0_n B_I}{(2\pi)^{1/2}} \frac{n}{2} (n+1)[(1+2\beta_1^2-\beta_2^2)\frac{\cos(n/2-1)\theta_1}{r_1^{1-n/2}} - 2h(n)\frac{\cos(n/2-1)\theta_2}{r_2^{1-n/2}}]$$
$$+ \sum_n \frac{K^*_n B_{II}}{(2\pi)^{1/2}} \frac{n}{2} (n+1)[(1+2\beta_1^2-\beta_2^2)\frac{\sin(n/2-1)\theta_1}{r_1^{1-n/2}} - 2h(\bar{n})\frac{\sin(n/2-1)\theta_2}{r_2^{1-n/2}}]$$

$$\sigma_y = \sum_n \frac{K^0_n B_I}{(2\pi)^{1/2}} \frac{n}{2} (n+1)[-(1+\beta_2^2)\frac{\cos(n/2-1)\theta_1}{r_1^{1-n/2}} + 2h(n)\frac{\cos(n/2-1)\theta_2}{r_2^{1-n/2}}]$$
$$+ \sum_n \frac{K^*_n B_{II}}{(2\pi)^{1/2}} \frac{n}{2} (n+1)[-(1+\beta_2^2)\frac{\sin(n/2-1)\theta_1}{r_1^{1-n/2}} + 2h(\bar{n})\frac{\sin(n/2-1)\theta_2}{r_2^{1-n/2}}]$$

$$\sigma_{xy} = \sum_n \frac{K^0_n B_I}{(2\pi)^{1/2}} \frac{n}{2} (n+1)\{-2\beta_1\frac{\sin(n/2-1)\theta_1}{r_1^{1-n/2}} + \frac{(1+\beta_2^2)}{\beta_2}h(n)\frac{\sin(n/2-1)\theta_2}{r_2^{1-n/2}}]$$
$$+ \sum_n \frac{K^*_n B_{II}}{(2\pi)^{1/2}} \frac{n}{2} (n+1)[+2\beta_1\frac{\cos(n/2-1)\theta_1}{r_1^{1-n/2}} - \frac{(1+\beta_2^2)}{\beta_2}h(\bar{n})\frac{\cos(n/2-1)\theta_2}{r_2^{1-n/2}}] \tag{14}$$

and

$$u = \sum_n \frac{K^0_n B_I}{2\mu}[\frac{2}{\pi}]^{1/2}(n+1)[r_1^{n/2}\cos\frac{n}{2}\theta_1 - h(n)r_2^{n/2}\cos\frac{n}{2}\theta_2]$$
$$+ \sum_n \frac{K^*_n B_{II}}{2\mu}[\frac{2}{\pi}]^{1/2}(n+1)[r_1^{n/2}\sin\frac{n}{2}\theta_1 - h(\bar{n})r_2^{n/2}\sin\frac{n}{2}\theta_2]$$

$$v = \sum_n \frac{K^0_n B_I}{2\mu}[\frac{2}{\pi}]^{1/2}(n+1)[-\beta_1 r_1^{n/2}\sin\frac{n}{2}\theta_1 + \frac{h(n)}{\beta_2} r_2^{n/2}\sin\frac{n}{2}\theta_2]$$
$$+ \sum_n \frac{K^*_n B_{II}}{2\mu}[\frac{2}{\pi}]^{1/2}(n+1)[\beta_1 r_1^{n/2}\cos\frac{n}{2}\theta_1 - \frac{h(\bar{n})}{\beta_2} r_2^{n/2}\cos\frac{n}{2}\theta_2] \tag{15}$$

In the above equations, the complex variables $z_j = x + i\beta_j y = r_j e^{i\theta_j}$ (j=1,2) were used, where $i = \sqrt{-1}$. Furthermore, n is the integer number($0 \le n < \infty$), n=n+1, and

$h(n)=\{2\beta_1\beta_2/(1+\beta_2{}^2):n \text{ odd}, (1+\beta_2{}^2)/2: n \text{ even}\}$. The undetermined parameters $K^0{}_n$, $K^*{}_n$ are related to the modes I and II deformations respectively. The functions of crack speed $B_M(C)$: M=I, II are defined by

$$B_I(C) = (1+\beta_2{}^2)/D(C) \tag{16.a}$$
$$B_{II}(C) = 2\beta_2/D(C) \tag{16.b}$$

and

$$D(C) = 4\beta_1\beta_2-(1+\beta_2{}^2)^2 \tag{17}$$

The equation $D(C)=0$ is the well-known Rayleigh equation and has the roots of $C=0$ and C_R(Rayleigh wave speed).

The general solutions given in Eqs.(14) and (15) contain the zero stress and rigid-body motion (n=0), the singular stresses and corresponding displacements (n=1), the constant stresses and linear displacements (n=2), and the higher-order terms (n≥3). Thus, the parameters $K^0{}_1$ and $K^*{}_1$ are equivalent to the dynamic stress-intensity factors K_I and K_{II} respectively.

For Mode I-problems, Nilsson[13] has shown that both the differential equations and the boundary conditions for an arbitrarily moving crack coincide with those for the problem of steady growth. From this it was concluded that the angular distribution of the singular stress field is only dependent on the instantaneous crack velocity. This conclusion can also be seen to be valid for the other fracture modes. By using the general solutions in Eqs.(10)∿(13), the Jacobian equation (see Eq.(13)) can be expressed by

$$J = \lambda^2+\lambda N(1-\beta_1{}^2)(\beta_1{}^2-\beta_2{}^2)$$
$$\cdot\sum_n[K^0{}_nC_nr^{n/2-3}\{-B_I\cos(\tfrac{n}{2}-3)\theta_1-\alpha_nB_{II}\sin(\tfrac{n}{2}-3)\theta_1\}]$$
$$-N^2\beta_1{}^2(\beta_1{}^2-\beta_2{}^2)^2\sum_n\sum_m[K^0{}_nK^0{}_mC_nC_mr_1{}^{(n+m)/2-6}$$
$$\cdot\{(B_I{}^2+\alpha_n\alpha_mB_{II}{}^2)\cos\tfrac{n-m}{2}\theta_1+(\alpha_n-\alpha_m)B_IB_{II}\sin\tfrac{n-m}{2}\theta_1\}] = 0 \tag{18}$$

where $\alpha_n=K_n{}^*/K_n{}^0$ and $C_n=n(1+n)(\tfrac{n}{2}-1)(\tfrac{n}{2}-2)$. The components of the position vector $\vec{W}$ at the screen can be expressed by

$$W_x = \lambda r_1\cos\theta_1-N(\beta_1{}^2-\beta_2{}^2)$$
$$\cdot\sum_n[K^0{}_nn(1+n)(\tfrac{n}{2}-1)r_1{}^{n/2-2}\{+B_I\cos(\tfrac{n}{2}-2)\theta_1+\alpha_nB_{II}\sin(\tfrac{n}{2}-2)\theta_1\}] \tag{19.a}$$

$$W_y = \lambda\tfrac{r_1}{\beta_1}\sin\theta_1-N\beta_1(\beta_1{}^2-\beta_2{}^2)$$
$$\cdot\sum_n[K^0{}_nn(1+n)(\tfrac{n}{2}-1)r_1{}^{n/2-2}\{-B_I\sin(\tfrac{n}{2}-2)\theta_1+\alpha_nB_{II}\cos(\tfrac{n}{2}-2)\theta_1\}] \tag{19.b}$$

If we use an appropriate variable transformation for r_1, Eq.(18) reduces to a polynomial equation with integer orders. For any given specific angle together with a given set of predetermined coefficients, the polynomial equation can be solved numerically by Bairstaw method. Substituting those numerical solutions in Eqs.(19.a and b), the caustic curve can be obtained.

By using only the singular terms (n=1), Nishioka and Kittaka[2] have derived completely analytical expressions for the caustic curves and for the initial curves, including crack-velocity effects.

LAGRANGEAN-ELEMENT MAPPING FOR FAST CURVING CRACK PROPAGATION

For the cases of mode I straight dynamic crack propagation, Nishioka and Atluri [10] have developed the moving isoparametric finite element method which gives accurate stress intensity factors from the numerical simulation. In order to simulate dynamic crack curving, the curved crack path should be modelled accurately by the finite element mesh pattern. Moreover, it is difficult to move a group of isoparametric elements around the crack-tip in accordance with the curved path. To overcome these difficulties, we introduce a mapping technique. The concept of this technique is illustrated in Fig.9. The procedures of the mesh movement and readjustment are controlled in the element-controlling plane. The mesh pattern in the real plane is created by the mapping from the element-controlling plane through a mapping function.

The mapping function for the entire region consists of several Lagrangean elements. The shape function for the i-th node of Lagrangean element is given by

$$N_i = N_{k\ell} = L_k^m(\xi)L_\ell^n(\eta) \tag{20}$$

where k and ℓ are the node numbers in the ξ and η directions, respectively; $L_k^m(\xi)$ and $L_\ell^n(\eta)$ are the Lagrangean polynomials. As shown in Fig.9, the Lagrangean-element mapping function can be obtained by using only the coordinates of the crack path and the external boundaries.

In the moving element procedure, the mesh pattern for the elements near the crack-tip translates along the curved crack path in each time step for which crack growth occurs as illustrated in Fig.10. Thus, the crack-tip always remains at the center of the moving elements throughout the analysis. The regular isoparametric elements surrounding the moving elements are continuously distorted. To simulate a large amount of crack propagation, the mesh pattern around the moving elements is periodically readjusted as is also shown in Fig.10.

The field variables, such as the displacements, velocities, and accelerations at the newly created nodes after shifting or readjusting the mesh pattern are required to be obtained through the interpolation of the field variables at the old nodes. An iterative numerical technique was used in the previous procedure to find the natural coordinates (ξ,η) at the newly created nodes. This technique requires additional computing time because of the nature of iteration. Recently, algebraic expressions for the transformation of the global coordinates (X,Y) to the natural coordinates (ξ,η) were obtained by using a computerized symbolic manipulation system (REDUCE)[11]. These algebraic expressions for the inverse mapping were incorporated in the moving element procedure. Such expressions are also very useful for the remeshing or zooming techniques often used in finite element analyses.

244

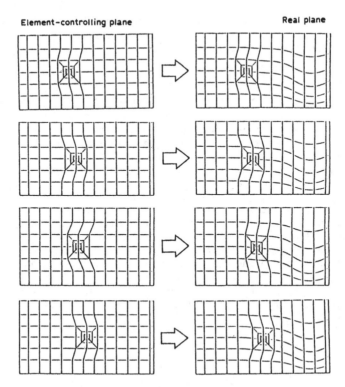

Fig.10 Moving isoparametric element procedure for dynamic crack curving

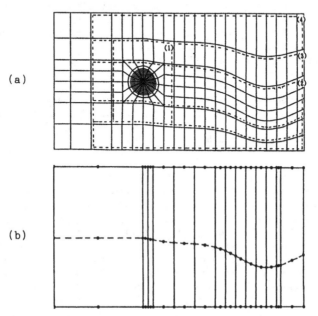

Fig.11(a) Finite element mesh pattern in the real plane
     (b) Lagrangean-element mesh pattern for the mapping from the element-controlling plane to the real plane

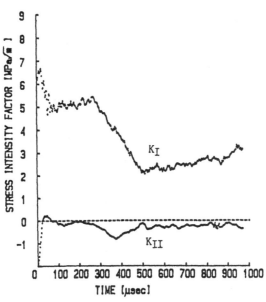

Fig.12 Stress intensity factors during fast curving crack propagation

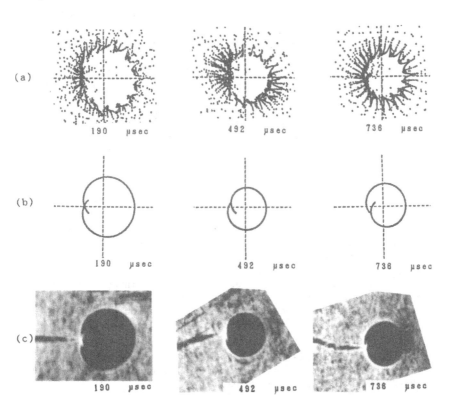

Fig.13(a) Numerically generated caustic patterns
    (b) Caustic patterns based on the higher-order theory of caustics
    (c) Caustic patterns in the high-speed photographs

NUMERICAL SIMULATION

Fast curving crack propagation in specimen 1 was simulated by using the currently developed procedure. The generation-phase simulation was carried out using the a_x-t and a_y-t curves. Figure 11(a) shows the finite element mesh pattern actually used in the simulation. Figure 11(b) shows the Lagrangean-element mesh pattern for the mapping function to obtain the actual mesh pattern (Fig.11(a)) from the element-controlling plane. To simulate the formation process of caustic pattern, a relatively fine mesh pattern was used around the crack-tip. The path independent dynamic J integral (J') was evaluated along the paths indicated by the broken lines in Fig.11(a). The numerical results for the K_I and K_{II} values are shown in Fig.12. As seen from the figure, K_I decreases rapidly after the crack extension until about 500 μsec. After certain crack extension (t > 500 μsec), the crack propagates with a nearly constant K_I value. The values of K_{II} are small comparing with those of K_I.

The numerically generated caustic patterns at t=190, 492 and 736 μsec are shown in Fig.13(a). The dots in the figure indicate the light beams at a screen, which are reflected from Gaussian integration points in each element. In the higher-order theory of caustics, the general solutions (including the higher-order terms) for stress and displacement around a dynamically propagating crack-tip were used. In the present simulation, the first twelve terms ($K^0{}_n, K^*{}_n; n=1{\sim}6$) including K_I and K_{II} in Eq.(15) were least-squares fitted to the numerical results of the displacement field. The caustic patterns based on the higher-order theory of caustics are also shown in Fig.13(b). The high-speed photographs at the corresponding instances in specimen 1 are shown in Fig.11(c). The photograph at 190 μsec shows the caustic pattern under mode I condition before the crack reaches at the point A indicated in Fig.3(a). The caustic pattern at 492 μsec was formed during clockwise curving along the path between the points A and B, whereas the caustic pattern at 736 μsec was formed by anti-clockwise curving between the points B and C. The both types of simulated caustic patterns agree well each other, and with the caustic patterns in the high-speed photographs.

CONCLUSIONS

(1) To overcome various difficulties in finite element modelling of fast curving crack propagation, a moving finite element method was developed, together with a concept of element-controlling plane based on Lagrangean-element mapping.
(2) A component separation method of the dynamic J integral was developed for the evaluation of mixed-mode stress intensity factors.
(3) It was found, from the generation phase simulation of the fast curving fracture test, that the K_{II} values during crack curving are small comparing with the K_I values.
(4) The caustic patterns obtained by the higher-order theory of caustics and the simulated caustic patterns by finite element method agree with the caustic patterns in the high-speed photographs.

ACKNOWLEDGEMENT

This study was partly supported by the Science and Technology Grant from Toray Science Foundation.

REFERENCES

1. Nishioka, T., Kittaka, H., Murakami, T. and Sakakura, K., A laser-caustic method for the measurement of mixed-mode dynamic stress intensity factors in fast-curving fracture tests. Int. J. Pres. Ves. & Piping., 1990, 44, 17-33.

2. Nishioka, T. and Kittaka, H., A theory of caustics for mixed-mode fast running cracks, Eng. Fract. Mech., 1990, 36, 987-98.

3. Nishioka, T., Murakami, R., Matsuo, S. and Ohishi, Y., A higher-order theory of caustics for fast running cracks under general loadings. Int. Conf. Photomechanics and Speckle Metrology, July 21-26, 1991, San Diego, California USA.

4. Nishioka,T., Atluri,S.N., Computational method in dynamic fracture. Computational Method in the Mechanics of Fracture, Elsevier Science Publishers, Chapter 10, 1986, 336-83.

5. Nishioka, T., and Atluri, S.N., Path-independent integrals, energy release rates, and general solutions of near-tip fields in mixed-mode dynamic fracture mechanics. Eng. Fract. Mech., 1983, 18, 1-22.

6. Nishioka,T., Murakami,R., Takemoto,Y., The use of the dynamic J integral (J') in finite-element simulation of mode I and mixed-mode dynamic crack propagation. Int. J. Pres. & Piping, 1990, 44, 329-52.

7. Nishioka, T., Uchiyama, H., Kittaka, H., Murakami, T., and Sakakura, K., Measurement of stress intensity factors for dynamic crack propagation and arrest using a laser caustic method. Transactions of the Japan Society of Mechanical Engineers, Series A.,1989, 55, No.520, 2410-15.

8. Nishioka,T., Invariance of the elastodynamic J integral (J'), with respect to the shape of an infinitesimal process zone. Eng. Fract. Mech., 1989, 32, 309-19.

9. Nishioka,T., Atluri,S.N., On the computation of mixed-mode K-factors for a dynamically propagating crack, using path-independent integrals J'_k. Eng. Fract. Mech., 1984, 20, 193-208.

10. Nishioka,T., Atluri,S.N., A path-independent integral and moving isoparametric elements for dynamic crack propagation. AIAA J., 1984, 22, 409-14.

11. Nishioka,T., Takemoto,Y., Moving finite element method aided by computerized symbolic manipulation and its application to dynamic fracture simulation. JSME International Journal, Ser. I, 1989, 32, 403-10.

12. Beinert, J., and Kalthoff, J.F., Experimental determination of dynamic stress intensity factors by shadow patterns. in Experimental Evaluation of Stress Concentration and Intensity Factors (G.C.Sih, Ed.), Martinus Nijhoff, 1981, pp.281-330.

13. Nilsson,F., A note on the stress singularity at a non-uniformly moving crack-tip. J. Elasticity, 1974, 4, 73-5.

STUDY OF DYNAMIC CRACK PROPAGATION IN POLYSTYRENE BY THE METHOD OF DYNAMIC CAUSTICS

GEORGE A. PAPADOPOULOS
The National Technical University of Athens,
Department of Engineering Science, Section of Mechanics,
5 Heroes of Polytechnion Av., GR-157 73, Zographou, Athens, Greece.

ABSTRACT

Dynamic crack propagation in thin, edge notched polystyrene specimens was studied by the method of dynamic caustics. During the crack propagation, an intensive zone of crazing surrounds and precedes the propagating crack. So, an active zone ahead of the crack tip is developed. This active zone is related to the velocity of crack propagation and the strain rate of loading. The velocity of the crack and the stress intensity factor K_I or the fracture surface energy γ were strongly influenced by the development of the active zone at the crack tip.

INTRODUCTION

Fatigue crack propagation and damage in thin, single edge notched polystyrene specimens was studied by Botsis et. al.[1,2]. According to this study, an intensive zone of crazing surrounds and precedes the propagating crack. The system of the crack and the craze zone constitute a crack layer [3]. The part of the crack layer ahead of the crack tip, where crazing accumulates prior to crack layer growth, is called the active zone. The width of the active zone was found to increase during slow crack layer propagation. Analysis of craze distribution within the active zone reveals that the craze density decreases away from the crack. Also, a considerable difference in the critical energy release rates was observed in specimens fatigued under different loading levels. This is attributed to the difference in the density of crazes at the critical crack tip.

Recently, experimental studies demonstrated that a zone, at the vicinity of the crack tip, precedes the propagating crack and this zone is usually termed as a plastic zone[4], process zone [5,6], dissipation zone [7] or deformation zone [8].

The objective of this study is to examine the crack propagation in polystyrene and the influence of the active zone on the crack velocity, the stress intensity factor and the fracture surface energy with parameter the strain rate. This study is taken out by means of dynamic caustics [9.10,11]. Also, the fracture surface energy is experimentally calculated from experimental values of the stress intensity factor.

THE EXPERIMENTAL METHOD OF DYNAMIC CAUSTICS

The experimental method of dynamic caustics was used to study the variation of the crack propagation velocity and the stress intensity factor [9,10,11]. According to the method of caustics, a convergent light beam impinges on the specimen in the vicinity of the crack tip and the transmitted rays are directed onto a reference plane parallel to the plane of the specimen. These rays are scattered and concentrated along a strongly illuminated curve, or **caustic**, on the reference plane located at distance z_0 from the specimen [9,10]. From the size and angular displacement ϕ of the axis of symmetry of the caustic relative to the crack axis [10], it is possible to calculate the stress intensity factors K_I and K_{II} for the case of mixed - mode conditions using the relations:

$$K_I = \frac{2(2\pi)^{\frac{1}{2}}}{3z_0 \, d\lambda_m^{3/2} c_t} \left[\frac{D_t^{max}}{\delta_t^{max}(\upsilon)} \right]^{5/2} \tag{1}$$

$$K_{II} = K_I \tan\left(\frac{\phi}{2} \right) \tag{2}$$

where d is thickness of the specimen, λ_m is the magnification ratio of the optical set-up, z_0 is the distance between the specimen and the reference plane, c_t is the optical constant of the material, D_t^{max} is the maximum transverse diameter of the caustic and $\delta_t^{max}(\upsilon)$ is a correction factor which depends on the crack velocity. This correction factor is given by nomograms in Ref.[9].

The elastic energy release rate G_I is calculated, for plane stress, by the relation:

$$G_I = K_I^2 \, / \, E \tag{3}$$

where E is the elastic modulus of the material.

The values of G_I at the crack initiation is defined as G_{Ic}, the critical strain energy release rate. The fracture surface energy γ as it is defined in the theory of Griffith [12] is equal to:

$$\gamma = G_I \, / \, 2 = K_I^2 \, / \, 2E \tag{4}$$

The stress intensity factor K_I is experimentally calculated by the relation (1) from the caustics. The critical stress intensity factor K_{Ic} or the corresponding γ is a useful index of the ultimate properties of the material. For the critical K_{Ic} the γ_c is given by the relation:

$$\gamma_c = K_{Ic}^2 \, / \, 2E \tag{5}$$

EXPERIMENTAL EVIDENCE

Notched polystyrene specimens of dimensions 220 x 50 x 0.25 mm were used for the experimental investigations. The specimens initially contained an edge transverse notch of 25 degree and length a_0 = 6 mm (Fig. 1).

The specimens were subjected to a dynamic tensile load to fracture using a

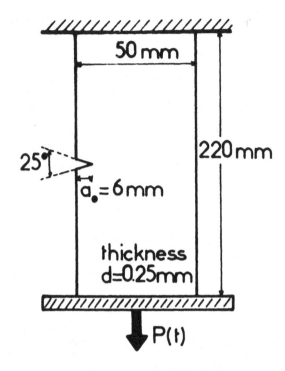

Figure 1. Geometry of cracked plate.

hydropulse high - speed testing machine (Carl - Schenk Co.) with a maximum strain rate $\dot{\varepsilon}$ = 80 s^{-1}. A Cranz - Schardin High - speed camera disposing 24 sparks with a maximum frequency of 10^6 frames per second was used to record the dynamic crack propagation. In the optical set-up used in the experiments z_0 = 0.80 m and λ_m = 0.77. The properties of polystyrene are E = 2.2 GPa , ν^0 = 0.3 and the stress optical constant c_t = 0.74 x 10^{-10} m^2/N. The loading strain rates $\dot{\varepsilon}$ in the present experiments were 10 and 20 s^{-1}.

RESULTS AND DISCUSSION

In order to study the influence of the active zone of the crack tip on the crack velocity, on the stress intensity factor K_I and on the fracture surface energy γ a number of specimens in the form shown in Fig. 1 was used. In all cases the specimens had an edge notch of 25 degree angle and initial length a_0 =6x10^{-3} m. The dynamic loading was applied with a strain rate $\dot{\varepsilon}$ = 10 s^{-1} and 20^{-1}.

The detailed crack propagation process may be studied from the series of photographs taken with a Cranz - Schardin high - speed camera. Fig. 2 presents a series of photographs showing the crack propagation process in a polystyrene plate with a strain rate $\dot{\varepsilon}$ = 10 s^{-1}. From these photographs we may see that the form of the caustic is not influenced by the crack layer which is developed at the crack tip. This means that the magnitude of the crack layer is smaller than the initial curve magnitude of the caustic. The magnitude of the initial curve is given by the relation [10]:

$$r_0 = D_t^{max}/\lambda_m \delta_t^{max}(\upsilon) = 1.6 \times 10^{-3} \text{ m} \qquad (6)$$

251

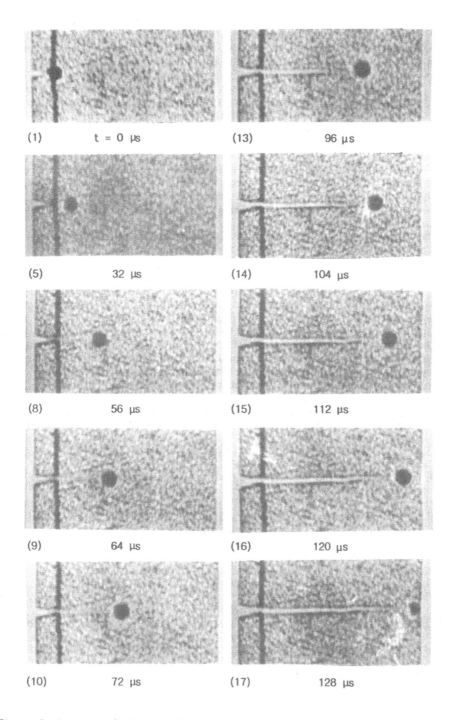

Figure 2. A series of photographs showing the crack propagation in a polystyrene plate with a strain rate $\dot{\varepsilon} = 10 \text{ s}^{-1}$.

where D_t^{max} is the mean value of the maximum diameter of the caustics, $D_t^{max} = 3.9 \times 10^{-3}$ m, λ_m is the magnificatio ratio of the set-up, $\lambda_m = 0.77$ and $\delta_t^{max}(\upsilon)$ is is the correction factor, which depends on the velocity of the crack and $\delta_t^{max}(\upsilon)$ equals $\delta_t^{max}(\upsilon) = 3.175$. So, the magnitude of the crack layer must be smaller than 1.6×10^{-3} m. The crack layer at the crack tip is presented in photographs of Figs 3 and 6. The crack layer, a region of crazes around the crack tip, increases as the strain rate increases. As the crack layer is developed, the crack propagation velocity decreases and so the magnitude of the crack layer decreases. After that, a new crack accelaration is observed and a new crack layer is developed. This phenomenon is presented in Fig. 3. In Fig. 3 the crack propagation velocity υ, versus the crack length a, is presented, for strain rate $\dot{\varepsilon} = 10$ s^{-1}. The photograph at the top of figure showing the crack propagation morphology with the craze regions. In this figure we can see a correspondence of the minima of the velocity with the crack layer, while the maxima of the velocity with the regions of the crack path without crack layers. From this is concluded that the crack velocity decreases by the crack layer.

The stress intensity factor K_I is calculated by the relation (1) from the diameter of the caustics. The variation of the K_I versus the crack length a for strain rate $\dot{\varepsilon} = 10$ s^{-1} is presented in Fig. 4. From this figure it may be observed that the variation of the K_I is approximately the same with those of the velocity.

Figure 5 presents a series of photographs showing the crack propagation process in a polystyrene plate with a strain rate $\dot{\varepsilon} = 20$ s^{-1}. From these photographs we may see the same phenomena as in Fig. 2, but most intense.

Figure 6 presents the variation of the crack propagation velocity υ, versus the crack length a, for strain rate $\dot{\varepsilon} = 20$ s^{-1}. The photograph at the top of figure showing the crack propagation morphology with the craze regions. In this figure we can see that the crack layers are wider than the crack layers of crack path in Fig. 3 and also, the crack velocities are greater than those in Fig. 3 for strain rate $\dot{\varepsilon} = 10$ s^{-1}. Also, in this figure it may be observed that the minima of the crack velocity correspond to the crack layers, while the maxima of the velocity correspond to the path regions without crack layers (regions between crack layers).

Figure 7 presents the variation of the stress intensity factor K_I, versus the crack length a, for strain rate $\dot{\varepsilon} = 20$ s^{-1}.

By comparison of the figures 3 and 6 we may be observed that the crack velocity increases as the strain rate increases. The crack velocity smoothly increases until the maximum value of 520 m/s at the position of crack path $a = 3.1$ cm for $\dot{\varepsilon} = 10$ s^{-1}, while for $\dot{\varepsilon} = 20$ s^{-1} the crack velocity rapidly increases until the maximum value of 580 m/s at the position of crack $a = 2.9$ cm. The same variation we can be observed for the stress intensity factor K_I.

Figure 8 presents the variation of the fracture surface energy γ, which was calculated by the relation (4), versus the crack length a, for strain rate $\dot{\varepsilon} = 10$ and 20 s^{-1}. From this figure it may be observed that the fracture surface energy γ for strain rate $\dot{\varepsilon} = 20$ s^{-1} is greater than the fracture surface energy for $\dot{\varepsilon} = 10$ s^{-1}. The critical fracture surface energy γ_c for $\dot{\varepsilon} = 10$ s^{-1} is $\gamma_c = 11.65$ KJ/m^2 and for $\dot{\varepsilon} = 20$ s^{-1} is $\gamma_c = 18.33$ KJ/m^2. Also, it may be observed that in the crack layer, where there is great crazes density, the fracture surface energy γ reduces considerably.

CONCLUSIONS

In the present work the effect of the crack layer on the dynamic crack propagation in polystyrene plates has been studied. The results may be summarized as follows:

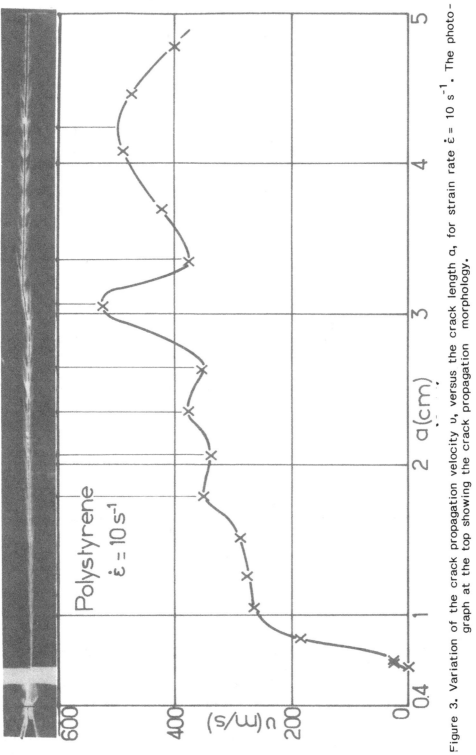

Figure 3. Variation of the crack propagation velocity υ, versus the crack length a, for strain rate $\dot{\varepsilon} = 10 \text{ s}^{-1}$. The photograph at the top showing the crack propagation morphology.

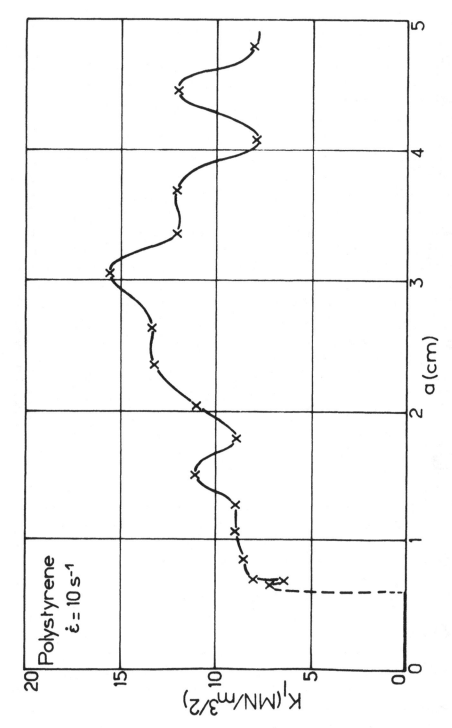

Figure 4. Variation of the stress intensity factor K_I, versus   the crack length a, for strain rate $\dot{\varepsilon} = 10$ s^{-1}.

255

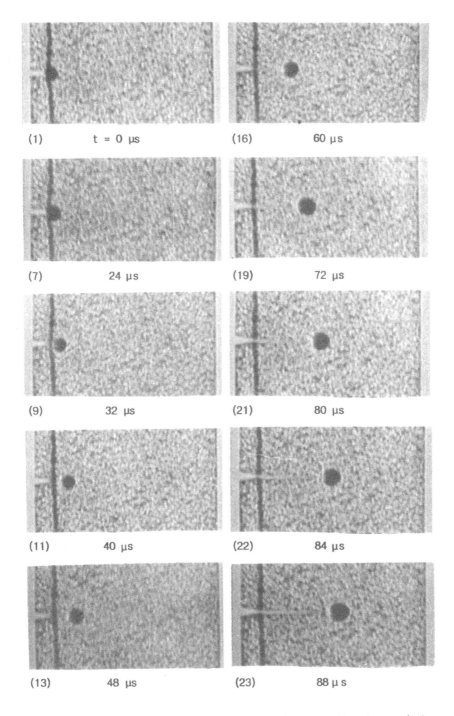

(1)	t = 0 μs	(16)	60 μs
(7)	24 μs	(19)	72 μs
(9)	32 μs	(21)	80 μs
(11)	40 μs	(22)	84 μs
(13)	48 μs	(23)	88 μ s

Figure 5. A series of photographs showing the crack propagation in a polystyrene plate with a strain rate $\dot{\varepsilon}$ = 20 s^{-1}.

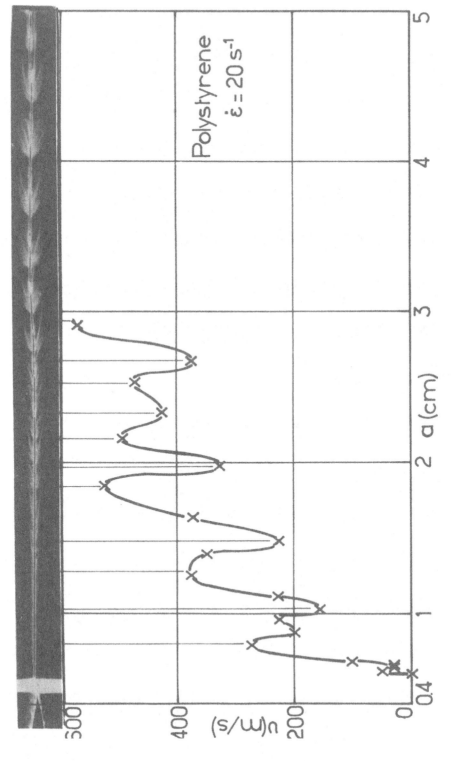

Figure 6. Variation of the crack propagation velocity υ, versus the crack length a, for strain rate $\dot{\varepsilon} = 20\ s^{-1}$. The photo-graph at the top showing the crack propagation morphology.

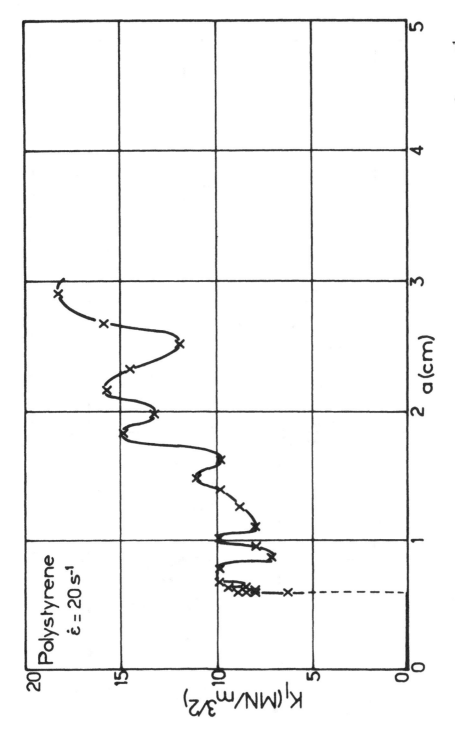

Figure 7. Variation of the stress intensity factor K_I, versus the crack length a, for strain rate $\dot{\varepsilon} = 20 \ s^{-1}$.

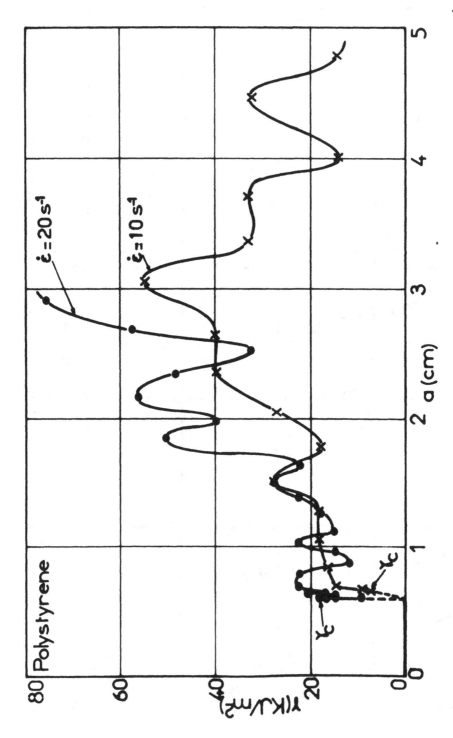

Figure 8. Variation of the fracture surface energy γ, versus the crack length α, for strain rate $\dot{\varepsilon}$ = 10 and 20 s^{-1}.

The shape and the magnitude of the active zone (crack layer) are strongly depended on strain rate.

During the crack propagation, the active zone is discontinuously developed at the crack tip.

The crack velocity and the stress intensity factor reduce when active zone is developed, while they increase in the regions where active zone is not developed.

The active zone is the area with intense damage. So, the modulus of elasticity E is reduced and then a decrease of the crack velocity is observed.

The active zone is formed ahead of crack propagating tip but when the crack velocity is great, the active zone remains back of the crack tip until a new active zone will be formed ahead of crack tip.

REFERENCES

1. Botsis, J., Chudnovski, A. and Moet, A., Fatigue crack layer propagation in polystyrene - Part I. Experimental observations. Int. J. Fracture, 1987, 33, 263 - 276.
2. Botsis, J., Chudnovski, A. and MOet, A., Fatigue crack layer propagation in polystyrene - Part II. Analysis. Int. J. Fracture, 1987, 33, 277 - 284.
3. Chudnovski, A. and Moet, A., Thermodynamics of translational crack layer propagation. J. Mater. Sci., 1985, 20, 630 - 635.
4. Rice, J.R., Mathematical analysis in mechanics of fracture. in Fracture, Ed. H. Liebowitz, Academic Press, NY 1964, Vol. II, 192 - 308.
5. Burech, F.E., About the process zones surrounding the crack tip in ceramics. in Fracture, Pergamon Press, London 1972, 3, 929 - 950.
6. Claussen, N., J. Amer. Ceram. Sco., 1976, 59, 49 - 60.
7. Pompe, H.A., Bahr, H.A., Gille, G. and Kreher, W., Increased fracture toughness of brittle materials by microcracking in a energy dissipative zone at the crack tip. J. Mater. Sci., 1978, 13, 2720 - 2723.
8. Donald, A.M. and Kramer, E.J., Micromechanics and kinetics of deformation zones at crack tips in polycarbonate. J. Mater. Sci., 1981, 16, 2977 - 2987.
9. Theocaris, P.S. and Papadopoulos, G.A., Elastodynamic forms of caustics for running cracks under constant velocity. Engng Fracture Mech., 1980, 13, 683 - 698.
10. Theocaris, P.S. and Papadopoulos, G., Mixed - mode dynamic stress-intensity factors from caustics. ASTM STP 791, 1983, II320 - 337.
11. Papadopoulos, G.A. and Theocaris, P.S., The variation of the dynamic modulus on crack propagation modes. Int. J. Fracture, 1987, 35, 195 - 219.
12. Griffith, A.A., The phenomenon of rupture and flow in solids. Phil. Trans. Roy. Soc., London, Ser. A, 1920, 221, 163 - 198.

RECENT ADVANCES IN DATA PROCESSING FROM CAUSTICS

H.P. ROSSMANITH and R.E. KNASMILLNER
Institute of Mechanics
University of Technology Vienna
Wiedner Hauptstraße 8-10/325, A-1040 Vienna, Austria

ABSTRACT

A technique for the determination of stress intensity factors from caustics by means of an interactive image processing system has been developed where a multipoint overdeterministic data reduction technique makes use of a number of data points along the recorded caustic. The method requires minimum manual input from the analyst. The selection of data points along the experimentally recorded caustic curve for the analysis is done automatically. Results of this procedure are checked by comparing for acceptable coincidence the numerically generated caustic on the basis of the result of the iterative data reduction technique with the experimentally recorded caustic.

INTRODUCTION

The method of caustics or shadow-spot technique originates in the fundamental work by Manogg [1][2] where the technique is developed for transparent materials. The associated optical system is termed transmitted light method. Theocaris and his coworkers have successfully revived and adapted this method to opaque materials by developing the reflected light method [3][4]. Within the framework of experimental fracture mechanics the method is utilized to investigate the highly strained regions surrounding the tip of a plane crack in brittle materials loaded in tension. From the geometrical dimensions of the shadow spot one can deduce the stress intensity factors K_1 and K_2 of the crack tip stress field.

The general equations of caustics for plane static and dynamic elasticity theory may be found in the review article by Beinert and Kalthoff [5] and Rossmanith [6].

The majority of papers in the literature pertain to idealized situations either mode-1 loading situations or cracks propagating at constant speed.

Recently, the importance of mixed-mode crack propagation, the transmission from plane strain to plane stress conditions, the influence of variable crack speed and the necessity to include higher order terms have been recognized [7, 8].

In addition, interactive computer-graphics-based techniques of caustic data reduction and analysis have been developed for various problems which, in terms of accuracy, applicability and versatility proved to be superior to classical methods, e.g. based on diameter measurements.

THE OPTICAL METHOD OF CAUSTICS

The physical principle of the method of caustics is the inhomogeneous deflection of parallel light rays during their passage through a plate specimen due to two effects: the reduction of the thickness of the specimen and the change of the refractive index of the material as a consequence of stress intensification. Several versions of the method of caustics are successfully employed in studying the behaviour of stress and strain around crack tips:

a) the transmitted light method,
b) the back face reflected light method and
c) the front face reflected light method.

The following analysis is valid for all types of caustic experiments upon replacing appropriate constants.

Consider a thin elastic isotropic plate containing a single internal crack or a semi-infinite crack arbitrarily oriented to the direction of application of in-plane biaxial tension. The stress field in the vicinity of the crack tip is regarded as homogeneous. In the transmitted light method a normally impinging light beam transverses an initially unstressed plate specimen of thickness h. When the specimen is subjected to external loading and it is state of plane stress, the lateral deformation causes a thickness reduction of the specimen and a change of the index of refraction (Fig. 1) and the direction of the incident light ray will be changed.

The direction of the light rays emerging from the plate is determined as the gradient of the light wave front. Because the magnitude and direction of light deflection are correlated with the principal stresses and stress intensification about the crack tip, the shadow pattern contains information about the stress-strain conditions in the illuminated region.

CAUSTIC ANALYSIS

When a normally impinging light beam transverses the unloaded specimen at the point $P(r, \theta)$ in the object plane its image $P'(r', \theta')$ on the shadow image plane is defined by the vector $\vec{r}'$ (Fig. 1). Load application induces deflection of the beam to the point $P''(x', y')$. The deflection vector is denoted by $\vec{w} = \vec{w}(r, \theta)$. The vector $\vec{W}$ of the image point P'' is

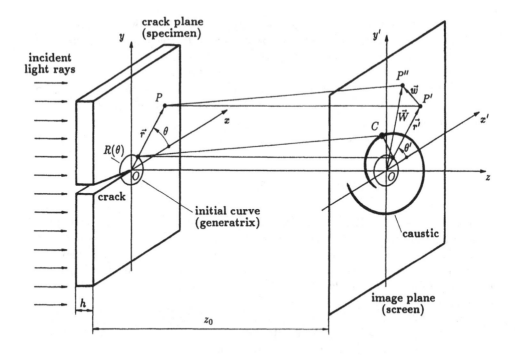

Figure 1: Stress-induced light deflection in a plate specimen with a crack subjected to mixed-mode loading conditions (transmitted light method)

given by

$$\vec{W} = \vec{r} + \vec{w}(r, \theta).\tag{1}$$

This deflection gives rise to the formation of a shadow space formed by the deflected light rays upon passage of the object. The caustic is a singular curve of the transformation equ(1) and is generated by the intersection of the image plane with the shadow space and ray field. A necessary condition for the existence of the singular caustic curve is that the Jacobian functional determinant J of equ(1) vanishes, i.e.

$$J = \frac{\partial x'}{\partial r}\frac{\partial y'}{\partial \theta} - \frac{\partial y'}{\partial r}\frac{\partial x'}{\partial \theta} = 0\tag{2}$$

for all points $C(R, \theta)$ of the caustic curve where $R(\theta)$ is the solution of equ(2). Hence, the fundamental problem of the method of caustics is to solve equ(2) under certain conditions imposed by the real physical problem.

An elasto-optical analysis [1][5] yields a relation between light deflection, the stress field characterized by the principal normal stresses σ_1, σ_2, the elasto-optical parameters and the geometry of the experimental set-up:

$$\vec{w}_{\pm} = -z_0 h c^* . grad\{(\sigma_1 + \sigma_2) \pm \varepsilon(\sigma_1 - \sigma_2)\}\tag{3}$$

where z_0 denotes the distance between model plane and image plane, h is the thickness of the model, c^* is an elasto-optical parameter and ε accounts for the optical anisotropy of the material. For optically isotropic or inert materials (e.g. Plexiglas) $\varepsilon = 0$ and one obtains one single caustic. Optically anisotropic or birefringent materials (e.g. Araldite B) give rise to a double caustic, an inner and an outer caustic [9, 10].

CRACK TIP STRESS FIELDS

The non-propagating crack

Consider a thin isotropic homogeneous elastic plate containing a semi-infinite crack or a central crack arbitrarily oriented to the direction of application of the tensile load. The plate is under conditions of plane stress, and only the opening mode (mode-1) and the shearing mode (mode-2) of crack deformations are present in various combinations depending on the angle of inclination of the orientation of the load applied.

Let the thin plate have its mid-plane coincide with the plane xy of a cartesian system of coordinates with its origin O at the crack tip. The x-axis coincides with the direction of the crack. For the symmetric and skew-symmetric plane extensional problems considered, the stress field around the crack tip is given by

$$
\begin{aligned}
\sigma_x &= \mathbf{Re}(Z_1 + 2Y) + 2\mathbf{Im}Z_2 + y[\mathbf{Re}Z_2' - \mathbf{Im}(Z_1' + Y')] \\
\sigma_y &= \mathbf{Re}Z_1 - y[\mathbf{Re}Z_2' - \mathbf{Im}(Z_1' + Y')] \\
\tau_{xy} &= \mathbf{Re}Z_2 - \mathbf{Im}Y - y[\mathbf{Re}(Z_1' + Y') + \mathbf{Im}Z_2']
\end{aligned}
\tag{4}
$$

where $z = x + iy = re^{i\theta}$ is a complex variable. The functions $Z_j(z)$, $(j = 1, 2)$ and $Y(z)$ are modified Westergaard type stress functions appropriately selected in the form

$$
Z(z) = Z_1(z) - iZ_2(z)
\tag{5}
$$

$$
Z_k(z) = \frac{K_k}{\sqrt{2\pi z}} \sum_{n=0}^{N} \beta_{kn}(\frac{z}{r_s})^n, \quad (k = 1, 2)
\tag{6}
$$

$$
Y(z) = \frac{K_1}{2\sqrt{2\pi z}} \sum_{n=0}^{N} \alpha_n(\frac{z}{r_s})^{n+\frac{1}{2}}
\tag{7}
$$

where α_n and β_{kn} are real coefficients. The complex stress intensity factor is defined as

$$
K^* = K_1 - iK_2,
\tag{8}
$$

where $k = 1$ and $k = 2$ correspond to opening and sliding modes, respectively.

The $K_k/\sqrt{2\pi z}$-term describes the crack tip singularity of any opening mode crack ($k = 1$) and shearing mode crack ($k = 2$) where the nonlinear zone containing the fracture process is relatively small, the plate exhibits uniform thickness and has no finite boundaries. The factors α_n and β_{kn} to some extend model the effect of the presence of near boundaries and/or loading configurations, and r_s is a reference length which can be selected to equal crack size or any other characteristic length associated with the specimen geometry.

Combinations of eqs(4-8) and taking the gradient yields

$$\nabla(\sigma_1 + \sigma_2) = 2\nabla\mathbf{Re}(Z + Y) = 2\frac{\partial}{\partial z}\overline{(Z + Y)} \tag{9}$$

$$\nabla(\sigma_1 - \sigma_2) = \overline{\Lambda}|\vartheta|^{-1} \tag{10}$$

with

$$\overline{\Lambda}(z,\overline{z}) = \vartheta_{\overline{z}}\,\overline{\vartheta} + \overline{\vartheta}_{\overline{z}}\,\vartheta \tag{11}$$

where $(\ )_z$ and $(\ )_{\overline{z}}$ denote partial differentiation with respect to z and $\overline{z}$, respectively, $\mathbf{Re}$ denotes the real part, and the derivatives of $\vartheta(z,\overline{z})$ are given in [6].

The deflection of a light ray defined by equ(3) is then given by

$$w_\pm = -z_0 hc^*\{2\frac{\partial}{\partial z}\overline{(Z + Y)} \pm \varepsilon\overline{\Lambda}|\vartheta|^{-1}\} \equiv -C\overline{\Gamma}_\pm(z,\overline{z}) \tag{12}$$

with $C = z_0 hc^*$ and the image equation of the caustic equ(1) takes the form

$$W_\pm(z) = z - C\overline{\Gamma}_\pm(z,\overline{z}). \tag{13}$$

The condition, equ(2), for the existence of a singular line, the caustic, in conjunction with eqs(12,13) renders the generatrix (initial curve) for the caustic:

$$1 - 2C\mathbf{Re}(\overline{\Gamma}_z) + C^2\{|\overline{\Gamma}_z|^2 - |\overline{\Gamma}_{\overline{z}}|^2\} = 0 \tag{14}$$

with the solution

$$R_\pm = R_\pm(\theta). \tag{15}$$

For a crack problem with known elasto-optical parameters, equ(12) and equ(14) permit the complete determination of the shape of the caustic generated about the crack tip(s). In a more general situation a numerical solution for the generatrix $R_\pm(\theta)$ is possible only. The governing equations for the caustic curve are then obtained upon substitution of $R_\pm(\theta)$ into equ(13), e.g. in Cartesian coordinates.

$$\begin{aligned} x'_{C\pm} &= R_\pm(\theta)\cos\theta - C\mathbf{Re}\overline{\Gamma}_\pm(R_\pm,\theta) \\ y'_{C\pm} &= R_\pm(\theta)\sin\theta - C\mathbf{Im}\overline{\Gamma}_\pm(R_\pm,\theta) \end{aligned} \tag{16}$$

For the case of optically isotropic materials ($\varepsilon = 0$) only one branch of the caustic is obtained and the subscripts $\pm$ become meaningless. The equation for the generatrix or initial curve for the caustic then reduces to:

$$1 - C^2|\overline{\Gamma}_{\overline{z}}|^2 = 0 \tag{17}$$

the solution of which is given by $R = R(\theta)$.

The caustic for the singular crack tip stress field governed by K_1 and K_2 is retained from equ(16) by dropping the high order terms in equ(6) and equ(7), i.e. for $N = 0$. Notice that the lowest order Z-related higher order term which corresponds to a homogeneous stress field parallel to the crack only appears in the term associated with the optical anisotropy, equ(10).

Very simple equations follow for the classical singular crack tip stress field caustic in optically isotropic materials, where the initial curve (generatrix) in the model plane degenerate to a circle:

caustic equations

$$x'_C = r_0 \cos\theta + C_{exp}r_0^{-3/2}\left[\cos\tfrac{3\theta}{2} - \mu\sin\tfrac{3\theta}{2}\right]$$

$$y'_C = r_0 \sin\theta + C_{exp}r_0^{-3/2}\left[\sin\tfrac{3\theta}{2} + \mu\cos\tfrac{3\theta}{2}\right]$$

(18)

$$\text{with} \qquad C_{exp} = K_1\frac{z_0 dc^{\tilde{}}}{\sqrt{2\pi}}, \qquad \mu = \frac{K_2}{K_1}$$

(19)

and initial curve

$$r_0 = \left[\tfrac{3}{2}C_{exp}\right]^{\tfrac{2}{5}}\left[1 + \mu^2\right]^{\tfrac{1}{5}}$$

(20)

where θ is the polar angle, r_0 is the radius of the initial curve defined in the model plane, μ denotes the mixed-mode index and the constant C_{exp} depends on the geometrical set-up and material parameters [11][12]. Although developed here for an infinite plate specimen with a single crack the method is generally applicable also for specimens of finite size with more complex crack geometries such as star cracks, interface cracks, etc. if the conformal mapping is introduced and selected properly [6].

Dynamic crack propagation

For a constant speed running crack in a finite body the stress field around a crack tip, equ(4) is replaced by [13][14][15].

$$\sigma_x = F[A\mathbf{Re}(Z_1 + Y_1) - B\mathbf{Re}Z_2 - C\mathbf{Re}Y_2]$$
$$\sigma_y = F[-C\mathbf{Re}(Z_1 - Y_1) + B\mathbf{Re}Z_2 - C\mathbf{Re}Y_2]$$
$$\tau_{xy} = F[-D\mathbf{Im}(Z_1 - Z_2 + Y_1) + E\mathbf{Im}Y_2]$$

(21)

with

$$
\begin{array}{ll}
A = 1 + 2\lambda_1^2 - \lambda_2^2 & B = (4\lambda_1\lambda_2)/(1 + \lambda_2^2) \\
C = 1 + \lambda_2^2 & D = 2\lambda_1 \\
E = (1 + \lambda_2^2)^2/2\lambda_2 & F = (1 + \lambda_2^2)/[4\lambda_1\lambda_2 - (1 + \lambda_2^2)^2]
\end{array}
$$

(22)

where

$$Z_k = Z(z_k) = \sum_{n=0}^{N}C_n Z_k^{n-\tfrac{1}{2}} \qquad (k = 1, 2)$$

$$Y_k = Y(z_k) = \sum_{n=0}^{N}D_n Z_k^{n}$$

(23)

$$z_k = x + i\lambda_k y$$

$$\lambda_k^2 = 1 - (c/c_k)^2$$

c is the crack velocity, c_1 is the longitudinal wave speed and c_2 is the shear wave speed in the material. The crack-tip coordinates x and y are oriented such that the negative branch of the x-axis coincides with the crack faces. C_n, D_n are unknown real coefficients to be determined for the problem of interest. The lowest order coefficient C_0 is proportional to the mode-1 stress intensity factor K_1.

Equ(9) will then be replaced by

$$\nabla(\sigma_1 + \sigma_2) = F\nabla \mathbf{Re}V = F(\mathbf{Re}V' - i\lambda_1 \mathbf{Im}V') \qquad (24)$$

with

$$\begin{aligned}
V &= (A - C)(Z_1 + Y_1) + 2C(Y_1 - Y_2) \\
&= 2(\lambda_1^2 - \lambda_2^2)(Z_1 + Y_1) + 2(1 + \lambda_2^2)(Y_1 - Y_2)
\end{aligned} \qquad (25)$$

and $(\ )' = d(\ )/dz$ [14].

For optically isotropic materials the vector of light ray deflection takes the form (cf. equ(12)):

$$w = -z_0 hc^* F\overline{V}^* \qquad (26)$$

The function $\overline{V}^* = \mathbf{Re}V - i\lambda_1 \mathbf{Im}V$ is the generalized dynamic conjugate of the generalized complex function $\overline{V} = \mathbf{Re}V + i\lambda_1 \mathbf{Im}V$. In the static case $\overline{V}^*$ reduces to the classical conjugate $\overline{V} = \mathbf{Re}V - i\mathbf{Im}V$ of the complex function V. The initial curve (generatrix) follows from equ(17) with $\overline{\Gamma} = F\overline{V}^*$ and represents a circle only if $c = 0$ and the first derivative of all higher order terms are identical to zero, i.e. $Y_k = const$.

For the generalization to transient asymptotic stress fields around the tip of a constant velocity crack see the contribution by Freund and Rosakis in this volume [8] and also [16].

INTERACTIVE DATA ANALYSIS

The hardware requirement for the technique introduced here consists of an image scanner (e.g. a CCD-camera with A/D-converter), an image storage device provided with a monitor and a "mouse". The digitized and stored caustic pattern displayed on the monitor is required to select interactively certain points on the screen. A multi-point overdeterministic method of data reduction for K-determination makes use of a large number of arbitrarily selected data points along the entire contour of the caustic.

For data selection, a coordinate system (x_D, y_D) is placed parallel to the crack with the origin within the caustic area as shown in Fig. 2. If the location of the crack tip were known the coordinate system could be placed appropriately, $\Delta x = \Delta y = 0$. In general, however, the exact site of the crack tip is not known a priori and the initial guess for Δx and Δy will be nonzero. The only requirement as to geometry in the following method of data reduction is the parallelism of the x-axis with the tangent to the crack at the crack tip.

Next, the region of the experimental caustic is selected where data points can easily be identified. In general, the part of the caustic adjacent to the faces of the crack is blurred and obscured (blurr zone **B**) and should therefore be discarded (Fig. 2).

The transformation

$$x_D = x' + \Delta x \qquad y_D = y' + \Delta y \qquad (27)$$

transforms the image equations of the method of caustics to the following expressions:

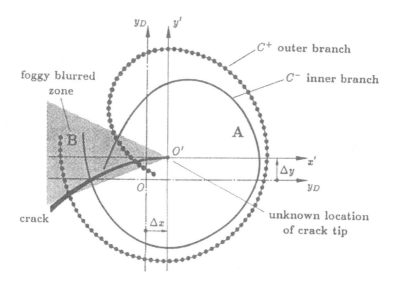

Figure 2: Caustic geometry, coordinate systems and data selection zones

- for a static crack in an optically isotropic material:

$$
\begin{aligned}
x_D &= f(K_1, \mu, \Delta x, \Delta y; \theta) \\
y_D &= g(K_1, \mu, \Delta x, \Delta y; \theta)
\end{aligned}
\tag{28}
$$

- for a static crack in an optically anisotropic material:

$$
\begin{aligned}
x_D^{\pm} &= f_-(\alpha_n, \beta_{kn}, \Delta x, \Delta y; \theta) \\
y_D^{\pm} &= g_-(\alpha_n, \beta_{kn}, \Delta x, \Delta y; \theta)
\end{aligned}
\tag{29}
$$

where the subscripts $+$ and $-$ hold for the inner and outer caustic branch, respectively;

- for a crack with constant propagation velocity c:

$$
\begin{aligned}
x_D &= f_c(C_n, D_n, \Delta x, \Delta y; \theta) \\
y_D &= g_c(C_n, D_n, \Delta x, \Delta y; \theta).
\end{aligned}
\tag{30}
$$

The as yet unknown quantities K_1, μ, ($(\alpha_n, \beta_{kn}$ or C_n, D_n, respectively), Δx and Δy will be determined by employing the method of least squares [17, 18].

The procedure is as follows (Fig. 3):

1. A caustic $C_j^{\pm}$ (initially $j = 0$) is constructed on the basis of the results of the preceding step of the successive iterative approximation procedure (for $j = 0$ estimated values) for K_{1j}, μ_j, $(\alpha_{nj}, \beta_{knj}$ or $C_{nj}, D_{nj})$, Δx_j and Δy_j, given by the equation

$$
\begin{aligned}
x_j^{\pm} &= f^{\pm}(K_{1j}, \mu_j, \ldots, \Delta x_j, \Delta y_j; \theta) \\
y_j^{\pm} &= g^{\pm}(K_{1j}, \mu_j, \ldots, \Delta x_j, \Delta y_j; \theta).
\end{aligned}
\tag{31}
$$

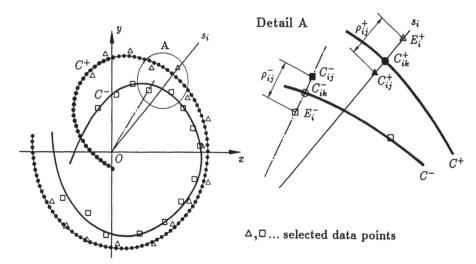

Figure 3: Iterative construction of best-fit caustic $C^{\pm}$.

2. A total of m data points $E_i^{\pm}$ $(i = 1,\ldots,m)$ on the experimentally recorded light intensity distribution which makes up for the inner and outer branch of the caustic is selected. For optically isotropic materials only one caustic curve is present and therefore the distinction $\pm$ becomes meaningless.

The proper selection of the position of the points $E_i^{\pm}$ can be achieved in several ways: they can be hand-picked or an interactive numerical procedure may be employed (see below).

3. Determine the argument $\alpha_i^{\pm}$ of $E_i^{\pm}$ by

$$\alpha_i^{\pm} = \arctan(y_{E_i}^{\pm}/x_{E_i}^{\pm}) \qquad (\pm; i = 1,\ldots,m) \tag{32}$$

4. Next, the value $\theta_{ij}^{\sim}$ of the parameter θ (as measured along the initial curve) associated with the radial s_i passing through $E_i^{\pm}$ must be determined from

$$\alpha_i^{\pm} = \arctan\frac{y_{ij}^{\pm}}{x_{ij}^{\pm}} = \frac{g^{\pm}(K_{1j},\mu_j,\ldots;\theta_{ij}^{\sim})}{f^{\pm}(K_{1j},\mu_j,\ldots;\theta_{ij}^{\sim})} \tag{33}$$

to yield $\theta_{ij}^{\sim}$ $(i = 1,\ldots,m)$. Each value $\theta_{ij}^{\sim}$ renders a point $C_{ij}^{\pm}$ by substituting in equ(31) $\theta_{ij}^{\sim}$ for θ.

5. The experimentally recorded and selected points $E_i^{\pm}(x_{E_i}^{\pm}, y_{E_i}^{\pm})$ and the points of the estimated caustic branches just determined, $C_{ij}^{\pm}(x_{C_{ij}}^{\pm}, y_{C_{ij}}^{\pm})$ claim the same radial $s_i^{\pm}$. Minimizing the sum S_j of the squares of the differences, $\rho_{ij}^{\pm}$,

$$S_j = \sum_{i=1}^{m}(\rho_{ij}^{\pm})^2 = \sum_{i=1}^{m}[(x_{E_i}^{\pm} - x_{C_{ij}}^{\pm})^2 + (y_{E_i}^{\pm} - y_{C_{ij}}^{\pm})^2] = \min \tag{34}$$

will yield improved estimates for the parameters K_{1j}, μ_j, $(\alpha_{nj}, \beta_{knj}$ or $C_{nj}, D_{nj})$, Δx_j and Δy_j $(j > 0)$ which will serve as new input data for the next step in this iterative procedure.

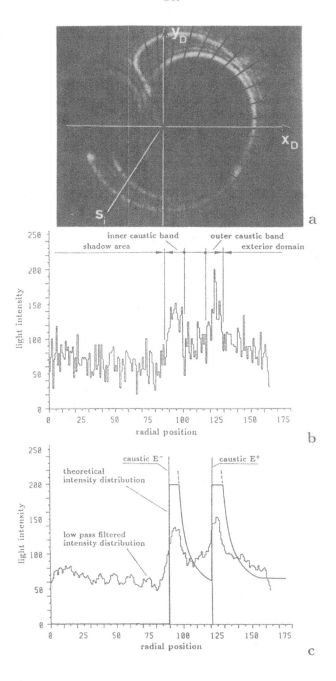

Figure 4: Density distributions for experimentally recorded crack tip
caustics obtained from an analysis with an image processing system:
   a) experimentally recorded and scanned caustic
   b) light distribution along the radial s_i
   c) low pass filtered light distribution and positions of
      caustic points E^- and E^+

6. Thus, a sequence of caustics $C_j^{\pm}$ $(j = 1, \ldots, k)$ is defined which converges to the best-fit caustic, $C_k^{\pm} \equiv C^{\pm}$.

In the interactive data reduction method a total of m radials s_i, homogeneously distributed within the zone of the clearly visible region of the caustic, $\mathbf{A}$, is automatically chosen for point identification (Fig. 4a). This selection leaves open the proper position of the data-points $E_i^{\pm}$ on the finite width of the experimentally recorded light intensity distribution which makes up for the caustic. Their proper positions are determined from an evaluation of the density distribution along the radials s_i within the caustic range. The experimentally recorded light intensity distribution (or film density distribution) across the caustic band is shown in Fig. 4b. Low pass filtering produces the smooth light intensity trace shown together with the ideal caustic intensity trace in Fig. 4c. The correct K-value would obviously be obtained from the proper selection of the points $E_i^{\pm}$ at distances $r_{E_i}^{\pm}$.

The automatic point selection method bases on the fact, that a variable relative grey level (intensity) serves as a degree of freedom. Upon interactive selection of a relative grey level estimate (0% and 100% correspond to grey minimum and grey maximum, respectively) radial positions $r_{E_i}^{\pm}$ for the data points $E_i^{\pm}$ will be determined. The magnitude of the relative grey level depends on the experimental set-up and the recording characteristics.

It is also found that for best fit the caustic branches do not necessarily have to coincide with the centerline of the caustic band. Further research is called for.

Once the data points $E_i^{\pm}$ have been identified the automatized method follows the multi-point overdeterministic data reduction technique outlined in the section before [19].

The last step in the K-determination procedure is the visual inspection of the result by plotting the caustic line and the associated experimental input points together with the stored image of the analyzed caustic on the monitor. If the coincidence of the analytically generated caustic line with the experimental recorded caustic is unsatisfactory, the filter characteristics, i.e. the relative grey level for the caustic line can be set to new values or some of the data points can be repositioned or deleted for a new analysis.

The method presented here converges very rapidly and even in cases where an appreciable part of the caustic is obscured or not fit for data analysis very reliable results have been obtained during an extensive test program, where caustics have been numerically generated, fuzzed, data-reduced and reconstructed on the basis of the calculated data.

ACKNOWLEDGEMENT

The authors would like to acknowledge the financial support granted by the *Austrian Science Foundation* under project numbers # P 5814 and # P 6632.

REFERENCES

[1] P. Manogg. *Anwendung der Schattenoptik zur Untersuchung des Zerreißvorganges von Platten.* PhD Thesis, Universität Freiburg i. Br., Freiburg, Germany, 1964.

[2] P. Manogg. Investigation of the rupture of a Plexiglas plate by means of an optical method involving high speed filming of the shadows originating around holes drilling in the plate. *Int.J. Fracture Mechanics,* 2:604–613, 1966.

[3] P.S. Theocaris. Local yielding around a crack tip in Plexiglas. *J. Applied Mechanics,* 37:409–415, 1970.

[4] P.S. Theocaris. Reflected shadow method for the study of constrained zones in cracked plates. *Applied Optics,* 10(10):2240–2247, 1971.

[5] J. Beinert and J.F. Kalthoff. Experimental determination of dynamic stress intensity factors by the method of shadow patterns. *Mechanics of Fracture,* 7:281–330, 1981.

[6] H.P. Rossmanith. The method of caustics for plane elasticity problems. *Elasticity,* 12:193–200, 1982.

[7] S. Suzuki and S. Fukuchi. Some experiments on measurement of dynamic stress intensity factor of fast propagating cracks. This volume.

[8] L.B. Freund and A.J. Rosakis. The influence of transient effects on the asymptotic crack tip field during dynamic crack growth. This volume.

[9] P. Manogg. Die Lichtablenkung durch eine elastisch beanspruchte Platte und die Schattenfiguren von Kreis und Rißkerbe. *Glastechnische Berichte,* 39:323–329, 1966.

[10] R. Podleschny and J.F. Kalthoff. Improved methodology for the determination of fracture mechanics shear-(mode-2)-stress intensity factors by means of the shadow optical method of caustics. In: *13.GESA-Symposium,* pages 323–335, Bochum, Germany, 1990.

[11] J.F. Kalthoff. The shadow optical methods of caustics. In: A.S. Kobayashi (Editor), *Handbook on Experimental Mechanics,* chapter 9, pages 430–500. Prentice Hall, Inc., 1987.

[12] H.P. Rossmanith. The method of caustics. In: Wang W.C. (Editor), *Advanced Photomechanics,* pages 1–450. Dynamic Engineering Institute, National Tsing-Hua University, Hsinchu, Taiwan, 1988.

[13] G.R. Irwin. Lecture Notes on Fracture Mechanics. Dept. of Mechanical Engineering, University of Maryland, 1979.

[14] H.P. Rossmanith. Determination of stress intensity factors by the dynamic method of caustics for optically isotropic materials. *Ingenieur-Archiv,* 48:363–381, 1979.

[15] H. Nigam and A. Shukla. Comparison of the techniques of transmitted caustics and photoelasticity as applied to fracture. *Experimental Mechanics,* pages 123–130, 1988.

[16] L.B. Freund. *Dynamic Fracture Mechanics*. Cambridge University Press, 1990.

[17] R.E. Knasmillner. Vielpunktmethode zur Bestimmung des Spannungsintensitätsfaktors mit Hilfe der Methode der Kaustik. *ÖIAZ*, 131:318–320, 1986.

[18] H.P. Rossmanith and R.E. Knasmillner. An interactive caustic-based method of stress intensity factor determination. In: D. Firrao (Editor), *Proc. ECF-8*, volume 2, pages 1093–1098. EMAS, October 1990.

[19] H.P. Rossmanith and R.E. Knasmillner. Interaction of Rayleigh waves with surface breaking and embedded cracks. In: Y. Kanto H. Homma (Editor), *Proc. of OJI Int. Seminar on Dynamic Fracture*, pages 236–262, Toyohashi University of Technology, Japan, 1989.

THE TRANSIENT ANALYSIS OF A SUBSURFACE INCLINED CRACK SUBJECTED TO STRESS WAVE LOADING

CHWAN-HUEI TSAI   and   CHIEN-CHING MA
Department of Mechanical Engineering
National Taiwan University
Taipei, Taiwan 10764, Republic of China

ABSTRACT

In order to understand the fracture phenomenon of an internal crack subjected to dynamic impact loading, a subsurface crack contained in a half-space is studied. The realistic impact loading is simulated by a Sine function with a duration period. The solutions are determined by linear superposition of the Green function obtained by Tsai and Ma [1]. The exact closed form solutions of stress intensity factor histories are obtained. These solutions are valid for the time interval from initial loading until the first wave scattered at the crack tip returns to the crack tip after being reflected by the free boundary. The probable crack propagation direction is also predicted from the fracture criteria of maximum circumferential tensile stress.

INTRODUCTION

The investigation of an idealized semi-infinite crack can provide some information for a realistic elastodynamic fracture problem. It is noted that while the analysis has been carried out assuming a semi-infinite crack, the results remain valid for a finite crack up until the time at which waves diffracted from the far tip reach the tip near the boundary. The incident wave generated by the normal impact on the half-space will be reflected from the crack surfaces and diffracted by the crack tip. From the study by Tsai and Ma [1], the transient response varies rapidly at the instant when the wave front passes the crack tip. The value of the peak usually is greater than the corresponding static value and might induce brittle fracture. The failure of a notched beam due to impact loading, called the dynamic tear test, was studied by Brock et al. [2]. They investigated the case in which the crack is normal to the half-plane surface and the point load is applied to the surface directly above the crack tip.

Freund [3] studied the problem of an elastic solid containing a semi-infinite crack subjected to concentrated impact loading on the faces of the crack. He proposed a fundamental solution arising from an edge dislocation climbing along the positive x_1 axis with a constant speed to overcome this difficulties of the case with characteristic length. The solution can be constructed by taking an integration over a climbing dislocation of different moving velocity. Basing his procedure on this method, Brock [4-5] and Ma and Hou [6,7] analyzed a series problems of a semi-infinite crack subjected to impact loading. Recently, Lee and Freund [8] analyzed fracture initiation of an edge cracked plate subjected to an asymmetric impact.

The problem to be considered in this study is the plane strain response of an elastic half-plane, with an inclined crack extending from infinity to the half-plane surface, subjected to the realistic impact loading of Sine function with a duration period. This problem involves multiple characteristic length which can not be solved by the methods proposed in [2,3]. Tsai and Ma [9] proposed a new fundamental solution to overcome these difficulties and applied sucessfully to solve the similar problem [1], but with Heaviside function dependence for dynamic loading. This alternative fundamental solution is represented by an exponentially distributed traction e^{pdx_1} applied at crack faces and expressed in the Laplace transform domain. The p is the Laplace transform parameter and d is a constant. The traction force can be divided into a normal force (mode I) and a tangential force (mode II). The corresponding stress intensity factors for the fundamental solution in the Laplace transform domain are

$$\bar{K}_I(p) = -\sqrt{\frac{2}{p}} K_I^F(d), \quad \bar{K}_{II}(p) = -\sqrt{\frac{2}{p}} K_{II}^F(d), \tag{1}$$

where

$$K_I^F(d) = \frac{(a+d)^{1/2}}{(c+d)S_+(d)}, \quad K_{II}^F(d) = \frac{(b+d)^{1/2}}{(c+d)S_+(d)},$$

$$S_+(d) = \exp\left(\frac{-1}{\pi} \int_a^b \tan^{-1}\left[\frac{4\lambda^2(\lambda^2-a^2)^{\frac{1}{2}}(b^2-\lambda^2)^{\frac{1}{2}}}{(b^2-2\lambda^2)^2}\right] \frac{d\lambda}{\lambda+d}\right),$$

in which $a = \sqrt{\rho/(\gamma+2\mu)}$, $b = \sqrt{\rho/\mu}$, a, b and c are the slownesses of the longitudinal wave, shear wave and Rayleigh wave, μ and ρ the shear modulus and mass density, and γ the Lame elastic constant.

The present study is an extension of the simple loading with Heaviside function dependence investigated in [1] which can be regard as a Green function. The more complicated loading condition can be obtained by superposition over the Green function. Final formulations are expressed explicitly and the dynamic effect of each wave is presented in a closed form. The results are valid before the first wave scattered from crack tip returns to the crack tip after being reflected by the free boundary. Finally, the fracture criteria of maximum circumferential tensile stress proposed by Erdogan and Sih [10] is used to determine the possible direction of crack propagation.

SUBSURFACE CRACK DUE TO SURFACE IMPACT

In the time period during which waves generated by the impact force and its diffractions at the notch end have not returned to the crack tip, the problem can be treated as a semi-infinite crack contained in an unbounded medium. The problem considered here is an inclined semi-infinite crack located under the surface of a half-plane as shown in Fig. 1. The origins of the two coordinate systems $(\bar{x}_1, \bar{x}_2)$ and (x_1, x_2) are located at the plane surface and crack tip, respectively. The planar crack lies in the plane $x_2 = 0$, $x_1 < 0$ and the inclined angle of the crack is θ.

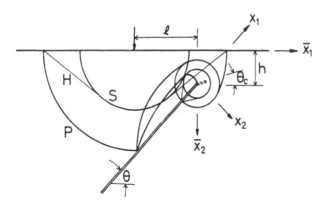

Fig. 1 Configuration, coordinate system and wave fronts of a subsurface crack subjected to stress wave loading.

The Green functions are generated by applying a point loading of Heaviside function time dependence. The stresses induced by the incident waves of Lamb's problem [11] can be represented by the integration of exponential functions $e^{p(\lambda \cos\theta + \alpha \sin\theta)x_1}$ and $e^{p(\lambda \cos\theta + \beta \sin\theta)x_1}$. It means that the traction that should be applied in the crack face to eliminate the stress induced by the incident wave can be represented by the integration of the above exponential function. Since the stress intensity factors resulting from applied tractions e^{pdx_1} are expressed in (1), the solution of this problem can be constructed by superposition of the fundamental solution. Replacing d with $\lambda \cos\theta + \alpha \sin\theta$ and

$\lambda \cos \theta + \beta \sin \theta$ and combining (1), the results for the mixed mode dynamic stress intensity factors in the Laplace transform domain can be obtained, and apply the Cagniard-de Hoop method to invert the Green functions to time domain, we get

$$\frac{\pi^{3/2}}{\sqrt{2}} K_I^G(t) = \int_{ar_0}^{t} \frac{1}{\sqrt{t-\tau}} \text{Im} \left[A_1(\lambda_1) K_I^F(\lambda_1 \cos \theta + \alpha \sin \theta) \frac{\partial \lambda_1}{\partial \tau} \right] d\tau$$

$$+ \int_{T_{H_0}}^{t} \frac{1}{\sqrt{t-\tau}} \text{Im} \left[A_2(\lambda_2) K_I^F(\lambda_2 \cos \theta + \beta \sin \theta) \frac{\partial \lambda_2}{\partial \tau} \right] H(br_0 - \tau) d\tau$$

$$+ \int_{br_0}^{t} \frac{1}{\sqrt{t-\tau}} \text{Im} \left[A_2(\lambda_3) K_I^F(\lambda_3 \cos \theta + \beta \sin \theta) \frac{\partial \lambda_3}{\partial \tau} \right] d\tau, \qquad (2)$$

$$\frac{\pi^{3/2}}{\sqrt{2}} K_{II}^G(t) = \int_{ar_0}^{t} \frac{1}{\sqrt{t-\tau}} \text{Im} \left[A_3(\lambda_1) K_{II}^F(\lambda_1 \cos \theta + \alpha \sin \theta) \frac{\partial \lambda_1}{\partial \tau} \right] d\tau$$

$$+ \int_{T_{H_0}}^{t} \frac{1}{\sqrt{t-\tau}} \text{Im} \left[A_4(\lambda_2) K_{II}^F(\lambda_2 \cos \theta + \beta \sin \theta) \frac{\partial \lambda_2}{\partial \tau} \right] H(br_0 - \tau) d\tau$$

$$+ \int_{br_0}^{t} \frac{1}{\sqrt{t-\tau}} \text{Im} \left[A_4(\lambda_3) K_{II}^F(\lambda_3 \cos \theta + \beta \sin \theta) \frac{\partial \lambda_3}{\partial \tau} \right] d\tau, \qquad (3)$$

where

$$A_1(\lambda) = -\frac{(b^2 - 2\lambda^2)^2}{R} \cos^2 \theta - \frac{(b^2 - 2\alpha^2)(b^2 - 2\lambda^2)}{R} \sin^2 \theta + \frac{2\alpha\lambda(b^2 - 2\lambda^2)}{R} \sin 2\theta,$$

$$A_2(\lambda) = \frac{4\alpha\beta\lambda^2}{R} \sin^2 \theta - \frac{4\alpha\beta\lambda^2}{R} \cos^2 \theta - \frac{2\alpha\lambda(b^2 - 2\lambda^2)}{R} \sin 2\theta,$$

$$A_3(\lambda) = \frac{(b^2 - 2\lambda^2)(a^2 - 2\lambda^2)}{R} \sin 2\theta + \frac{2\alpha\lambda(b^2 - 2\lambda^2)}{R} \cos 2\theta,$$

$$A_4(\lambda) = \frac{4\alpha\beta\lambda^2}{R} \sin 2\theta - \frac{2\alpha\lambda(b^2 - 2\lambda^2)}{R} \cos 2\theta, \qquad (4)$$

$$\alpha = \sqrt{a^2 - \lambda^2}, \quad \beta = \sqrt{b^2 - \lambda^2}, \quad R = (b^2 - 2\lambda^2)^2 + 4\alpha\beta\lambda^2$$

$$\lambda_1 = \frac{\tau}{r_0} \cos \theta_0 + i\sqrt{\frac{\tau^2}{r_0^2} - a^2} \sin \theta_0, \quad \lambda_2 = \frac{\tau}{r_0} \cos \theta_0 - \text{sgn}(l)\sqrt{b^2 - \frac{\tau^2}{r_0^2}} \sin \theta_0,$$

$$\lambda_3 = \frac{\tau}{r_0}\cos\theta_0 + i\sqrt{\frac{\tau^2}{r_0^2} - b^2\sin\theta_0}, \quad T_{H_0} = r_0(a|\cos\theta_0| + \sqrt{b^2 - a^2}\sin\theta_0),$$

$$r_0^2 = l^2 + h^2, \quad \cos\theta_0 = \frac{l}{r_0}, \quad \sin\theta_0 = \frac{h}{r_0},$$

$$\operatorname{sgn}(l) = 1, \qquad l > 0,$$

$$= -1, \quad l < 0.$$

We consider then a concentrated point impact loading of Sine function $\sigma_0\sin(\pi t/T_0)$ applying at the position $\bar{x}_1 = l$ at time $t = 0$ and with duration time T_0. The corresponding stress intensity factor histories can be determined exactly by linear superposition of the Green function $K_{I,II}^G$. Suppose that at time $t = 0$, the pulse is applied suddenly and the magnitude of the pressure increases according to the function $f(t)$, the corresponding stress intensity factors can be obtained by the superposition method

$$K_{I,II}(t) = \int_0^t \frac{df(\tau)}{d\tau} K_{I,II}^G(t - \tau)d\tau. \tag{5}$$

In this study, $f(\tau) = \sigma_0\sin(\pi\tau/T_0)$ and $0 < \tau < T_0$, here the stress intensity factors can be expressed as follows

$$K_{I,II}(t) = \int_0^t \sigma_0\frac{\pi}{T_0}\cos\frac{\pi\tau}{T_0}K_{I,II}^G(t - \tau)d\tau, \qquad \text{for} \quad 0 \le t \le T_0,$$

$$\tag{6}$$

$$K_{I,II}(t) = \int_0^{T_0} \sigma_0\frac{\pi}{T_0}\cos\frac{\pi\tau}{T_0}K_{I,II}^G(t - \tau)d\tau, \qquad \text{for} \quad T_0 < t.$$

In mixed mode experiments, it is usually observed that crack extension takes place at an angle with respect to the original crack. When the mixed stress intensity factors have been obtained, the criteria of maximum circumferential tensile stress proposed by Erdogan and Sih [10] is introduced to examine the crack growth direction. The maximum circumferential tensile stress criterion postulates that the crack will grow in a direction determined by the condition that when the circumferential tensile stress within the asymptotic field is at a maximum, the angle θ_2 $(= \theta - \theta_c)$ between the crack line and the direction of crack growth satisfies

$$\sin\theta_2 K_I + (3\cos\theta_2 - 1)K_{II} = 0. \tag{7}$$

NUMERICAL RESULTS

For numerical calculation of the mixed dynamic stress intensity factors, Poisson's ratio ν is assumed to be equal to 0.25. In this case, the ratios of the slownesses are $b = \sqrt{3}a$ and $c = 1.88a$. The point impact loadings of Sine function $\sigma_0 \sin(\pi t/T_0)$ are applied at the positions $l = -2h$, 0 and $2h$. The inclined angles θ of the crack are 0° and 90°. There are five combinations of loading position and inclined crack angle. The duration period is chosen for $T_0/ah = 1$ and 5. The dynamic stress intensity factors for the time interval of interest are shown in Figs 2 to 5. The crack propagation angle θ_c is shown in Fig. 6.

We consider first for the case of an applied loading at the left (right) hand side of crack tip which makes $l = -2h$ ($l = 2h$). As indicated previously, the results obtained in this study are valid for the time period in which waves generated at the impact point and its diffractions from the notch end have not yet returned to the crack tip, that is $\sqrt{5} \le t/ah \le \sqrt{5}+2$. There are incident P, H, and S waves from the applied loading point on the semi-infinite surface. The normalized arrival times t/ah of incident wave fronts at the crack tip are $\sqrt{5}$ for the P wave, $2+\sqrt{2}$ for the H wave, and $\sqrt{15}$ for the S wave. If loading is applied directly above the crack tip (i.e. $l = 0$), the valid time interval in this case is $1 \le t/ah \le 3$. The normalized arrival times are 1 for the P wave and $\sqrt{3}$ for the S wave. It is shown that the arrival of the stress waves instantaneously places the crack edge in compression for vertical crack $l = 2h$ and horizontal crack $l = 0$. The other case are all suffered in tensile state during the whole valid time history. Unlike the rapid change for short duration period case, the long duration period $(T_0/ah = 5)$ behaves a smooth change of the stress intensity factor.

Without considering the feasibility when the compression state is generated in the near tip, the crack propagation angle θ_c predicted by maximum circumferential tensile stress criteria for the problem analyzed here is shown in Fig. 6. It can be concluded that the most dangerous case will be one of a loading applied directly above the crack tip. Because, in this case, the crack propagation angle θ_c is always greater than zero, meaning that the crack will propagate toward the half-plane surface.

CONCLUSION

In the previous sections, a subsurface inclined crack subjected to impact loading of Sine function dependence on a half-plane surface is investigated. The net result of this loading will induce a mixed mode field at the crack tip. Exact mixed mode I and II stress intensity factors are obtained in an explicit form. The exact solution to this configuration can provide a valuable examination in the experimental analysis or numerical methods such as the finite element, finite difference or boundary element methods in solving more realistic consideration. The maximum circumferential tensile stress criteria is used to predict the direction of crack propagation. It is found in this study that the crack will extrude out of the half-plane surface if the impact loading is applied in the region above the crack tip.

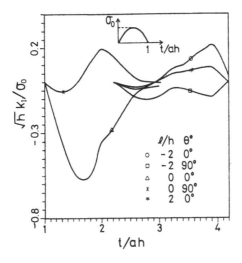

Fig. 2 Stress intensity factors K_I for duration period $T_0 = 1$.

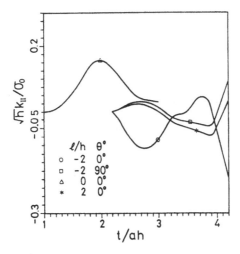

Fig. 3 Stress intensity factors K_{II} for duration period $T_0 = 1$.

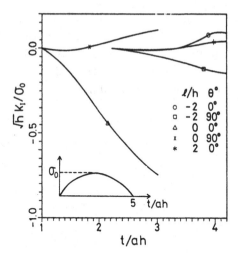

Fig. 4 Stress intensity factors K_I for duration period $T_0 = 5$.

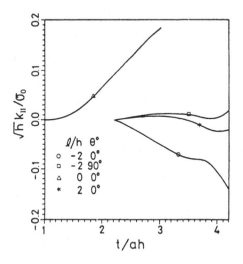

Fig. 5 Stress intensity factors K_{II} for duration period $T_0 = 5$.

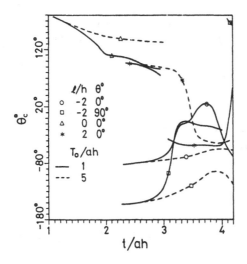

Fig. 6 Prediction of crack propagation direction.

ACKNOWLEDGMENT

The work described here was supported by the National Science Council (Republic of China), through grant NSC 79-0401-E-002-36 to National Taiwan University. This research support is gratefully acknowledged.

REFERENCES

1. Tsai, C. H. and Ma, C. C., The stress intensity factor of a subsurface inclined crack subjected to dynamic impact loading. submit to ASME Journal of Applied Mechanics, 1991.

2. Brock, L. M., Jolles, M. and Schroedl M., Dynamic impact over a subsurface crack : applications to the dynamic tear test. ASME Journal of Applied Mechanics, 1985, 52 287-290.

3. Freund, L. B., The stress intensity factor due to normal impact loading of the faces of a crack. International Journal of Engineering Science, 1974, 12, 179-189.

4. Brock, L. M., Shear and normal impact loadings on one face of a narrow slit. International Journal of Solids and Structures, 1982, 18, 467-477.

5. Brock, L. M., Stresses in a surface obstacle undercut due to rapid indentation. Journal of Elasticity, 1984, 14, 415-424.

6. Ma, C. C. and Hou, Y. C., Transient analysis for antiplane crack subjected to dynamic loadings. ASME Journal of Applied Mechanics, 1991, in press.

7. Ma, C. C. and Hou, Y. C., Theoretical analysis of the transient response for a stationary inplane crack subjected to dynamic impact loading. International Journal of Engineering Science, 1991, in press.

8. Lee, Y. J. and Freund, L. B., Fracture initiation due to asymmetric impact loading of an edge cracked plate. ASME Journal of Applied Mechanics, 1990, 57, 104-111.

9. Tsai, C. H. and Ma, C. C., Transient analysis of a semi-infinite crack subjected to dynamic concentrated forces. submit to ASME Journal of Applied Mechanics, 1991.

10. Erdogan, F. and Sih, G. C., On the crack extension in plates under plane loading and transverse shear. Journal of Basic Engineering, 1963, 85, 519-527.

11. Lamb, H., On the propagation of tremors over the surface of an elastic solid. Philosophical Transactions of the Royal Society (London), 1904, A 203, 1-42.

Numerical Modeling of Dynamic Failure of Elastic-Plastic Materials

by Godunov's Method

XIAO LIN[+] and JOSEF BALLMANN[‡]

[+]Post-doctoral fellow of the AvH Foundation and [‡]Professor
Lehr- und Forschungsgebiet für Mechanik der RWTH Aachen
Templergraben 64, 5100 Aachen, FRG

Abstract — In this paper, we deal with the application of Godunov's method for the computational modeling of dynamic fracture. First the numerical scheme is presented. The material is assumed to obey Hooke's law in the elastic region, and to have a linearly work-hardening behaviour in the plastic region with the von Mises yielding condition. Three numerical examples are presented to demonstrate the capability of the method, and to show the time history of forthcoming plastic zones and their growing size in a plate with a crack or a rectangular hole when the plate is loaded by a given dilatational pulse. Varied parameters are the work-hardening parameter and the strength of the loading pulse

1. INTRODUCTION

The dynamic failure of materials related to the stress waves is one of the most important topics in the study of dynamic fracture. Many researchers have been working on elastic materials with cracks, and got a lot of exciting achievements. Thau and Lu[1] studied the transient stress intensity factors caused by a dilatational wave. Sih et al. [2] analyzed similar problems using an integral formulation. Freund [3] has given the exact solution of the stress intensity factor for a semi-infinite crack in an infinite plate. There exist also many publications in the field of numerical computation, e.g. Chen [4] who investigated crack problems by a finite-difference method (HEMP code). Fan and Hahn [5] dealt with a solution using the boundary integral equation method, where the time was eliminated by Laplace transformation. Recently, Nicholson and Mettu [6] have developed a time domain boundary integral equation method to treat the related problem.

Major difficulties arise in dealing with elastic-plastic materials since they behave differently in the loading and unloading phase. Godunov's method has been demonstrated to be very successful in dealing with shock problems in fluid dynamics by many authors, e.g. [7]. This method can also be used for solids. In this paper, we will present first results of our research about using Godunov's method for the modeling of dynamic failure in the two-dimensional dynamical stress problems of elastic-plastic plates.

In this method, a computational grid is introduced which subdivides the plate, i.e. the spatial part of the solution domain, into a limited number of cells. The components of stresses, strains and velocities are assumed to be constant in each cell. Since they have different values in different cells, there a Riemann problem can be formulated at every interface of the cells. Solving these Riemann problems in a step by step procedure the integration with respect to time can be carried out to give the time rates of changes for the stresses, strains and velocities. To get the informations about the materials failure the elastic-plastic material law is involved.

The paper concentrates on two-dimensional plates with a crack or a notch. The constitutive equations are assumed to be of a linearly work-hardening form and for yielding processes in the plastic region the von Mises yield condition is adopted. For unloading linear elastic behaviour, i.e. Hooke's law, is assumed. Computational results for three examples are presented to demonstrate the efficiency of the method and to show the formation of the plastic region in a plate loaded by a tension shock.

2. GOVERNING EQUATIONS AND GODUNOV'S SCHEME

Let us consider a two-dimensional plate with a propagating elastic-plastic plane wave. The governing equations in matrix form are

$$\underset{\sim}{w}_t = \underset{\sim}{F}_x + \underset{\sim}{G}_y \tag{2-1}$$

where $\underset{\sim}{w} = (u, v, \varepsilon_1, \varepsilon_2, \gamma)^t$, $\underset{\sim}{F} = (\sigma_1/\rho, \tau/\rho, u, 0, v)^t$, $\underset{\sim}{G} = (\tau/\rho, \sigma_2/\rho, 0, v, u)^t$; u and v are the particle velocities in x- and y-direction, σ_1, σ_2 and τ are the components of the plane stress tensor, ε_1, ε_2 and γ the components of the related strain tensor. When the material is an isotropic work-hardening one, the constitutive relations are

$$\left.\begin{aligned}
d\varepsilon_1 &= (1/E + Hf_1^2)d\sigma_1 + (\nu/E + Hf_1 f_2)d\sigma_2 + Hf_1 f_3 d\tau \\
d\varepsilon_2 &= (\nu/E + Hf_1 f_2)d\sigma_1 + (1/E + Hf_2^2)d\sigma_2 + Hf_2 f_3 d\tau \\
d\gamma &= Hf_1 f_3 d\sigma_1 + Hf_2 f_3 d\sigma_2 + (1/G + Hf_3)d\tau
\end{aligned}\right\} \tag{2-2a}$$

or, in a matrix form

$$d\underset{\sim}{\varepsilon} = \underset{\approx}{A}(\sigma) \, d\underset{\sim}{\sigma} \tag{2-2b}$$

There

$$f_1 = (\sigma_1 - \frac{\sigma_2}{2})/\vartheta, \qquad f_2 = (\sigma_2 - \frac{\sigma_1}{2})/\vartheta, \qquad f_3 = \vartheta\,\tau \tag{2-3}$$

with Young's modulus E, the shear modulus G and Poisson's ratio ν, the hardening modulus H is determined by the von Mises yield condition

$$H = \frac{1}{\varkappa^2}\left[\frac{1}{g(\vartheta\varkappa)} - \frac{1}{E}\right] \tag{2-4}$$

$$\varkappa^2 = (\sigma_1^2 + \sigma_2^2 - \sigma_1\sigma_2)/\vartheta^2 + \tau^2 \tag{2-5}$$

where $\vartheta = \sqrt{3}$ and $g(\sigma)$ is the slope of the stress-strain curve for a simple tension test, namely

$$g(\sigma) = d\sigma/d\varepsilon \tag{2-6}$$

So, H may have two values: Starting in the elastic region with an incremental step from $\underset{\sim}{\sigma}$ to $\underset{\sim}{\sigma} + d\underset{\sim}{\sigma}$ still remaining in the elastic region, $g(\vartheta\varkappa) = E$, and then $H = 0$; if $\underset{\sim}{\sigma} + d\underset{\sim}{\sigma}$ tends to go outside of the yield surface determined by $\underset{\sim}{\sigma}$, $H \neq 0$. For linear isotropic work-hardening materials $g(\vartheta\varkappa) = E_p$, where E_p is constant, H has a simple form.

For the numerical integration of eq. (2-1) the solution domain is subdivided into finite cells by two families of lines, e. g. parallel to the cartesian coordinates. If necessary grid cells adjusted to curved boundaries or in elastic-plastic domains an adaptive grid may be introduced. In general case the grid cells for the plane problem shall have four boundaries. Using a cartesian grid the first-order Godunov scheme for a local time step in eq. (2-1) will have the following integral form

$$\underset{\sim}{w}^{n+1} = \underset{\sim}{w}^n + \frac{\Delta t}{\Delta A} \oint \underset{\sim}{F}\,dy - \underset{\sim}{G}\,dx \tag{2-7}$$

where ΔA is the area of the cell and Γ its boundary. For a cell with indices ij the above formula can be rewritten in the following algebraic form

$$\underset{\sim}{w}_{ij}^{n+1} = \underset{\sim}{w}_{ij}^n + \frac{\Delta t}{\Delta x}\left(\underset{\sim}{F}_{i+1/2,j}^n - \underset{\sim}{F}_{i-1/2,j}^n\right) + \frac{\Delta t}{\Delta y}\left(\underset{\sim}{G}_{i,j+1/2}^n - \underset{\sim}{G}_{i,j-1/2}^n\right) \tag{2-8}$$

The stress components are updated by integration of eq. (2-2) in the space of strains,

$$\underset{\sim}{\sigma}^{n+1} = \underset{\sim}{\sigma}^n + \int_{\underset{\sim}{\varepsilon}^n}^{\underset{\sim}{\varepsilon}^{n+1}} \underset{\sim}{A}^{-1}(\underset{\sim}{\sigma})\,d\underset{\sim}{\varepsilon} \tag{2-9}$$

Therein the local increments $d\varepsilon_1$, $d\varepsilon_2$ and $d\gamma$ are used, since the complete path of elastic-plastic loading is not unique. After updating the stress components, the new yield surface can be determined by eq. (2-5).

3. RIEMANN PROBLEM

Like in fluid dynamics, the fluxes F and G through the cell boundaries shall be treated by coupled one-dimensional Riemann problems. Therefore, first the one-dimensional Riemann problem is considered, where the x-axis is supposed to be in the normal direction of the boundary. Then in eq. (2-1) all flux functions have vanishing gradients in y direction, and all components related to this direction are to be ignored. And in eq. (2-2) $d\varepsilon_2=0$. Then eq. (2-1) and eq. (2-2) are written as a system of first order pde's,

$$\rho\frac{\partial u}{\partial t}=\frac{\partial \sigma_1}{\partial x}$$

$$\rho\frac{\partial v}{\partial t}=\frac{\partial \tau}{\partial x}$$

$$(1/E+Hf_1^2)\frac{\partial \sigma_1}{\partial t}+(\nu/E+Hf_1 f_2)\frac{\partial \sigma_2}{\partial t}+Hf_1 f_3\frac{\partial \tau}{\partial t}=\frac{\partial u}{\partial x}$$

$$(\nu/E+Hf_1 f_2)\frac{\partial \sigma_1}{\partial t}+(1/E+Hf_2^2)\frac{\partial \sigma_2}{\partial t}+Hf_2 f_3\frac{\partial \tau}{\partial t}=0$$

$$Hf_1 f_3\frac{\partial \sigma_1}{\partial t}+Hf_2 f_3\frac{\partial \sigma_2}{\partial t}+(1/G+Hf_3^2)\frac{\partial \tau}{\partial t}=\frac{\partial v}{\partial x}$$

(3-1)

The system of (3-1) is hyperbolic and its characteristic wave speeds in the elastic region ($H=0$) are

$$c_1=c_0/\sqrt{1-\nu^2}, \quad c_2=\sqrt{G/\rho}, \quad 0 \tag{3-2}$$

where $c_0 = \sqrt{E/\rho}$, and in the plastic region c has also three values: c_f, c_s and 0, which are related to the stress state $(\sigma_1, \sigma_2, \tau)$ and satisfy

$$0\le c_s\le c_2\le c_f\le c_1 \quad . \tag{3-3}$$

A simple wave solution can be found, which results in the following stress changing path in $(\sigma_1, \sigma_2, \tau)$ space

$$\frac{d\sigma_1}{\left(\frac{c}{c_0}\right)^2\left[\left(\frac{c}{c_2}\right)^2-1\right](f_1+\nu f_2)}=\frac{d(\sigma_2-\nu\sigma_1)}{\left[\left(\frac{c}{c_1}\right)^2-1\right]\left[\left(\frac{c}{c_2}\right)^2-1\right]f_2}=\frac{d\tau}{\left(\frac{c}{c_0}\right)^2\left[\left(\frac{c}{c_1}\right)^2-1\right]f_3}$$

(3-4)

and the stress-velocity relations in the direction defined by $dx=cdt$ in the x,t- plane are

$$du=-\frac{d\sigma_1}{\rho c}, \quad dv=-\frac{d\tau}{\rho c} \tag{3-5}$$

It is evident that the stress loading path described by eq. (3-4) will be a curve in the space of stresses σ_1,σ_2,τ. However, for any given initial and final states of (σ_1,σ_2,τ) there exists always a loading path to connect them. An interesting fact is that from (3-4) one derives for the characteristic wave speed $c=0$

$$d\sigma_1=0, \qquad d\sigma_2=1, \qquad d\tau=0 \qquad\qquad (3\text{-}6)$$

It turns out that the stress paths with $c=c_1, c_f, c_2$ and c_s will first take σ_1 and τ to the final state, and then σ_2. This fact permits to construct a Riemann solution by the linear tangential approximation method[8] for eq. (3-5), where only (σ_1, τ) and (u,v) at both sides of the boundary are required to be equal. There is no postulate for σ_2, the last part of loading for changing σ_2 can be ignored, and its values can be different there.

On the other hand, since the stress path is always a curve in the 3-D space of stresses, it is not only time consuming, but also difficult to find an exact path to process the iteration in the Riemann solver. To overcome this difficulty, we assume here that the stress loading paths related to $c=c_1$, c_f, c_2 and c_s can be restricted to a plane given by

$$\sigma_2 - \nu\sigma_1 = const \qquad\qquad (3\text{-}7)$$

where the constant at the r.h.s. follows from the initial loading state. For example, suppose a loading path related to $c= c_f$, c_s, 0, then the projection of first and second parts of exact paths on (σ_1, σ_2) plane will be curves, while that of simplifying will be a straight line, see Fig.1.

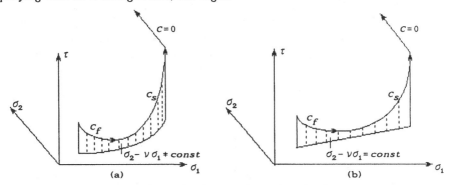

Fig.1. Elastic-plastic loading paths in $(\sigma_1, \sigma_2, \tau)$ stress space
(a) True path; (b) Simplifying path

Next, we give a simple outline of the construction of a Riemann solver by the linear tangential approximation method. Suppose the initial state variables in two elements are known, see Fig.2. We want to calculate the result at the boundary between the domains L and R for $t>0$. The iteration steps are as follows:

(i) Suppose the result in m-th iteration step $(\sigma_1, \sigma_2, \tau)^{(m)}$ is known. (In the first step it can be taken as the elastic result).

(ii) Find a loading path from $(\sigma_1, \sigma_2, \tau)_L$ to $(\sigma_1, \sigma_2, \tau)^{(m)}$, namely, to work out the sound speed values $c=c(\underset{\sim}{\sigma})$ along the curve connecting the point L to (m). As mentioned above, this shall be done in the plane defined by eq. (3-7).

(iii) Integrate eq. (3-5) from $(\sigma_1, \sigma_2, \tau)_L$ to $(\sigma_1, \sigma_2, \tau)^{(m)}$ to get $(u,v)_L^{(m)}$.

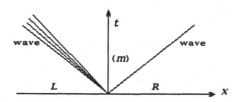

Fig.2. Riemann problem in (x,t) space

(iv) For the right side, $(u,v)_R^{(m)}$ is determined in the analogous way.

(v) If $|u_L^{(m)} - u_R^{(m)}|$ and $|v_L^{(m)} - v_R^{(m)}|$ keep within a certain limit, the iteration can be stopped. If not, the $(\sigma_1, \sigma_2, \tau)^{(m+1)}$ are assumed at the cross points of the tangential lines of eq. (3-5). Then return to (i).

4. EXAMPLES

We want to show that the Godunov's method can be used in solid mechanics.

4-1. As the first example, we consider a simple elastic problem which was a testcase also discussed by Raatschen[9]. He considered a rectangular plate with free boundaries where a shock-like pulsating load $P(t)$ was applied at the right side of the boundary. For our case, the zoning of the plate and the function $P(t)$ are shown in Fig.3. The plate has the following parameters: E=21000 N/mm^2, v =0.2, ρ =7.85 gm/cm^3, and the grid size $\Delta x = \Delta y$=1 mm. The Courant number is taken $c_1 \Delta t / \Delta x$=1. Two views of the σ_x distribution at different times are shown in Fig.4. The result is in good agreement with that of Raatschen calculated by the method of characteristics.

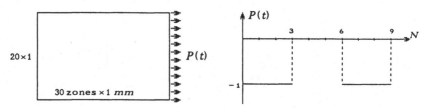

Fig.3. Zoning of plate and $P(t)$ curve

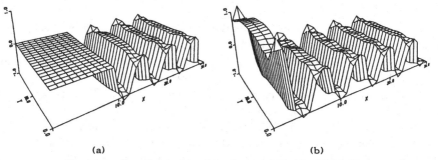

(a)   (b)

Fig.4 σ_x distribution in the plate. (a) N=20, (b) N=33

4-2 As the second example, we consider the forming of a plastic zone at a crack when the plate is dynamically loaded by a dilatational pulse. The zoning of the plate is shown in Fig.5, where only half of the plate is taken into account because of symmetry to the x-axis. The numbers N_1, N_2 of grid cells in x and y direction are chosen big enough to avoid influences of wave reflections from the outer boundaries interfering with the stress field at the crack in the considered intervall of time.

As a test for the capability of the present method an elastic case is treated first to make a comparison with analytical results from Thau, Lu [1] and Kim [10]. The solid line and the dotted line of Fig. 6 show the time history of the stress intensity factor at the crack tip from these authors. In the present calculations the data of Kim were used for the material parameters. In Fig.6 the curve indicated by circles shows the numerical results where the stress intensity factor was deduced from the crack op ning displacements, which were calculated by integration of the particle velo cities. The results are in good agreement with those derived analytically.

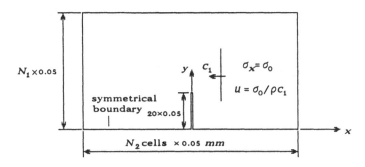

Fig.5. Zoning of the crack problem

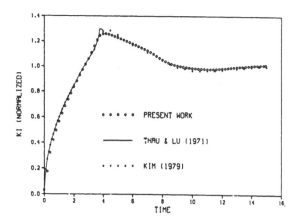

Fig.6 Normalized dynamic stress intensity factor at a crack (elastic case)

Now, a problem of an elastic-plastic crack is dealt with and some graphical illustrations will be presented to show how the plastic zone is formed near a crack tip. The material is taken as the 304 stainless steel, where, $\rho=7.85\ gm/cm^3$, $c_L=5.77\ km/s$ (speed of P-wave), $c_2=3.12\ km/s$. From c_L and c_2 we know $c_1=5.249\ km/s$, $\nu=0.2934$. The initial yield surface is given by $x_0=\sigma_s/\vartheta=0.1/\sqrt{3}$. Since the Hugoniot yield stress in a plate is $\sigma_h=\sigma_s/\sqrt{(1-\nu+\nu^2)}$, we choose a limited initial dilatational pulse, $\sigma_o\leq\sigma_s<\sigma_h$, to insure the initial state being elastic. The initial data belong to a give n wave field with a shock-like wave front. The computation is started just when the wave front has reached the crack.

Calculations were carried out for different linearly work-hardening materials, where the length of the half crack A = 1 mm and the parameter $c_p=\sqrt{(E_p/\rho)}$ comes from eq. (2-6). Fig.7 shows the elastic/plastic fronts $N=120$ time steps after the passing of the wave front which corresponds to a travel distance of 6 mm. The critical condition for the elastic/plastic fronts is $x=1.05x_0$, i.e. inside of the plastic zone $x\geq1.05x_0$. Fig.8 shows the growth of the yielding zones with time. The different loading pulses will have also different influences on the yield history. Fig.9 shows the elastic-plastic boundaries, i.e. contour lines with $x=1.05x_0$ for different values of the pulse σ_0, and the related yield area histories can be seen in Fig.10.

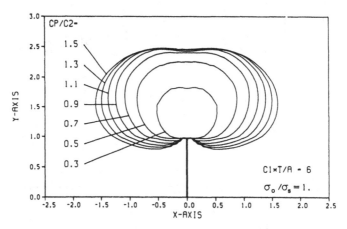

Fig.7 Elastic-plastic fronts ($x=1.05x_0$) for different linearly work-hardening materials after the diffraction of a mode I dilatational pulse at the crack tip (positive half plane).

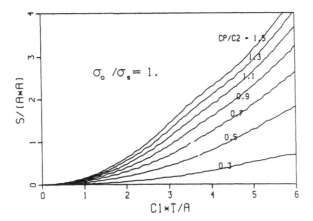

Fig.8 Time history of the yielding area at the crack for different
linearly work-hardening materials.

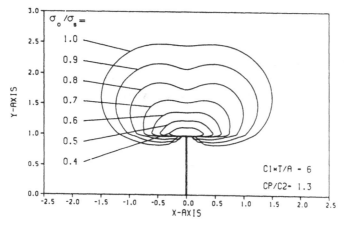

Fig.9 Elastic-plastic fronts ($x=1.05x_0$) for different loading pulses
after the diffraction of a mode I dialational pulse at the crack
tip (positive half plane).

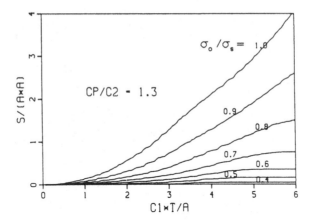

Fig.10 Time history of the yielding area at the crack for different
loading pulses.

4-3 The last example concerns the diffraction of a dilatational pulse at a
rectangular hole, see Fig.11. The material parameters are taken from 304 stainless
steel, as in the example 4-2. But the initial time $t=0$ is chosen as the instant
when the wave front just has reached the right corner of the hole, i.e. $x=1$ mm.
For this calculation the cells at the boundary of the hole are chosen to have
triangular form. The computational results can be seen from Fig.12-15, where
again the elastic-plastic fronts are the contour lines $x=1.05x_0$, and the varied
parameters are work-hardening factor and the strength of the loading pulse.

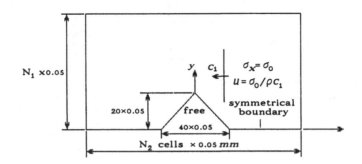

Fig.11 Zoning of the rectangular hole problem

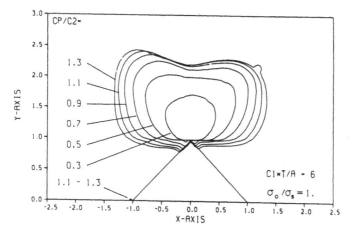

Fig.12 Elastic-plastic fronts ($x=1.05x_0$) for different linearly work-
hardening materials after the diffraction of a dilatational pulse
at the corner of a rectangular hole (positive half plane).

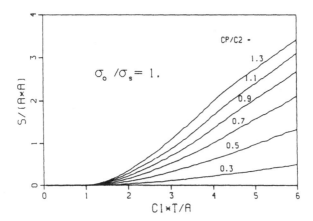

Fig.13 History of the yielding area at the corner of a rectangular
hole for different linearly work-hardening materials.

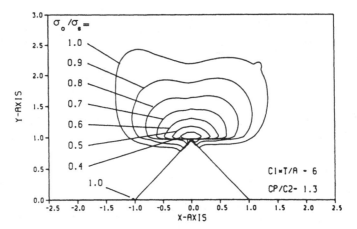

Fig.14 Elastic-plastic fronts (x=1.05x₀) for different loading pulses
after the diffraction of a dilatational pulse at the corner of
a rectangular hole (positive half plane).

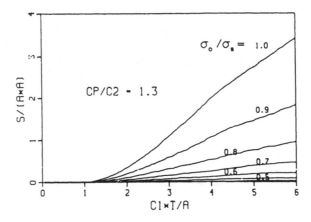

Fig.15 Time history of the yielding area at the corner of a rectangular
hole for different loading pulses.

5. CONCLUSIONS

This paper concerns the application of Godunov's method to compute elastic-plastic wave problems. Though there are always two wave speeds in solids, they can easily be dealt with by Godunov's method. There are two difficulties to overcome. One is the construction of a Riemann solver, for which a linear tangential approximation method can be used. The other is the updating of the stress components which can be treated by the proportional deformation loading path. Examples are presented to demonstrate the correct description of finite amplitude wave propagation and the capability to simulate the elastic- plastic deformation.

From the plastic region formed at a crack tip or at a corner of a rectangular hole one can see, the lower the value of c_p is, the smaller the plastic area will be. Of course the strains will be greater in yielding domains for smaller values of c_p. Therefore, the materials are more receptive to large local deformations.

Acknowledgements: The authors wish to express their gratitude to the Alexander von Humboldt Foundation for the research fellowship of Xiao Lin, and to the East China Institute of Technology for the providing the opportunity to realize this cooperation.

REFERENCES

1. S. A. Thau and T. H. Lu, Transient stress intensity factors for a finite crack in an elastic solid caused by a dilatational wave, *Int. J. Solids Structure*, 7, 731- 750 (1971).

2. G. C. Sih, G. T. Embley and R. S. Ravera, Impact response of a finite crack in plane extension. *Int. J. Solids Structure*, **8**, 977-993 (1972).

3. L. B. Freund, The stress intensity factor due to normal impact loading of the faces of a crack. *Int. J. Engng Sci.* **12**, 179-189 (1974)

4. Y. M. Chen, Numerical computation of dynamic stress intensity factors by a lagrangian finite-defference method (the HEMP code), *Engineering Fracture Mechanics* **7**, 653-660 (1975)

5. T. Y. Fan and H. G. Hahn, An application of the boundary integral equation method to dynamic fracture mechanics, *Engineering Fracture Mechanics*, **21**, 307-313 (1985).

6. J. W. Nicholson and S. R. Mettu, Computation of Dynamic stress intensity factors by the time domain boundary integral equation method-I. Analysis; -II. Examples. *Engineering Fracture Mechanics* **31**, 759-782 (1988).

7. S. K. Godunov, A finite difference method for the numerical computation of discontinuous solutions of the equations of fluid dynamics, *Mat. Sb.* **47**, 271 (1959)

8. B. van Leer, Towards the ultimate conservative difference scheme., V. a second -order sequel to Godunov's Method, *J. of Computational Physics*, **32**, 101-136 (1979)

9. H.-J. Raatschen, Ein Bicharakteristikenverfahren zur Berechnung von Spannungswellen in krummlinig berandeten Scheiben, doctor thesis, Lehr und Forschungsgebiet Mechanik der, RWTH Aachen, 1986.

10. K.S. Kim, Dynamic propagation of a finite crack, *Int. J. Solids Structructures*, **15**, pp685-699 (1979).

TRANSIENT VISCOELASTODYNAMIC RESPONSE OF CRACKED BODIES

H.G. GEORGIADIS
Physics and Mechanics Division, Box 422, School of
Technology, The Aristotle University of Thessaloni-
ki, 540 06 THESSALONIKI, GREECE

ABSTRACT

The transient elastodynamic stress intensity factor was determined for a cracked linearly viscoelastic body under impact.
The body is of infinite extent and contains a finite length
crack. A plane or a torsional step pulse strikes the crack
and stress wave diffraction takes place. The solution was obtained from Sih's elastodynamic analysis through the correspondence principle and the use of the Dubner-Abate-Crump Laplace
transform inversion technique. Numerical results were given
for the standard linear solid.

INTRODUCTION

In dynamic fracture mechanics problems, one is usually interested in obtaining the stress intensity factor (SIF) as a function of time and of geometrical and material parameters. For
stationary cracks the latter information is desirable because of
the well-known *dynamic overshoot* of the SIF in impacted bodies.
Determining the SIF dynamic overshoot answers to the question
whetther the fracture toughness of the material will be exceeded and catastrophic crack propagation will follow. For *moving*
cracks the information about dynamic SIF when combined with an
appropriate fracture criterion may answer to the question of
crack *arrest* or *propagation*.
   The specific problems considered here are the plane and
torsional impact of a *viscoelastic* plate containing a central
crack (see in [1,2] for brief reviews on viscoelastodynamic
crack problems).   The solution is obtained by utilizing integral transforms, a numerical solution of integral equations
and a *numerical* technique for the Laplace-transform *inversion*
due to Dubner and Abate [3] and Crump [4] (DAC technique).
One of the goals of the paper is to emphasize the suitability
of the DAC technique for inertial viscoelastic problems. In

fact, we propose the replacement of the old-fashioned *Orthogonal Polynomials* method by the DAC method for reasons explained below.

Indeed, the most widely utilized Laplace-transform inversion method in mixed BVPs in elastodynamics is the orthogonal polynomials method. This was introduced by Papoulis [5] and modified by Miller and Guy [6] and Bellman et al. [7] . The research group of G.C. Sih has made an extensive use of the Miller/Guy technique for crack problems [8].

However, the orthogonal polynomials technique suffers from serious drawbacks. First, there is a "non-uniqueness" in the results since these are strongly dependent upon the choice of the parameters β and δ (in the notation of Miller and Guy [6]). Thus, in order to adjust these parameters, independent analytical results should be available (e.g. already known static results can be used for large times and some classical solutions by Baker [9] , Kostrov [10] and Thau and Lu [11] for small time). However, in more advanced fracture mechanics problems, such as viscoelastic or composite-bodies situations, analytical results are not usually available. So, we need a more reliable inversion method.

Second, by increasing the number of terms N in a certain series expansion, a matrix becomes more and more ill-conditioned leading to instability. Thus, one gets inaccuracy by using few terms and instability by using a lot of terms.

Third, a slight error in the values of the transformed function has a great effect on the values of the original function (see in Cost [12] for a relative investigation).

In a recent survey by Davies and Martin [13] , over twenty methods for Laplace-transform inversion were tested and compared. One of the best techniques was found to be the DAC technique. We have adopted this method here and we recommend its further use for mixed BVPs in elastodynamics.

Finally, although our procedure is fairly general, numerical results were obtained here only for the case of the *standard linear solid*. The numerical study revealed some interesting features of the problem. For instance, several combinations of material constants may have as a result severe dynamic SIF overshoots or *unexpected* overshoots. Furthermore, a purely elastic analysis based on the steady-state (long time) material response may give *conservative* estimates of the SIF values. Therefore, our results suggest the necessity of an analysis similar to the present one for other viscoelastodynamic crack problems too.

BASIC RELATIONS AND BOUNDARY CONDITIONS

Two separate geometries were treated.

In the first case, the body was in plane strain and the finite-length crack was subjected to a step-type loading. Then, in respect to a Cartesian system (x_1, x_2), the equations of motion are written as [14].

$$\frac{\partial \sigma_{ij}}{\partial x_j} = \rho \frac{\partial^2 u_i}{\partial t^2} \quad , \tag{1}$$

where ρ is the mass density and the indices i,j, take the values 1 and 2. The stress-strain relations for this type of material can be written in a *convolution* form as

$$s_{ij} = 2 \int_0^t G_1(t-z) \frac{d\, e_{ij}(z)}{dz} dz \quad , \qquad (2.1)$$

$$\sigma_{kk} = \int_0^t G_2(t-z) \frac{d\, \varepsilon_{kk}(z)}{dz} dz \quad , \qquad (2.2)$$

where

$$s_{ij} = \sigma_{ij} - \frac{1}{3} \delta_{ij} \sigma_{kk} \quad , \qquad s_{ii} = 0 \quad , \qquad (3.1)$$

$$e_{ij} = \varepsilon_{ij} - \frac{1}{3} \delta_{ij} \varepsilon_{kk} \quad , \qquad \varepsilon_{ii} = 0 \quad , \qquad (3.2)$$

and ε_{ij} is the strain tensor, k takes the values 1 and 2, whereas $G_1(t)$ and $G_2(t)$ are the relaxation functions of the material.

The crack of length 2a contained in the viscoelastic body is acted upon by *impact* normal tractions of amplitude σ_0. The boundary conditions are written as (see Fig. 1)

$$\sigma_{22}(x_1,0^+,t) = -\sigma_0 H(t) \quad , \qquad -a<x_1<a \quad , \qquad (4.1)$$

$$\sigma_{12}(x_1,0^+,t) = 0 \quad , \qquad -\infty<x_1<\infty \quad , \qquad (4.2)$$

$$u_2(x_1,0^+,t) = 0 \quad , \qquad |x_1|>a \quad , \qquad (4.3)$$

$$u_i = 0 \quad , \qquad (x_1^2 + x_2^2)^{1/2} \to \infty \quad , \qquad (4.4)$$

where H(t) is the Heaviside step function. Quiescent initial conditions are relevant for this problem.

In the second case, the body was in an axisymmetric torsional state and the penny-shaped crack was subjectd to a step-type twist. Then, in respect to a cylindrical polar coordinate system (r,ϑ,x_3), the only nonzero displacement component is $u_\vartheta = u_\vartheta(r,x_3,t)$, whereas the equation of motion is written as [14]

$$\frac{\partial^2 u_\vartheta}{\partial r^2} + \frac{\partial u_\vartheta}{\partial r} \frac{1}{r} - \frac{u_\vartheta}{r^2} + \frac{\partial^2 u_\vartheta}{\partial x_3^2} = \frac{\rho}{\mu(D)} \frac{\partial^2 u_\vartheta}{\partial t^2} \quad , \qquad (5)$$

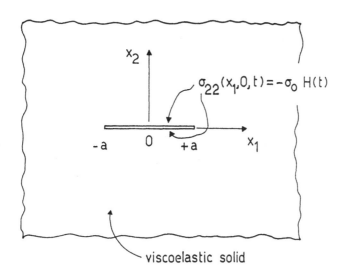

Figure 1.   Viscoelastic sheet with a stationary crack under
plane-strain impact.

where $\mu(D)$ is a differential operator containing the time-de-
rivative operator $D=\partial/\partial t$.   The stress-strain relations of the
viscoelastic material can be written in a *differential-operator*
form (such a form is simply an alternative to the convolution
form utilized in (2)) as

$$\tau_{r\vartheta} = \mu(D) \left(\frac{\partial u_\vartheta}{\partial r} - \frac{u_\vartheta}{r}\right) \quad , \tag{6.1}$$

$$\tau_{\vartheta x_3} = \mu(D) \frac{\partial u_\vartheta}{\partial x_3} \quad . \tag{6.2}$$

The crack of diameter 2a is acted upon by *impact* torsional tractions of amplitude τ_0 .   The boundary conditions
are written as

$$\tau_{\vartheta x_3}(r,x_3=0^+,t) = -\tau_0 \left(\frac{r}{a}\right) H(t) \quad , \quad r<a \quad , \tag{7.1}$$

$$u_\vartheta(r,x_3=0^+,t) = 0 \quad , \quad r>a \quad , \tag{7.2}$$

$$u_\vartheta = 0 \quad , \quad (r^2+x_3^2)^{1/2} \to \infty \quad . \tag{7.3}$$

Results were obtained for a typical viscoelastic model,
namely the *standard linear solid*.   The relaxation functions for
this model are given as

$$G_1(t) = \frac{\mu_\infty}{1+f} \left[1 + f \exp\left(-\frac{1+f}{\tau} t\right)\right] \quad, \tag{8.1}$$

$$G_2(t) = \frac{2(1+\nu)}{1-2\nu} G_1(t) \quad, \tag{8.2}$$

where $f=\left[(\mu_\infty/\mu_0)-1\right]$ is a measure of the difference between the short-time, μ_∞ and the long-time, μ_0, shear moduli and $\left[\tau/(1+f)\right]$ is the relaxation time. Moreover, identical viscoelastic behavior in bulk and in shear was assumed in arriving at (8.2). On the other hand, the viscoelastic shear "modulus", which is relevant to the second configuration, is given by

$$\mu(D) = \mu_\infty \frac{\dfrac{\partial}{\partial t} + \dfrac{1}{\tau}}{\dfrac{\partial}{\partial t} + \dfrac{1+f}{\tau}} \quad. \tag{9}$$

Laplace-and Fourier-transform pairs are also needed for the analysis as these are defined in [16].

The object of the present work is to determine the *viscoelastodynamic* SIFs for the problems defined by (4) and (7).

ANALYTICAL FORM OF SOLUTION

The respective *purely elastic* problem was treated in [16]. Therefore, by invoking the *correspondence principle* [14,15], we can obtain a formal solution in the Laplace-transform domain. However, this is not of too much help from the numerical point of view, since the Laplace transformed intensity function, $F^*(1,p)$ below, depends already upon the material constants.

The viscoelastodynamic SIFs for the plane and torsional case are defined as

$$k(t) = \lim_{x_1 \to +a} \left[\left[2(x_1-a)\right]^{1/2} \sigma_{22}(x_1,0,t)\right] \quad, \tag{10.1}$$

$$k(t) = \lim_{r \to +a} \left[\left[2(r-a)\right]^{1/2} \tau_{\vartheta x_3}(r,0,t)\right] \quad. \tag{10.2}$$

The analytical form of solution for the plane case is found as

$$k(t) = \sigma_0 a^{1/2} \frac{1}{2\pi i} \int_{Br} \frac{F^*(1,p)}{p} e^{pt} \, dp \quad, \tag{11}$$

where the intensity function $F^*(1,p)$ results as the solution of the Fredholm integral equation

$$F^*(\xi,p) + \int_0^1 K(\xi,\eta,p)\ F^*(\eta,p)\ d\eta = \xi^{1/2} . \qquad (12)$$

The kernel in (12) is given in [16] but the longitudinal- and shear-wave speeds are now complex functions of the Laplace-transform variable p

$$c_1 = \left(\frac{\lambda^* + 2\mu^*}{\rho}\right)^{1/2} , \quad c_2 = \left(\frac{\mu^*}{\rho}\right)^{1/2} , \qquad (13)$$

where

$$\mu^* = p\ G_1^*(p) , \quad 3\lambda^* + 2\mu^* = p\ G_2^*(p) . \qquad (14)$$

This solution holds for a general linearly viscoelastic solid having arbitrary relaxation functions $G_1(t)$ and $G_2(t)$. The formal solution for the torsional case has an analogous form but it involves only c_2.

LAPLACE TRANSFORM INVERSION

The inversion of the Laplace transform in (11) was carried out numerically by following the DAC technique [3,4]. If we write p=c+iu, express the original function as a Fourier cosine and sine integral and apply the trapezoidal rule for integrals over semi-infinite intervals, we find an approximate expression for f(t) as a Fourier series

$$f(t) \simeq (e^{ct}/T) \left[1/2\ f^*(c) + \sum_{k=1}^{\infty} \left[\mathrm{Re}\ f^*(c+ik\pi/T)\ \cos(k\pi t/T) - \right.\right.$$

$$\left.\left. -\ \mathrm{Im}\ f^*(c+ik\pi/T)\ \sin(k\pi t/T)\ \right]\right] . \qquad (15)$$

Crump [4] has presented a systematic analysis of errors in the above procedure, from which one can compute f(t) to a predetermined accuracy. It is possible also to increase the rate of convergence of (15) and thus reduce the truncation error by using a suitable series transformation.

NUMERICAL PROCEDURE

First, the integral equation (12) has to be solved numerically. Since this equation contains complex functions, we have to separate real from imaginary parts in the following manner

$$F^*(\xi,p) = F_1^*(\xi,\mathrm{Re}\ p,\mathrm{Im}\ p) + i\ F_2^*(\xi,\mathrm{Re}\ p,\mathrm{Im}\ p), \quad (16.1)$$

$$K(\xi,\eta,p) = K_1(\xi,\eta,\mathrm{Re}\ p,\mathrm{Im}\ p) + i\ K_2(\xi,\eta,\mathrm{Re}\ p,\mathrm{Im}\ p) . $$

$$(16.2)$$

Then, eqn (12) is equivalent to the following coupled Fredholm integral equations

$$F_1^* + \int_0^1 (K_1 F_1^* - K_2 F_2^*) \; d\eta = \xi^{1/2} \qquad , \qquad (17.1)$$

$$F_2^* + \int_0^1 (K_2 F_1^* + K_1 F_2^*) \; d\eta = 0 \qquad . \qquad (17.2)$$

The above two-dimensional system was solved for $F^*(\xi, \text{Re } p,$ Im $p)$ by using the Gauss quadrature in conjunction with *complex algebra* on the computer. Since $p = c + i(k\pi/T)$ is a parameter in (17), these equations have to be solved a number of times equal to the number of terms considered in the Laplace transform inversion (15).

Since $\xi, n < 1$, the integral giving the kernel $K(\xi, n, p)$ in (12) is not highly oscillatory, so the standard *Gauss rule* can be applied. The semi-infinite integration interval was normalized to $[-1, +1]$ by the Stieltjes transformation [17]. The above procedure was checked both by the *IMT rule* (which is convenient for integrands with end-point singularities) [17] and the *Longman rule* (which is convenient for oscillatory integrals over infinite intervals) [17] and found to be competitive. Thus, we utilized the simple *Gauss rule* throughout the numerical procedure.

After determining the complex *intensity* function $F^*(1,p)$, the next step is the numerical inversion of the Laplace transform in (11). An algebraic inspection was made and we found that the function $F^*(1,p)$ and the respective one for the torsional case are always *analytic* in Re $(p) > 0$ for the material constants used (standard linear solid). Branch cuts for the complex radicals $\gamma_j = [s^2 + (p/c_j)^2]^{1/2}$, $(j=1,2)$, are extending only in the half-plane Re $(p) < 0$. Before applying however the DAC technique, we normalize both the SIF and the time in the following way

$$m(t) = k(t)/\sigma_0 a^{1/2} \; , \; m(t) = 3\pi \; k(t)/4\tau_0 a^{1/2}, \qquad (18)$$

for the plane-strain and torsional case, respectively, and

$$p = \frac{c_2^\infty}{a} \; w \quad , \quad t = \frac{a}{c_2^\infty} \; T_d \quad , \qquad (19)$$

where $c_2^\infty = (\mu_\infty/\rho)^{1/2}$ is the *short-time* shear-wave velocity. Then, eqn (11) becomes

$$m(t) = (2 \pi \; i)^{-1} \int_{c-i\infty}^{c+i\infty} L(1,w) \; e^{wT} d \; dw \quad , \qquad (20)$$

and further is approximated according to (15) as

$$m(t) \simeq (e^{cT}d/T) \; [1/2 \; L(c) + \sum_{k=1}^{N} \; [\text{Re } L(c+ik\pi/T) \; \cos(k\pi T_d/T) -$$

$$- \; \text{Im } L(c+ik\pi/T) \; \sin(k\pi T_d/T) \;] \;] \quad , \qquad (21)$$

where

$$L(1,w) = \frac{F^*(1,\frac{c_2^\infty}{a} w)}{w} \quad , \tag{22}$$

and N is the number of terms required to obtain convergence. In this study, sixty to seventy terms were usually enough to get very accurate results. Of course, the accelerating epsilon algorithm was also employed.

RESULTS AND CONCLUDING REMARKS

Numerical results for the dynamic SIF were obtained for: a) PMMA whose viscoelastic material contants were found experimentally by Rosenfield (see in [18] as μ_∞=1690 MN/m^2 , ν=0.345 , f=0.71 , τ=20 sec , ρ=1200 kg/m^3 , and b) a fictitious material with constant short-time modulus μ_∞=1268 MN/m^2 , ν=0.345 and ρ=1200 kg/m^3 and variable f, τ in each case. The latter groups of material constants were considered in order to examine the effect of the viscoelastic behavior.

In both cases, relations (8) and (9) pertinent to the *standard linear solid* were considered. Then, the complex Lamé functions are given as

$$\mu^*(p) = \frac{\mu_\infty(p + \frac{1}{\tau})}{p + \frac{1+f}{\tau}} \quad , \tag{23}$$

$$\lambda^*(p) = \frac{2\nu}{1 - 2\nu} \mu^*(p) \quad . \tag{24}$$

Clearly, the *purely elastic* case can be recovered from the above material model by letting either f→0 or τ→∞ .
Fig. 2 shows the dynamic SIF time history in the case of a stationary crack under plane impact of an *elastic* (f=0.0) plate having shear modulus μ=1690 MN/m^2 and ν,ρ as in case (a) above. The static limit for this problem is m(t=∞)=1 and so this may serve for checking the numerical results. For the purpose of comparison, both the DAC and Miller-Guy techniques were utilized for this particular case. It can be clearly seen the strong dependence of the Miller-Guy results on the choice of the parameters contained in this technique.
Figure 3 shows the plane dynamic SIF time history in a PMMA plate. We can observe that there is no appreciable change from the elastic case of Fig. 2. This should be expected since PMMA presents no strong viscous behavior. The peak SIF value is slightly lower now, whereas the steady-state elastic SIF value (m(t=∞)=1) is never reached.
Figs. 4,5 and 6 show the dynamic SIF time history for the fictitious model of case (b). The first two correspond to the case of plane impact and the third to the torsional-impact case.

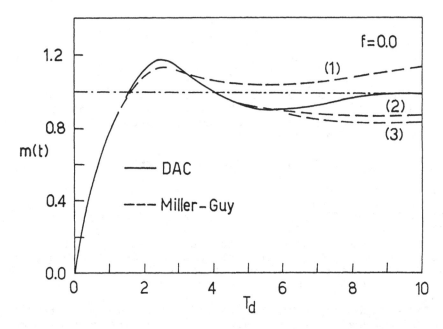

Figure 2. Elastodynamic SIF time history by the DAC and Miller-
-Guy inversion techniques. Curve (1): β=1.0, δ=1.0 and N=16,
Curve (2): β=0.2, δ=0.9 and N=15, and Curve (3): β=1.0,
δ=1.0 and N=15.

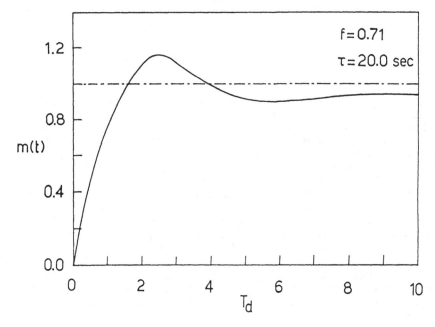

Figure 3. Viscoelastodynamic SIF time history for a PMMA
cracked plate.

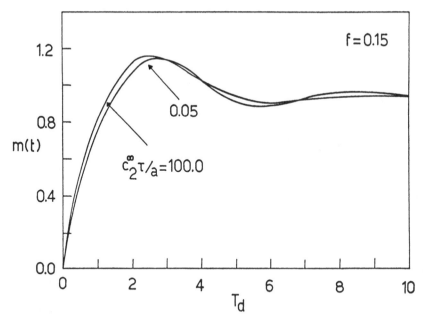

Figure 4. Viscoelastodynamic SIF time history for $\mu_\infty=1268\,MN/m^2$, f=0.15 and $c_2^\infty \tau/a$ = 100.0 and 0.05.

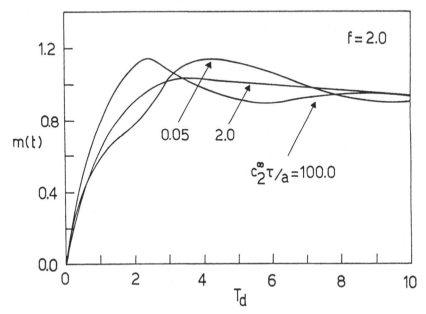

Figure 5. Viscoelastodynamic SIF time history for $\mu_\infty=1268\,MN/m^2$, f=2.0 and $c_2^\infty \tau/a$ = 100.0, 2.0 and 0.05.

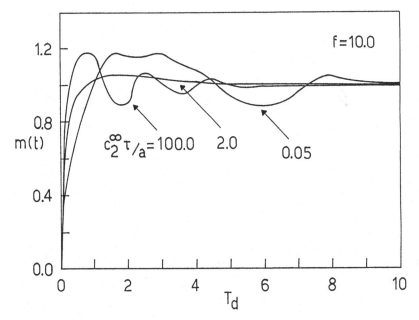

Figure 6. Viscoelastodynamic SIF time history for μ_∞=1268 MN/m^2,
f=10.0 and $c_2^\infty\tau/a$ = 100.0, 2.0 and 0.05 (torsional case).

From our numerical study, the following conclusions can be reached (some of these are also observable in the figures presented):

1. When f is *small*, even radical changes of the relaxation time produce *no appreciable* viscous effect. *Strong* viscous effects are observed, when *medium* and *large* f is considered.

2. When the values G(t=0) and G(t=∞) are fixed, in ranges where the G(t) curve is almost *flat* and *slowly decreasing*, one should expect to observe instantaneously, an *almost elastic* material response (in respect to the overshoot of SIF), whereas in ranges of *continuously varying* G(t) the *viscous* effect is more pronounced.

3. Performing a viscoelastodynamic analysis for cracked materials is useful, since *unexpected* dynamic SIF overshoots may occur.

4. A *purely elastic* analysis based on the steady-state material response may underestimate the SIF values.

REFERENCES

[1] Popelar, C.H. and Atkinson, C., Dynamic crack propagation in a viscoelastic strip. J. Mech. Phys. Solids, 1980, 28, 79-92.

[2] Georgiadis, H.G. and Mouskos, S.C., Integral transform ap-

proach to an inertial viscoelastic problem involving moving singularities. Proceedings of the 12th Canadian Congress of Applied Mechanics, Ottawa, 1989, 204-205.

[3] Dubner, H. and Abate, J., Numerical inversion of Laplace transforms by relating them to the finite Fourier Cosine transform. J. Assoc. Comp. Machinery, 1968, 15, 115-123.

[4] Crump, K.S., Numerical inversion of Laplace transforms using a Fourier Series approximation. J. Assoc. Comp. Machinery, 1976, 23, 89-96.

[5] Papoulis, A., A new method of inversion of the Laplace transform. Quart. Appl. Math., 1957, 14, 405-414.

[6] Miller, M.K. and Guy, W.T., Numerical inversion of the Laplace transform by use of Jacobi Polynomials. SIAM J. Numer. Analysis, 1966, 3, 624-635.

[7] Bellman, R., Kalaba, R. and Lockett, J., Numerical Inversion of the Laplace Transform, American Elsevier, New York, 1966.

[8] Sih, G.C. and Chen, E.P., Cracks in Composite Materials, Nijhoff Publ., The Hague, 1981.

[9] Baker, B.R., Dynamic stresses created by a moving crack. J. Appl. Mech., 1962, 29, 449-458.

[10] Kostrov, B.V., Unsteady propagation of longitudinal shear cracks. J. Appl. Math. Mech. (PMM), 1966, 30, 1241-1248.

[11] Thau, S.A. and Lu, T.H., Transient stress intensity factors for a finite crack in an elastic solid caused by a dilational wave. Int. J. Solids Structures, 1971, 7, 731-750.

[12] Cost, T.L., Approximate Laplace transform inversions in viscoelastic stress analysis. AIAA Journal, 1964, 2, 2157-2166.

[13] Davies, B. and Martin, B., Numerical inversion of the Laplace transform: A survey and comparison of methods. J. Comp. Physics, 1979, 33, 1-32.

[14] Fung, Y.C., Foundations of Solid Mechanics, Prentice Hall, Englewood Cliffs, 1965.

[15] Christensen, R.M., Theory of Viscoelasticity, Academic Press, New York, 1971.

[16] Chen, E.P. and Sih, G.C., Elastodynamic Crack Problems, Noordhoff, The Netherlands, 1977.

[17] Davis, P.J. and Rabinowitz, P., Methods of Numerical Integration, Academic Press, New York, 1984.

[18] Popelar, C.H. and Kanninen, M.F., A dynamic viscoelastic
     analysis of crack propagation and crack arrest in a double
     cantilever beam test specimen. ASTM STP 711, 1980, 5-23.

DISLOCATION GENERATION BY DYNAMICALLY-LOADED CRACKS AND ITS EFFECTS ON FRACTURE CHARACTERIZATION

L. M. Brock
Engineering Mechanics
University of Kentucky
Lexington, Kentucky 40506, USA

ABSTRACT

Brittle fracture involves micromechanical processes, including the generation of dislocations by the crack edge. Here, dislocation generation under dynamic loading is analyzed for insight into its effects on fracture characterization.

INTRODUCTION

Evidence suggests that nominally brittle fracture involves micromechanical processes such as dislocation generation at the crack edge [1-3]. Studies [4] also suggest that dislocations shield cracks by lowering stress intensity factors. Most analyses of crack-dislocation interaction are, however, static. Here, therefore, some idealized 2D cases of dislocation generation at cracks dynamically loaded by stress wave diffraction are studied. The cracks in each case are semi-infinite, and the solids are unbounded, linearly elastic, isotropic and homogeneous. On the basis of exact transient solutions for each case, a dislocation force criterion [1] and a stress intensity factor-based brittle fracture criterion [5], insight is possible into several features of dislocation generation. In the next section, we begin the studies with cases of screw dislocation generation as a precursor to fracture.

STATIONARY MODE III CRACK

Consider the crack defined by the Cartesian coordinates (x,y) as $y=0, x<0$. Prior to $s=0$, where $s=(\text{time})\text{x}(\text{rotational wave speed})$, the crack is undisturbed, but a plane SH-wave, traveling at right angles to it, reaches the crack plane at $s=0$, and is diffracted at the crack edge. This process places the crack in a Mode III state, and at $s=s_o>0$, a right-handed screw dislocation is generated and glides out ahead of the crack for a distance d^*, where is arrests. Its position is given, therefore, by $y=0, x=d(s)$, where $d\equiv 0(s\leq s_o)$, $d=d^*(s\geq s^*)$, and $0<v<1(s_o<s<s^*)$. Here s^* gives the arrest instant, and v is the dislocation speed non-dimensionalized by the rotational wave speed, so that the inequality imposes subsonic dislocation motion.

The diffraction-generation process and the associated elastic waves are shown in Fig. 1. An exact solution for the process is readily obtained [6-8] in terms of (s_o,s^*) and the function d. These must be determined, of course, by imposing a criterion for dislocation motion. It is generally held [9-11] that dislocation motion by glide cannot occur until the dislocation force overcomes lattice friction, which can be identified [1-3,12] with the yield stress. This criterion is adopted here, under an assumption [1,3] that the dislocation does not move very far. Because a characteristic length may not arise in this transient problem, we require that

$$d^*/s_o \ll 1 \qquad\qquad (1)$$

That is, the distance traveled by the dislocation is much less than the

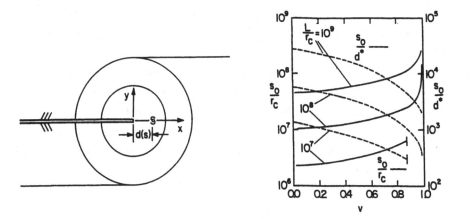

Figure 1. Generation process.   Figure 2. Generation parameters.

distance traveled by a rotational wave in the interval between diffraction and generation.

The dislocation force on the glide plane per unit length of dislocation edge, F, can be obtained from the formula [9]

$$F = \int [(\tau_{yz}\dot{w})_+ - (\tau_{yz}\dot{w})_-]dx \qquad (2)$$

where $(\ )_{\pm}$ denotes $y=0\pm$, $(\dot{\ })$ denotes s-differentiation, w is the anti-plane displacement, and integration is over the glide plane. By following [9], this integration produces a finite result which can, in view of (1), be written for the interval of motion as

$$\frac{\pi}{b}F = \frac{H}{\sqrt{d}} - \frac{\mu b B}{4d} \quad (s_o < s < s^*) \qquad (3)$$

Here b is the Burgers vector magnitude, μ is the shear modulus, and

$$H = P(s_o), \quad P(s) = \int_0^s \frac{g(u)}{\sqrt{(s-u)}}du, \quad B = \sqrt{(\frac{1+v}{1-v})} > 1 \qquad (4)$$

where g(u) is the traction τ_{yz} induced by the SH-wave at the distance u behind the wavefront. The dislocation motion criterion requires that F/b not fall below the yield stress Y during the motion.

Constant Dislocation Speed

If, following [13], dislocation speed is taken to be constant, then setting F/b equal to Y gives a quadratic equation for the values of d at which motion ceases:

$$\sqrt{d_{\pm}} = \frac{H}{2\pi Y} \pm \sqrt{[(\frac{H}{2\pi Y})^2 - \frac{\mu b B}{4\pi Y}]} \qquad (5)$$

The smaller value, d_-, cannot exceed the dislocation core radius r_c; to do so would imply that dislocation generation occurs away from the crack edge [1]. The larger value d_+, therefore, is the arrest distance d^*. It follows that

$$d_- = \lambda r_c \quad (0 < \lambda \leq 1) \qquad (6)$$

where λ is dimensionless, whereupon (5) gives the relations

$$\frac{1}{J(\lambda r_c)} \frac{H}{\pi Y} = 1 + \frac{\mu b}{Y r_c} \frac{B}{4\pi\lambda}, \quad \lambda \frac{d^*}{r_c} = (\frac{\mu b}{Y r_c} \frac{B}{4\pi})^2 \qquad (7a,b)$$

Equation (7b) is derived under the generally valid assumptions [1,14] that $b/r_c \sim 0(1)$, $\mu/Y > 0(10^2)$. Equation (7a) clearly cannot hold if $H \leq 0$. In view of (4), this implies that, while the instantaneous value of the SH-wave stress g is not restricted, its time history must give $H > 0$ in order to produce a right-handed screw dislocation. An analogous result $H < 0$ follows for left-handed dislocations, because (b,Y) would merely change sign.

Because v is constant, it is clear that (s_o, d^*, v) completely characterize the generation process in terms of the given parameters (g, μ, Y, b, r_c). However, (7) gives only two equations, and these involve the parameter λ. Therefore, the force criterion is not sufficient; additional requirements are needed. Another difficulty becomes clear when the dislocation force after arrest is written:

$$\frac{\pi}{b} F = \frac{H}{Jd^*} - \frac{\mu b}{4} \frac{B^*}{s-s^*} \quad (s^* < s \leq s^* + 2d^*) \qquad (8a)$$

$$\frac{\pi}{b} F = \frac{H}{Jd^*} - \frac{\mu b}{4d^*} \quad (s \geq s^* + 2d^*) \qquad (8b)$$

Here the second term in (8b) is the quasi-static result [3], while

$$B^* = 2[J(\frac{1}{v}+1)J(\frac{s-s_o}{d^*})-1] \qquad (9)$$

and it is noted that $B^*(s^*) < 0$, $B^*(s^*+2d^*) = 2$. Equation (8a) is valid during the interval $2d^*$ between dislocation arrest and the arrival at the dislocation of the arrest signal diffractions from the crack edge, while (8b) holds thereafter. Clearly, $F/b \to \infty$ just after arrest, then drops to an essentially constant value. Even if it is argued that $2d^*$ is "small" and (8a) can be ignored, (8b) itself gives an F/b that exceeds Y, because $B > 1$. Therefore, regeneration of dislocation motion should, strictly speaking, follow.

Incident Wavelength Dependence

In spite of these difficulties, the results for constant v do allow insight

into the influence of the incident wave stress g. In particular, because
of Fourier superposition, it is useful to consider the sinusoidal incident
wave

$$g(u) = \tau \sin\pi\frac{u}{L} \quad (0<\tau<<Y) \tag{10}$$

where τ is the stress amplitude and L is the wavelength. Equation (4)
gives

$$P(s) = \tau G(\frac{s_o}{L})JL, \quad G(x) = Jx\int_0^{\pi/2}J(1+\sin u)\sin(\pi x \sin u)du \tag{11}$$

whereupon (7a) becomes

$$G(\frac{s_o}{L}) = \mu J(\lambda\frac{r_c}{L})\frac{Y}{\tau}(1+\frac{\mu b}{Yr_c} \frac{B}{4\pi\lambda}) \tag{12}$$

For various values of L/r_c and

$$\frac{Y}{\mu} = 0.001, \frac{\tau}{\mu} = 0.0001, \frac{b}{r_c} = 0.5, \lambda = 1.0 \tag{13}$$

dimensionless parameters involving (s_o,d^*) are obtained from (12) and
plotted vs. allowable values of v in Fig. 2. There it is seen that
assumption (1) is verified, and that the generation instant (s_o) varies
directly with L. Indeed, no data for $L/r_c<0(10^6)$ appears because (12)
cannot then be satified for $0<v<1[7]$. Because $r_c\sim0(10^{-4})\mu m[1]$ and
rotational wavespeeds are $o(10^3)$m/sec [15], this limitation implies that
$L>0(10^{-1})$mm - and, therefore, incident wave frequency $<0(10)$MHz - for
generation to occur. Again in view of rotational wave speed values, Fig. 2
also implies that generation occurs $0(10^{-1})-0(10)\mu sec$ after diffraction,
and that the arrest distance d^* is $0(10^{-2})-0(1)\mu m$. Thus, the generation
characterization parameters (s_o,s^*,d^*) are on a micromechanical scale,
while L is not. The scale of the former clearly justifies the use of semi-
infinite cracks and unbounded solids in the model.

Mode III Dynamic Stress Intensity Factor

The Mode III dynamic stress intensity factor k_3 can also be extracted from
the analytical solution [6-8] to the Fig. 1 problem as

$$k_3 = J(\tfrac{2}{\pi})P(s) = k_3^o \quad (0 < s < s_o) \tag{14a}$$

$$k_3 = k_3^o - \mu b J(\tfrac{1+\nu}{2\pi d}) \quad (s_o < s < s^* + d^*) \tag{14b}$$

$$k_3 = k_3^o - \frac{\mu b}{J(2\pi d^*)} \quad (s > s^* + d^*) \tag{14c}$$

Equation (14a) is valid before dislocation generation, (14b) governs between generation and reception of the arrest signal at the crack edge, and (14c) holds thereafter. These formulas show that the crack edge stress field can actually be enhanced, not relaxed, if the history term $P(s) < 0$ for $s > s_o$. This behavior is in contrast to non-transient results [4], which generally predict relaxation.

Equations (14) assume, as does the Fig. 1 problem statement, that the crack remains stationary during dislocation generation. The question of whether this is valid or not can be examined in light of a standard stress intensity factor criterion [5]. That is, purely brittle fracture will occur if k_3 given by (14a) reaches a critical value k_3^c at some $s = s_c < s_o$. For instance, the sinusoidal wave (10) gives

$$G(\frac{s_c}{L}) = J(\frac{2r_c}{\pi L})^{\frac{1}{2}} k_c, \quad k_c = \frac{k_3^c}{Y J r_c} \tag{15}$$

for (14a). If we again consider the situation (13), then it can be shown that $k_c = 29.6$ and (15) gives the data in Table 1.

TABLE 1

Brittle fracture initiation instants

L/r_c	10^8	10^9	10^{10}	10^{11}	10^{12}
s_c/r_c	$-$	$5.4(10^8)$	$2.1(10^9)$	$9.3(10^9)$	$4.3(10^{10})$

Comparison of Fig. 2 with Table 1 shows that $s_c > s_o$, indicating that fracture here does not occur in a purely brittle manner. It is also noted that no data for $L/r_c < 0(10^8)$ appears in Table 1. This is because (15) cannot then be satified, which indicates that wavelength/frequency cutoffs exist for both dislocation generation and brittle fracture.

Dislocation Acceleration

Along with its post-arrest behavior, another feature of constant-speed dislocation generation is that (3) gives an F/b that exceeds Y during motion. That is, micromechanical strain-hardening [11] occurs. Indeed, F/b in (3) can remain fixed at Y during motion only when v is allowed to vary. Thus, dislocation acceleration from the crack edge implies micromechanical perfect plasticity. If v is assumed to be continuous once motion has begun, then (3) is valid for all $s>s_o$, and a stop-start motion may not arise. To examine this possibility, we treat the case of a step-stress SH-wave

$$g(u) = \tau \quad (0<\tau<<Y) \tag{16}$$

Setting F/b in (3) equal to Y then gives

$$v = d = \frac{d(2\alpha - Jd)^2 - \beta^2}{d(2\alpha - Jd)^2 + \beta^2}, \quad \alpha = \frac{1}{\pi}\frac{\tau}{Y}Js_o, \quad \beta = \frac{1}{4\pi}\frac{\mu}{Y}b \tag{17a-c}$$

Equation (3) has, under the criterion, now provided a true equation of motion for the dislocation - a nonlinear differential equation for d. Because s does not appear explicitly, (17a) can be integrated by separation of variables to give [8] an inverse relation

$$\frac{1}{\lambda r_c}(s-s_o) = D^2 - 1 + M\ell n(\frac{D-e_2}{1-e_2})^{k_2}(\frac{e_3-1}{e_3-D})^{k_3}(\frac{e_4-D}{e_4-1})^{k_4}(\frac{1+e_1}{D+e_1})^{k_1} \tag{18}$$

Here

$$D = J(\frac{d}{r_c}), \quad Q_+k_i = e_i \ (i=1,4), \quad Q_-k_i = e_i \ (i=2,3) \tag{19a}$$

$$e_1 = Q_+ - N, \quad e_2 = N - Q_-, \quad e_3 = N + Q_-, \quad e_4 = N + Q_+ \tag{19b}$$

$$Q_\pm = J(N^2 \pm M), \quad M = \frac{1}{4\pi\lambda}\frac{\mu b}{Y r_c}, \quad 2N = 1 + J(\frac{1+v_o}{1-v_o})M \tag{19c}$$

where v_o is v at $s=s_o$. Study [8] of (17a) and (18) for (13) gives the values in Table 2.

TABLE 2

Dislocation acceleration parameters

τ/μ	v_o	v_m	s_o/r_c	s_m/r_c	d_m/r_c	d^*/r_c
10^{-4}	$0(10^{-8})$	0.9819	$4.105(10^5)$	$4.11(10^5)$	$4.16(10^2)$	$1.583(10^3)$

Here v_m is the maximum value of v and (s_m,d_m) give, respectively, the instant at which the dislocation achieves the maximum speed, and the distance traveled by that instant. Table 2 shows that the maximum speed is almost critical, and that the dislocation instantaneously achieves a finite speed, but one that is almost negligible. Study of (17a) and (18) also shows that d^* is attained only as $s \to \infty$, but that v itself has become almost negligible when $s \sim 3s_m$. Thus, dislocation arrest does occur, but as an asymptotic process. It should also be noted that values of d^* are essentially identical to those that would be calculated in a quasi-static analysis [3]. That is, micromechanical perfect plasticity gives a dynamic, high-rate process, but one that is, nevertheless, quasi-static in regard to arrest distance.

STATIONARY MODE III INTERFACE CRACK

Cracks often lie at the interface of rigidly bonded, dissimiliar elastic materials, and dislocation motion may well occur along such interfaces. The dissimilarities may be small but, in a micromechanical process, even grain boundaries [14] may qualify as interfaces. We now, therefore, consider the situation displayed in Fig. 3: In Fig. 3a a plane SH-wave travels through a material #2 ($y<0$), thereby generating both a reflection and a transmitted wave into material #1 ($y>0$). This wave ensemble approaches the undisturbed interface crack $y=0, x<0$ and, at $s=0$, is diffracted at its edge. Now $s=(\text{time})\text{x}(\text{rotational wave speed in material}$ #1), and Fig. 3b shows the wave diffraction pattern, where the dimensionless constant m is the ratio of the rotational wave speeds in materials #1 and #2. For purposes of illustration, we have chosen $m>1$. By Snell's law [15], the angles of incidence of the plane waves shown in Fig. 3 must satisfy the conditions

$$\cos\phi_1 = m\cos\phi_2, \quad \cos^{-1}(1/m) < \phi_2 < \pi/2, \quad 0 < \phi_1 < \pi/2 \qquad (20)$$

At $s = s_o > 0$ a right-handed screw dislocation is generated at the crack edge, and glides along the interface, as it did in the Fig. 1 problem. Here, however, the dislocation speed non-dimensionalized with respect to the rotational wave speed in material #1, v, must be bounded as $0 < v < 1/m$ in order to maintain subsonic dislocation motion.

The analytical solution for this problem has been derived [16], and, again, the output parameters (s_o, s^*, v) must be obtained from a dislocation force [1-3,9-12] criterion. Now, (2) is generalized to

$$F = \int [(\tau_{yz}\dot{w})_1 - (\tau_{yz}\dot{w})_2] dx \qquad (21)$$

and, under the reasonable assumption (1), can be made to yield the results

$$\pi\frac{F}{b} = \frac{H}{Jd} - \frac{\mu_2 bB}{4d} \quad (s_o < s < s^*) \qquad (22)$$

where μ_i is the shear modulus in material #i. For the constant v-case,

$$Q_+(\cos\phi_1)H = \frac{2mn\sin\phi_1\sin\phi_2}{\sin\phi_1 + mn\sin\phi_2} P(s_o) \qquad (23a)$$

$$\frac{1}{4}B = \frac{1}{vK^2}\left(\frac{a^2}{b} + \frac{b^2}{a}\right) + \frac{ab}{K}\left[\int_1^m \frac{\Omega}{(u+\frac{1}{v})^2} du - \frac{1}{2}\frac{v}{1+v}\right] \qquad (23b)$$

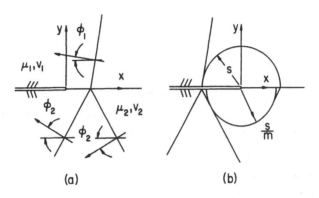

(a)   (b)

Figure 3. Diffraction process at interface crack.

where

$$K = a+nb, \quad \mu_1 n = \mu_2 \tag{24a}$$

$$\ln\left[\frac{Q_+(x)}{a_1^{\pm}(x)}\right] = \frac{1}{\pi}\int_1^m \frac{\Omega}{u\pm x}du, \quad \Omega = \tan^{-1}\frac{1}{n}\sqrt{(\frac{m^2-u^2}{u^2-1})} \tag{24b}$$

$$a_1^{\pm}(x) = \sqrt{(1\pm x)}, \quad a = \sqrt{(\frac{1}{v^2} - 1)}, \quad b = \sqrt{(\frac{1}{v^2} - m^2)} \tag{24c}$$

Equation (22) governs during dislocation motion, and equations analogous to (8a,b) hold for the post-arrest period. Indeed, the F/b-behavior noted for this problem is strictly analogous to (3) and (7a,b). Specifically, (22) gives a quadratic equation for d. Under the condition (6), this quadratic gives results identical in form to (7a,b), with μ replaced by μ_2, (B,H) now defined by (23), and Y now defined as the interface bond yield stress. Similarly, the post-arrest response of F/b is the same as that noted for the Fig. 1 problem. However, in obtaining the analogy to (8b), it is found [16] that $d^*=\lambda r_c$, i.e. the dislocation never leaves the crack edge, unless

$$1 < \frac{\mu_2 b}{Y r_c}\frac{B}{4\pi\lambda} \tag{25}$$

Because $b/r_c \sim 0(1)$ and $\mu_2/Y \geq 0(10^2)$[14], satisfaction of (25) falls largely on n. In particular, if it is too large, indicating a substantial shear modulus mismatch, then (25) cannot be satisfied. As an illustration of possible dislocation motion, we choose a step-stress pulse for the incident SH-wave, and

$$m = 1.2, \quad n = 1/\sqrt{m}, \quad \frac{\tau}{\mu_2} = 0.0001, \quad \frac{Y}{\mu_2} = 0.001, \quad \frac{b}{r_c} = 0.5, \quad \lambda = 1.0 \tag{26}$$

where τ is the anti-plane shear traction τ_{yz} induced by the incident SH-wave. For different angles of incidence ϕ_2, the dimensionless parameters $(s_o/r_c, s_o/d^*)$ are plotted vs. allowable v in Fig. 4. There it is noted that (s_o/d^*) are sensitive to both ϕ_2 and to v. In particular, the dislocation generation process takes an infinite time to initiate if it is to proceed at the subsonic (rotational wave) limit, but a preferred quasi-static limit also exists. That is, s_o achieves a minimum value when v→0.

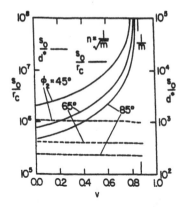

Figure 4. Generation parameters for interface crack.

MULTIPLE GENERATION AT A GROWING MODE III CRACK

We have heretofore treated dislocation generation as a precursor to
fracture, and assumed that glide takes place in the crack plane. For
insight into alternative possibilities, we return to the crack of Fig. 1,
but now assume that the plane SH-wave diffraction has triggered fracture at
$s = s_c > 0$. As indicated in Fig. 5, the crack edge is located at $y = 0, x = C(s)$,
where $C \equiv 0 (s < s_c)$, $0 \leq c < 1 (s \geq s_c)$, and c is the non-dimensionalized crack speed.
Dislocation generation again occurs at $s = s_o$, but now $s_o > s_c$ and glide is at
an angle $\phi (0 \leq \phi \leq \pi/2)$ with respect to the crack plane. In Fig. 5, $C_o = C(s_o)$
locates the crack edge at the dislocation generation instant, and we treat
the dislocation speed as constant and subsonic ($0 < v < 1$).

While more complicated, this problem can also be solved analytically
[17] and, under the assumption (1), an expression for the dislocation force
derived. For a constant crack speed and a step-stress SH-wave, this
expression prior to dislocation arrest takes the form

$$\frac{\pi}{b} F = 2\tau I(v,c,\phi) J(\frac{s_o}{d}) - \frac{\mu b}{4} \frac{B(v,c,\phi)}{d} \qquad (27)$$

where τ is the stress wave magnitude, and

$$I = J(\frac{S}{2v}) \frac{R + c(1 - cv\cos\phi)}{J(1-c)(1+c)R} \qquad (28a)$$

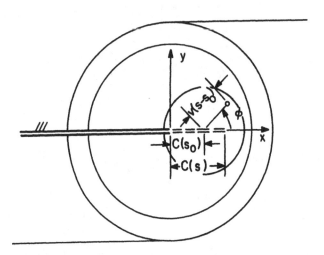

Figure 5. Generation process at growing crack.

$$B = \frac{2(1-cv\cos\phi)}{(1+c)R\sqrt{(1-v^2)}}(R+c+cv\cos\phi) - \frac{vK}{SR}\sqrt{(1+c)}\,\text{Re}(Q) \qquad (28b)$$

$$Q = (\frac{K_+ - iK_-}{\Delta})\sqrt{(\frac{\omega-v}{\omega+v})}(2 - \frac{ivsin\phi}{\sqrt{(1-v^2)}})(\cos2\phi - \frac{isin2\phi}{\sqrt{(1-v^2)}}) \qquad (28c)$$

$$K = \sqrt{(1-R-cv\cos\phi)}, \quad K_\pm = \sqrt{[R\pm(v-c\ \cos\phi)]} \qquad (28d)$$

$$S = \sqrt{(R+v\cos\phi-c)}, \quad R = \sqrt{[(c-v\cos\phi)^2+v^2(1-c^2)\sin^2\phi]} \qquad (28e)$$

$$\Delta = \sqrt{(\omega-v)} + \frac{K_+K}{\sqrt{(1+c)}S} + i\sqrt{(\frac{1-c}{1+c})}\frac{K_-}{S}[\sqrt{(\omega+v)} + \frac{K}{\sqrt{(1-c)}}] \qquad (28f)$$

$$\omega = \cos\phi + i\sqrt{(1-v^2)}\sin\phi \qquad (28g)$$

Under the criterion used throughout this paper, (27) gives the results

$$\frac{s_o}{\lambda r_c} = (\frac{\pi}{2\tau})^2(\frac{Y}{\tau} + \frac{\mu b}{4\pi}\frac{B}{\tau\lambda r_c})^2, \quad \lambda\frac{d^*}{r_c} = (\frac{\mu b}{4\pi}\frac{B}{Yr_c})^2 \qquad (29a,b)$$

For the values in (13), various values of v, c=0.3 and a range of ϕ, equations (29a,b) give the plots in Fig. 6. It is seen that the ϕ-dependence is pronounced, and gives minimum s_o-values when glide is directly ahead of the crack.

The force criterion does not specifically forbid simultaneous generation of more than one dislocation [1-3]. Therefore, we consider Fig.

321

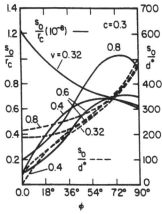

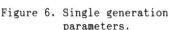

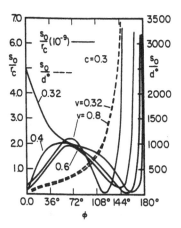

Figure 6. Single generation
parameters.

Figure 7. Multiple generation
parameters.

5 again, but visualize a second screw dislocation which moves as a crack
plane mirror image to the first. Analytical solutions to this problem are
again possible [17], and expressions identical to (29) result, except that
B is replaced by B_2, where,

$$B_2 = 2[B- \frac{1}{\sqrt{(1-v^2)}}]+ \frac{\sqrt{(1-v^2)}}{1-v^2\cos^2\phi} \tag{30}$$

A plot similar to Fig. 6 is now made in Fig. 7, but for $0<\phi<\pi$. Figure 7
shows a substantially different behavior when $\phi>\pi/2$. In particular, $s_0\to\infty$
as $\phi\to\pi$, indicating that glide ahead of the crack is preferred, even when a
dislocation pair is involved. Comparison of Figs. 6 and 7 also shows,
however, that in the sense of minimum generation time (s_0), single
generation is the preferred mechanism.

To close examination of this case, we make a direct comparison with
the Fig. 1 case by setting $\phi=0$ and considering only single generations.
The dimensionless quantities $(s_0/r_c,s_0/d^*)$ are plotted vs. v in Fig. 8,
where v>c in order to insure that the crack edge does not overtake the
dislocation. Figures 2 and 8 shows that, nevertheless, the crack edge
speed is not the preferred dislocation speed, in the sense of requiring a
minimum generation time. That is, dislocation generation is more.likely to
occur at speeds greater than the crack edge speed. It is noted, however,

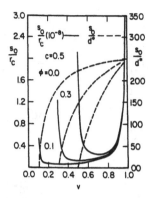

Figure 8. Parameters for generation ahead of moving crack.

that this disparity vanishes in the limit as the crack is stationary, so that Figs. 2 and 8 are consistent.

SUMMARY

This paper has studied a variety of cases involving dislocation generation at a crack edge under stress wave diffraction. The results of the studies suggest that the generation process is on a micromechanical scale, and that neither dislocation generation nor purely brittle fracture may occur for certain incident wavelength/frequency ranges. Moreover, dislocation generation is preferred over brittle fracture, in the sense of needing a smaller generation time after diffraction.

For dislocations moving at constant speeds, the dislocation force behavior suggests that the motion is stop-start. Indeed, an analogy with stick-slip friction would be apt, since dislocation glide is viewed as the overcoming of lattice friction. Moreover, constant-speed motion seemed to require micromechanical strain-hardening.

Dislocation acceleration, on the other hand, is a direct consequence of micromechanical perfect plasticity, and arrest in this case is an asymptotic process that, interestingly enough, involves very high speeds, yet an arrest distance that can be predicted quasi-statically.

Dislocation glide after generation at constant speeds along interfaces was found to have many features of such glide in homogeneous solids. However, results did indicate that a substantial elastic constant mismatch across the interface could prevent a dislocation from actually leaving the crack edge.

Analysis of generation after fracture occurs showed that dislocations moving at constant speeds in the crack plane prefer, in the sense of minimum generation time, to leave the crack edge at a greater speed than that of the crack. A consideration of multiple generation and non-crack plane glide showed that glide in the crack plane and single, as opposed to multiple, generation events, require smaller generation times.

Finally it was found that incident wave history, not its instantaneous behavior, determines whether or not dislocation generation can occur and that, if it does, lowering of the resulting stress intensity factor for a stationary crack is not necessarily guaranteed. Indeed, the factor may actually be enhanced.

In closing, it is noted that the dislocation force criterion alone could not completely specify the dislocation motion. Generation instant minimization seemed to be a good candidate for incorporation into the criterion - especially for constant dislocation speeds. Moreover, idealized models were used, although the semi-infinite crack and the unbounded solid are apparently justified by the aforementioned micromechanical scale of the generation process. In particular, Mode III/screw dislocation mechanisms were treated, owing to the simplicity of their mathematical solution forms. Work for transient cases of Mode I, II/edge dislocation mechanisms is available [7,18] and, while some distinctive features arise there, the overall results are similar to those discussed here. One feature common to all such mechanisms is the singular nature of the stress fields imposed by the classical [19] dislocation model; it should be noted that dislocation generation studies are being performed [20] using models which do not cause such severe singular behavior.

ACKNOWLEDGEMENT

This work was partially supported by NSF Grants MEA 8319607 and MSM 8917944, and was carried out in part under the auspices of the Navy/ASEE Summer Faculty Research Program at the Naval Research Laboratory, Washington, DC.

REFERENCES

1. Rice, J. R. and Thomson, R., Ductile versus brittle behavior of crystals, Phil. Mag., 1982, 29, 73-97.

2. Li, J.C.M., Dislocation sources, in Dislocation Modelling of Physical Systems, Pergamon Press, New York, 1981.

3. Ohr, S. M., An electron microscope study of crack tip deformation and its impact on the dislocation theory of fracture, Matl. Sci. Engng., 1985, 72, 1-35.

4. Lin, I.-H. and Thomson, R., Cleavage, dislocation emission and shielding for cracks under general loading, Acta Metl., 1985, 34, 187-200.

5. Rice, J. R., Mathematical analysis in the mechanics of fracture, in Fracture, Vol. II, Academic Press, New York, 1968, pp. 191-311.

6. Brock, L. M., An exact transient analysis of dislocation emission and fracture, J. Mech. Phys. Solids, 1989, 37, 47-69.

7. Brock, L. M. and Wu, J.-S., An analysis of incident wave behavior required for screw and edge dislocation emission from cracks, Int. J. Engng. Sci., 1989, 27, 1223-1239.

8. Brock, L. M. and Wu, J.-S., An equation of motion for the emission of accelerating dislocations from a crack, J. Mech. Phys. Solids, 1990, 38, 273-285.

9. Dundurs, J., Elastic interaction of dislocations with inhomogeneities, in Mathematical Theory of Dislocations, ASME. New York, 1968, pp. 70-115.

10. Hirth, J. P. and Lothe, J., Theory of Dislocations, 2nd edn., Wiley-Interscience, New York, 1982.

11. Mura, T., Micromechanics of Defects in Solids, Martinus Nijhoff, The Hague, 1982.

12. Shaw, M. C., A critical review of mechanical failure criteria, J. Engng. Matl. Tech., 1984, 106, 219-226.

13. Johnston, W. G. and Gilman, J. J., Dislocation velocities, dislocation densities, and plastic flow in lithium fluoride crystals, J. Appl. Phys., 1959, 30, 129-144.

14. Guy, A. G., _Elements of Physical Metallurgy_, Addison-Wesley, Reading, MA, 1959.

15. Achenbach, J. D., _Wave Propagation in Elastic Solids_, North-Holland/American-Elsevier, Amsterdam, 1973.

16. Brock, L. M. and Wu, J.-S., Transient emission of screw dislocations from a dynamically-loaded interface crack, University of Kentucky College of Engineering Technical Report, October 1989.

17. Brock, L. M. and Wu, J.-S., Transient analysis of dislocation emission from a growing crack, _J. Mech. Phys. Solids_, 1989, **37**, 583-601.

18. Brock, L. M. and Wu, J.-S., Transient analyses of dislocation emission in the three modes of fracture, _Int. J. Engng. Sci._, 1989, **27**, 1479-1495.

19. Love, A.E.H., _A Treatise on the Mathematical Theory of Elasticity_, Dover, New York, 1944.

20. Wu, J.-S. and Brock, L. M., Transient analysis of dislocation emission by a crack using a modified dislocation model, in _Developments in Mechanics, Vol. 14b, Proceedings, 20th Midwestern Mechanics Conference_, Purdue University, 1987.

DYNAMIC INTERACTION BETWEEN DISLOCATION AND MODE-III CRACK

M. K. Kuo

Institute of Applied Mechanics

National Taiwan University

Taipei, Taiwan 10764, R. O. C.

ABSTRACT

The transient elastodynamic stress intensity factors of a sub-surface semi-infinite crack in an elastic half plane are analyzed. The crack is subjected to an incident SH wave initially, and some time later a screw dislocation emitting from a point located away from the crack tip. The crucial steps in the analysis are the direct application of integral transforms together with the Wiener-Hopf technique. Exact expressions are obtained for the resulting mode-III stress intensity factors as functions of time. The solution is constructed as a series. Each term in the solution series can be interpreted as the contribution of waves that have reflected at the free surface of half-plane different times. The first two terms in the series for the stress intensity factor history are then computed. The results are exact for the time interval from initial loading until the first wave scattered at the crack tip is reflected twice at the free surface of the half-plane and returns to the crack tip. Numerical results show that the smaller the emitting angle (and/or the smaller the emitting speed) of the dislocation is, the larger shielding effect to the crack.

INTRODUCTION

Rice and Thomson [1] and Ohr [2] have discussed criteria for fracture, and definitions of its brittleness and ductility, in terms of the emission of dislocations from crack edges. These studies have emphasized quasi-static or steady-state results, so that rates of emission are hard to determine. Moreover, dynamic overshoot effects must be neglected. Recently, Brock [3]-[5] and Brock and Wu [6]-[10] studied the transient interaction between a crack and individual dislocation in a material. They developed a model for dislocation emission which allowed the prediction of important quantities such as the times of emission and the distance traveled by an emitted dislocation prior to arrest. In particular, Brock [3], Brock and Wu [6]-[9] considered only interaction between a mode-III crack and screw dislocations; while Brock [4]-[5] and Brock and Wu [10] considered either only mode-I crack with edge dislocation or both mode-I-II and mode-III cracks with edge and screw dislocations. In this above analyses, the material is an unbounded, linearly elastic, isotropic homogeneous solid, and the crack is semi-infinite either stationary or growing with constant speed.

In this paper the effects of free surface on the transient interactions between a mode-III crack and screw dislocations are investigated. In particular, the model assumes a half-plane contained a stationary semi-infinite sub-surface crack which is parallel to the free surface of the half-plane. The problem can be regarded as a sub-surface crack version of the one considered by Brock and Wu [10].

PROBLEM STATEMENT

In the Cartesian coordinate system, two-dimensional anti-plane wave motions of homogeneous, isotropic, linearly elastic solids are governed, in terms of the out-of-plane displacement $w(x, y, t)$, by

$$\frac{\partial^2 w}{\partial x^2} + \frac{\partial^2 w}{\partial y^2} + \frac{F}{\mu} = s^2 \frac{\partial^2 w}{\partial t^2} \tag{1}$$

where F is the body force per unit volume and s is the slowness of the shear wave, which is related to the mass density ρ and shear modulus μ by $s = (\rho/\mu)^{1/2}$. The relevant stress component is

$$\sigma_{yz} = \mu \frac{\partial w}{\partial y} \quad . \tag{2}$$

We will also be particularly interested in the elastodynamic mode-III stress intensity factor which is defined as

$$K_{III}(t) = \lim_{x \to 0+} [2\pi x]^{1/2} \sigma_{yz}(x, 0, t) \quad . \tag{3}$$

Consider a sub-surface semi-infinite crack along $y = 0$, $x < 0$, which is parallel to the free surface of a half-plane. The half-plane, $y \le a$, containing the crack is at rest initially except for a stress pulse traveling as a horizontally polarized shear (SH) wave at an angle α to the crack plane (and hence to the free surface of the half plane). The incident wave, indicated with superscript in, is given by, for $y \le a$,

$$\sigma_{yz}^{in} = \sigma_0 \cos \alpha \, f(t - sy \cos \alpha + sx \sin \alpha) \tag{4}$$

where f is identically zero when its argument is negative, but otherwise is an arbitrary waveform. Thus the wave reaches the crack tip at time $t = 0$. Ahead of the advancing incident wavefront the medium is undisturbed. In the latter numerical examples, f is taken as the unit step function for illustration.

At $t = t_d > 0$, a single right-handed screw dislocation emits from a point located a distance d away from the crack tip. The dislocation is assumed to travel rectilinearly along the X-axis, which is making an angle ϕ ($0 \leq \phi \leq \pi$) with respect to the crack plane, with a constant subsonic speed v

$$w|_{Y=0+} - w|_{Y=0-} = bH(X - d)\, H[v(t - t_d) - X + d] \qquad (5)$$

where b and $H(\ )$ are the z-component of the Burger's vector and the Heaviside step function, respectively, and the coordinate system (x, y) and (X, Y) are shown in the Fig. 1. The delay time t_d, the emitting angle ϕ and the emitting speed v are supposed to be determined from criteria such as dislocation force argument etc., which we will not discuss in this paper. The traction free boundary condition at the surface of half-plane is

$$\sigma_{yz}(x, y, t) = 0 \quad \text{along} \quad y = a \quad . \qquad (6)$$

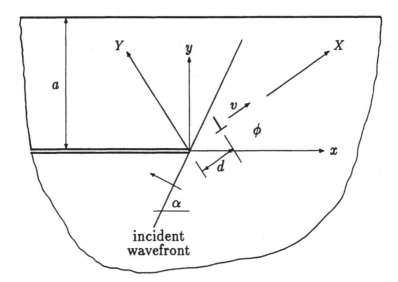

Figure 1. Geometry of the sub-surface crack, the incident wave and the dislocation.

The process itself can be treated as one of anti-plane strain, so that only the out-of-plane displacement $w(x, y, t)$ exists. By linear superposition, this can be written as

$$w = w^{in} + w^r + w^{sc} + w^d \qquad (7)$$

where w^{in} and w^r are the displacements of the incident and the reflected SH-wave in the half-plane, respectively, as if there were no crack; w^{sc} is the displacement generated in order to cancel the w^{in}- and w^r-induced stresses from the crack surface; and w^d is the displacement induced by the movement of the dislocation in the half-plane with crack present. The solution $w^{in} + w^r + w^{sc}$ is in fact the solution of for the sub-surface crack subjected to the incident wave loading.

INCIDENT WAVE ON THE SUB-SURFACE CRACK

The problem of incident wave striking on a semi-infinite stationary crack in an un-bounded medium has been treated by Achenbach [11], and expressions for both dis-placement and stresses directly ahead of the crack derived. In the case of sub-surface crack, the wave will multiply reflect and diffract among the crack tip, the crack surfaces and the free surface of the half-plane. We intend only to find the early time solutions where only few of reflection and diffraction have been encountered.

The solution to the reflection of the incident SH wave in the uncracked half-plane is trivial. It can be deduced directly by the method of image as

$$\sigma^r_{yz}(x,y,t) = -\sigma_0 \cos\alpha \; f[t + s(y-2a)\cos\alpha + sx\sin\alpha] \tag{8}$$

The scattered waves which are added to the incident and reflected waves to form the total field for the sub-surface crack problem without dislocation are determined from (1) with F vanishing, and boundary conditions (6) and

$$\sigma^{sc}_{yz} = -\sigma^{in}_{yz} - \sigma^r_{yz} \quad \text{along} \quad y = 0, \quad x < 0 \tag{9}$$

which specifies that the crack surfaces are free of traction.

To solve (1) for the scattered waves we employ the integral transforms and the Wiener-Hopf method. First, the one-sided Laplace transform in t and two-sided Laplace transform in x with transform parameter p and $p\xi$, respectively, are applied to the complete set of governing equations. The appropriate doubly transformed solutions of (1) (with $F = 0$) satisfying the boundary condition (6) are

$$\tilde{\bar{w}}^{sc} = \begin{cases} A_s \cosh[p\gamma(y-a)] & , \; 0 \le y \le a \\ B_s \exp(p\gamma y) & , \; y \le 0 \end{cases} \tag{10}$$

where $\gamma = (s^2 - \xi^2)^{1/2}$, and A_s and B_s are arbitrary functions of ξ and p. The branch cuts have been taken in the complex ξ-plane along the real axis, from $\xi \to -\infty$ to $-s$ and $\xi = s$ to ∞, such that $\text{Re}(\xi) \ge 0$ in the whole complex ξ-plane. The symbol 'Re' denotes the real part of a function, while the superposed bar and tilde denotes the one-sided and two-sided Laplace transforms.

From (10) the double integral transformed displacement and stress along the crack line are expressed in terms of A_s and B_s as

$$\tilde{\bar{w}}^{sc}_- + \tilde{\bar{w}}^{sc}_+ = \begin{cases} A_s \cosh(p\gamma a) & , \; y = 0^+ \\ B_s & , \; y = 0^- \end{cases} \tag{11}$$

$$\tilde{\bar{\sigma}}^{sc}_{yz} = \begin{cases} -\mu p\gamma A_s \sinh(p\gamma a) & , \; y = 0^+ \\ \mu p\gamma B_s & , \; y = 0^- \end{cases} \tag{12}$$

In (11) $\tilde{\bar{w}}^{sc}_-(\xi, 0^\pm, p)$ and $\tilde{\bar{w}}^{sc}_+(\xi, 0, p)$ are the double integral transformed unknown crack-face-displacements $w^{sc}(x, 0^\pm, t)$ of the (upper and lower) crack faces and displacements ahead of the crack tip $w^{sc}(x, 0, t)$, respectively. Since w^{sc}_+ vanishes for $x > 0$ by defini-tion, moreover, because of anticipated wavefronts it is zero for $x < -t/s$, $y = 0^\pm$. By virtue of the theory of Laplace transform, $\tilde{w}_-^{sc}$ is analytic and goes to zero as $|\xi| \to \infty$ in the half complex ξ-planes $\text{Re}(\xi) < s$.

Next, the doubly transformed boundary condition of crack surfaces, from (9), is found to be along $y = 0$

$$\tilde{\tilde{\sigma}}_{yz}^{sc} = \frac{\sigma_0 \cos \alpha \, \bar{f}(p)}{p \, (\xi - s \sin \alpha)} \left(1 - e^{-2psa \cos \alpha}\right) + \tilde{\tilde{\Sigma}}_+^{sc} \quad , \quad \text{Re}(\xi) < s \sin \alpha \qquad (13)$$

where $\tilde{\tilde{\Sigma}}_+^{sc}$ is the double transform of the σ_{yz}^{sc} along $y = 0$, $x > 0$, and $\bar{f}(p)$ is the one-sided Laplace transform of $f(t)$. The subscript plus signs, as of $\tilde{\tilde{\Sigma}}_+^{sc}$ and $\tilde{w}_+^{sc}$, are intended to denote functions that are analytic in the right (plus) half plane of the transform parameter ξ. Likewise, the subscript minus signs, as of $\tilde{w}_-^{sc}$, are for functions that are analytic in the left (minus) half plane of the transform parameter ξ. In relating transformed crack line stresses and displacements through (11)-(13) by eliminating A_s, B_s and $\tilde{w}_+^{sc}$, one has

$$-\mu p \gamma L(\xi, p) \tilde{\tilde{W}}_-^{sc} = \tilde{\tilde{\Sigma}}_+^{sc} + \frac{\sigma_0 \cos \alpha \, \bar{f}}{p \, (\xi - s \sin \alpha)} \left(1 - e^{-2psa \cos \alpha}\right) \qquad (14)$$

where

$$L(\xi, p) = [1 - \exp(-2p\gamma a)]/2 \qquad (15)$$

and $\tilde{\tilde{W}}_-^{sc} \equiv \tilde{w}_-^{sc}(\xi, 0^+, p) - \tilde{w}_-^{sc}(\xi, 0^-, p)$ denotes the double integral transform of the crack opening displacement. By (14) the problem has been reduced into a form suitable for the application of the Wiener-Hopf technique. The solution of (14) then depends upon being able to write those related functions as the product and the sum, respectively, of regular functions.

Since the function $L(\xi, p)$ has infinite numbers of zeros in the complex ξ-plane, the direct application of the formula of sum splitting to the logarithm of $L(\xi, p)$ can only be evaluated numerically in the complex ξ-plane. It is also possible to introduce a new function which has only simple poles and zeros but no branch cuts in the ξ-plane. It can be then expressed as an infinite product and the quotient splitting is immediately followed. Unfortunately, though the formal expression of the one-sided Laplace transformed stress intensity factor can be found explicitly, the inversion of the integral transform can then only be performed numerically. Moreover, the convergence of the infinite product expressions are usually very slow, as pointed out in Knauss [12].

Instead of finding the solution by performing the quotient splitting directly to L, let us construct series for the crack opening displacement $\tilde{\tilde{W}}_-^{sc}$ and the crack-line stress $\tilde{\tilde{\Sigma}}_+^{sc}$ of general forms

$$\tilde{\tilde{W}}_-^{sc}(\xi, p) = \tilde{\tilde{W}}_{s-}^{(0)}(\xi, p) + \tilde{\tilde{W}}_{s-}^{(1)}(\xi, p) + \tilde{\tilde{W}}_{s-}^{(2)}(\xi, p) + \cdots \qquad (16)$$

$$\tilde{\tilde{\Sigma}}_+^{sc}(\xi, p) = \tilde{\tilde{\Sigma}}_{s+}^{(0)}(\xi, p) + \tilde{\tilde{\Sigma}}_{s+}^{(1)}(\xi, p) + \tilde{\tilde{\Sigma}}_{s+}^{(2)}(\xi, p) + \cdots \qquad (17)$$

with $\tilde{\tilde{W}}_{s-}^{(j)}$ and $\tilde{\tilde{\Sigma}}_{s+}^{(j)}$ being solutions of the following problems

$$-\frac{1}{2}\mu p \gamma_- \tilde{\tilde{W}}_{s-}^{(0)} = \frac{\tilde{\tilde{\Sigma}}_{s+}^{(0)}}{\gamma_+} + \frac{\sigma_0 \cos \alpha \, \bar{f}}{p} D_s(\xi) \qquad (18)$$

$$-\frac{1}{2}\mu p\gamma_-\tilde{\tilde{W}}_{s-}^{(1)} \;=\; \frac{\tilde{\tilde{\Sigma}}_{s+}^{(1)}}{\gamma_+} - \frac{1}{2}\mu p\gamma_-\,\tilde{\tilde{W}}_{s-}^{(0)}e^{-2p\gamma a} - \frac{\sigma_0\cos\alpha\,\bar{f}}{p}e^{-2psa\cos\alpha}D_s(\xi) \quad (19)$$

$$-\frac{1}{2}\mu p\gamma_-\tilde{\tilde{W}}_{s-}^{(2)} \;=\; \frac{\tilde{\tilde{\Sigma}}_{s+}^{(2)}}{\gamma_+} - \frac{1}{2}\mu p\gamma_-\,\tilde{\tilde{W}}_{s-}^{(1)}e^{-2p\gamma a} \qquad\qquad\qquad\quad (20)$$

$$\cdots$$

where

$$D_s(\xi) = \frac{1}{(\xi - s\sin\alpha)\gamma_+} \;. \qquad\qquad\qquad (21)$$

It is easy to verify that the infinite series of (16) and (17) constructed in this manner certainly satisfies (14). By using the definition of stress intensity factor given as (3), it is apparent that the series of stress intensity factor corresponding to (16) and (17) is

$$K_s = K_s^{(0)} + K_s^{(1)} + K_s^{(2)} + \cdots \;. \qquad\qquad\qquad (22)$$

Notice that the zeroth order equation (18) is exactly the corresponding Wiener-Hopf equation for a cracked infinite plane subjected to expanding loadings on the crack faces. We can now interpret physically results corresponding to each order. From point of view of wave propagation, the zeroth order solutions correspond to the diffracted field of incident wave from the crack tip $x = 0$ starting at time $t = 0$. Thus, in (19) where the zeroth order crack opening displacement is multiplied by $\exp(-2p\gamma a)$, it represents the disturbance of the wave which has been reflected again at the surface of the half-plane and finally come back to the crack tip at $t = 2sa$. The last term in (19) with $\exp(-2psa\cos\alpha)$ represents the disturbance due to the reflection of the incident wave at the surface of the half plane which arrives at the crack tip at $t = 2sa\cos\alpha$. Similarly, in (20) where the first order crack opening displacement is multiplied by $\exp(-2p\gamma a)$, it represents the disturbance due to the reflection of the first order crack opening displacement at the surface of the half-plane and coming back to the crack tip. Consequently, the second order solutions occur at the time $2sa$ after the zeroth order solutions have been in effect.

$\tilde{\tilde{W}}_{s-}^{(0)}$ and $\tilde{\tilde{\Sigma}}_{s+}^{(0)}$

As mentioned above, (18) corresponds to the problem of a cracked infinite solid subjected to expanding loadings on the crack faces. It, as one would expect, implies that in the short time range the Mode-III elastodynamic responses for a cracked half plane can be computed for a crack in an infinite solid, provided that the crack face tractions are the ones of the original half plane problem. The cracked infinite solid problem has been discussed by Achenbach [11], and expressions for both displacement and stresses directly ahead of the crack derived. Here we list only the needed results.

$$\frac{1}{2}\mu p\gamma_-\tilde{\tilde{W}}_{s-}^{(0)} \;=\; -\frac{\sigma_0\cos\alpha\,\bar{f}}{p}D_{s-}(\xi) \qquad\qquad\qquad (23)$$

$$\tilde{\tilde{\Sigma}}_{s+}^{(0)} \;=\; -\frac{\sigma_0\cos\alpha\,\bar{f}}{p}\gamma_+D_{s+}(\xi) \qquad\qquad\qquad (24)$$

$$K_s^{(0)}(t) = \left(\frac{2}{\pi}\right)^{1/2} \frac{\sigma_0 \cos \alpha}{s^{1/2}(1+\sin \alpha)^{1/2}} H(t) \int_0^t f(t-\eta)\eta^{-1/2}d\eta \qquad (25)$$

with

$$D_{s-}(\xi) = \frac{1}{s^{1/2}(1+\sin \alpha)^{1/2}} \frac{1}{\xi - s\sin \alpha} \qquad (26)$$

$$D_{s+}(\xi) = \frac{1}{s(1+\sin \alpha)} \left[\frac{(\xi+s)^{1/2} - s^{1/2}(1+\sin \alpha)^{1/2}}{\xi - s\sin \alpha} - \frac{1}{(\xi+s)^{1/2}}\right] \cdot \qquad (27)$$

$\tilde{\tilde{W}}_{s-}^{(1)}$ and $\tilde{\tilde{\Sigma}}_{s+}^{(1)}$

From (19) and (23), the equation governing $\tilde{\tilde{W}}_{s-}^{(1)}$ and $\tilde{\tilde{\Sigma}}_{s+}^{(1)}$ is obtained as

$$-\frac{1}{2}\mu p\gamma_- \tilde{\tilde{W}}_{s-}^{(1)} = \frac{\tilde{\tilde{\Sigma}}_{s+}^{(1)}}{\gamma_+} + \frac{\sigma_0 \cos \alpha \bar{f}}{p}\left[\frac{G_s(\xi, 2a)}{s^{1/2}(1+\sin \alpha)^{1/2}} - e^{-2psa\cos \alpha}D_s(\xi)\right] \cdot \qquad (28)$$

where

$$G_s(\xi, \Lambda) = e^{-p\gamma\Lambda}/(\xi - s\sin \alpha) \quad . \qquad (29)$$

The solution for the problem described by (28) can be obtained again by the Wiener-Hopf technique. The last term in the right hand side of (28) is exactly the same as the nonhomogeneous term in (18) except a sign difference and an extract factor $e^{-2psa\cos \alpha}$, hence in the first order solution due to this term is simply the negative of the zeroth order solution with a time shifting $2sa\cos \alpha$.

Since $G_s(\xi, \Lambda)$ has only a simple pole and branch cuts, it can be expressed as the sum of two regular functions $G_{s+}(\xi, \Lambda)$ and $G_{s-}(\xi, \Lambda)$ by the direct application of the formula of sum splitting

$$G_{s+}(\xi, \Lambda) = -\frac{1}{2\pi i}\int_{-c-\infty i}^{-c+\infty i} \frac{e^{-p(s^2-u^2)^{1/2}\Lambda}}{(u-\xi)(u-s\sin \alpha)} du = G_s(\xi, \Lambda) - G_{s-}(\xi, \Lambda)$$
$$, \quad c > 0 \quad . \qquad (30)$$

As in the last section, the uses of the analytic continuation and the Liouville's theorem conclude

$$\tilde{\tilde{\Sigma}}_{s+}^{(1)} = -\frac{\sigma_0 \cos \alpha \bar{f}}{p}\gamma_+ \left[\frac{G_{s+}(\xi, 2a)}{s^{1/2}(1+\sin \alpha)^{1/2}} - D_{s+}(\xi)e^{-2psa\cos \alpha}\right] \qquad (31)$$

$$\frac{1}{2}\mu p\gamma_- \tilde{\tilde{W}}_{s-}^{(1)} = -\frac{\sigma_0 \cos \alpha \bar{f}}{p}\left[\frac{G_{s-}(\xi, 2a)}{s^{1/2}(1+\sin \alpha)^{1/2}} - D_{s-}(\xi)e^{-2psa\cos \alpha}\right] \cdot \qquad (32)$$

The double integral transform of (31) can now be inverted. Notice that $|G_{s+}(\xi)| \sim O(|\xi|^{-1})$, as $|\xi| \to \infty$ in the right half plane. From Abelian theorem [13] concerning asymptotic relations between functions and their Laplace transforms, one has

$$\bar{K}_s^{(1)}(p) = \left(\frac{2}{\pi}\right)^{1/2} \frac{\sigma_0 \cos \alpha \bar{f}}{s^{1/2}(1+\sin \alpha)^{1/2}} \bar{K}_{s1}(p, 2a) - \bar{K}_s^{(0)}e^{-2psa\cos \alpha} \qquad (33)$$

where

$$\bar{K}_{s1}(p,\Lambda) = -\int_0^\infty w^{1/2} \left\{ \frac{1}{2\pi i} \int_{Br} \frac{e^{-p[w+(s^2-u^2)^{1/2}\Lambda]}}{u-s\sin\alpha} du \right\} dw \quad . \tag{34}$$

Finally, by letting $w + (s^2 - u^2)^{1/2}\Lambda = t$ and using the Cagniard-de Hoop method, the one-sided Laplace transform in (34) can be inverted by inspection. Hence the first order correction to the stress intensity factor is

$$K_s^{(1)}(t) = \left(\frac{2}{\pi}\right)^{1/2} \frac{\sigma_0 \cos\alpha}{s^{1/2}(1+\sin\alpha)^{1/2}} f(t) * K_{s1}(t,2a) - K_s^{(0)}(t-2sa\cos\alpha) \tag{35}$$

where

$$K_{s1}(t,\Lambda) = \frac{\Lambda}{\pi} H(t-s\Lambda) \int_{s\Lambda}^t \frac{ws\sin\alpha}{[w^2-(s\Lambda)^2]^{1/2}(t-w)^{1/2}(w^2-s^2\Lambda^2\cos^2\alpha)} dw \tag{36}$$

and the asterisk symbol indicates the convolution. The convolution integral in (36) can be simplified, when $f(t) = H(t)$ as used in the numerical examples, as

$$f(t) * K_{s1}(t,\Lambda) = (2\Lambda/\pi)H(t-s\Lambda) \int_{s\Lambda}^t \frac{w(t-w)^{1/2}s\sin\alpha \, dw}{[w^2-(s\Lambda)^2]^{1/2}(w^2-s^2\Lambda^2\cos^2\alpha)} \quad . \tag{37}$$

DISLOCATION EMITTING FROM THE SUB-SURFACE CRACK

The dislocation described in (5) can be replaced by the equivalent body forces F as suggested by Burridge and Knopoff [14], where

$$F = -\mu b\delta'(Y)H(X-d)H[v(t-t_d)-X+d] \tag{38}$$

and $\delta'()$ is the derivative of the Dirac delta function. The appropriate doubly transformed solutions of (1) satisfying the boundary condition (6), with F describing as (38), can easily be obtained. As the procedure in the previous section, after relating transformed crack line stresses and displacements, one has

$$- \mu p\gamma L\tilde{\tilde{W}}_-^d = \tilde{\tilde{\Sigma}}_+^d - \frac{\mu b}{2p} e^{-p(t_d+\xi d\cos\phi)} \left[ge^{-p\gamma d\sin\phi} + he^{-p\gamma(2a-d\sin\phi)} \right] \tag{39}$$

where

$$g(\xi) = \frac{\xi\sin\phi - \gamma\cos\phi}{s_1 + \xi\cos\phi + \gamma\sin\phi}, \quad h(\xi) = \frac{\xi\sin\phi + \gamma\cos\phi}{s_1 + \xi\cos\phi - \gamma\sin\phi}, \quad \mathrm{Re}(\xi) > -s_1|\cos\phi| \tag{40}$$

with $s_1 = 1/v$, and $\tilde{\tilde{W}}_-^d \equiv \tilde{\tilde{w}}_-^d(\xi,0^+,p) - \tilde{\tilde{w}}_-^d(\xi,0^-,p)$ being the slowness of the dislocation emitting speed and the double integral transform of the crack opening displacement, respectively. Notice that in (39) a term corresponding the contribution of the reflection of the dislocation motion from the free surface of the half plane has been discarded, since we consider only the early time solution. Again, let us construct series for the crack opening displacement and the crack-line stress of general forms as (16) and (17), except that the superscript sc and subscript s are replaced by the superscript d and subscript d, respectively, and

$$-\frac{1}{2}\mu p\gamma_- \tilde{\tilde{W}}_{d-}^{(0)} = \frac{\tilde{\tilde{\Sigma}}_{d+}^{(0)}}{\gamma_+} - \frac{\mu b}{2p} D_d(\xi) \tag{41}$$

$$-\frac{1}{2}\mu p\gamma_- \tilde{\tilde{W}}_{d-}^{(1)} = \frac{\tilde{\tilde{\Sigma}}_{d+}^{(1)}}{\gamma_+} - \frac{1}{2}\mu p\gamma_- \tilde{\tilde{W}}_{d-}^{(0)} e^{-2p\gamma a} - \frac{\mu b}{2p} G_d(\xi) \tag{42}$$

$$-\frac{1}{2}\mu p\gamma_- \tilde{\tilde{W}}_{d-}^{(2)} = \frac{\tilde{\tilde{\Sigma}}_{d+}^{(2)}}{\gamma_+} - \frac{1}{2}\mu p\gamma_- \tilde{\tilde{W}}_{d-}^{(1)} e^{-2p\gamma a} \tag{43}$$

$$\cdots$$

where

$$D_d(\xi) = (g/\gamma_+) \, e^{-p[(\xi \cos\phi - \gamma \sin\phi)d - t_d]} \tag{44}$$

$$G_d(\xi) = (h/\gamma_+) \, e^{-p[(\xi \cos\phi - \gamma \sin\phi)d + 2\gamma a - t_d]} \tag{45}$$

The corresponding series of stress intensity factor is

$$K_d = K_d^{(0)} + K_d^{(1)} + K_d^{(2)} + \cdots \quad . \tag{46}$$

Since the methods of the solution are again the Wiener-Hopf technique together with the Cagniard-de Hoop method as discussed in the last section, we will list only the brief results.

Notice that (41) is exactly the corresponding Wiener-Hopf equation for a cracked infinite plane subjected to the moving dislocations. The physical interpretation of results corresponding to each order are as follows. In (42) where the zeroth order crack opening displacement is multiplied by $\exp(-2p\gamma a)$, it represents the disturbance due to the reflection of the zeroth order crack opening displacement at the surface of the half-plane and come back to the crack tip at time $2sa$ after the zeroth order solutions have been in effect. The last term in (42) with $\exp(-2p\gamma a)$ represents the disturbance which emits from the moving dislocation, reflects at the surface of the half plane and finally arrives at the crack tip. Similarly, in (43) where the first order crack opening displacement is multiplied by $\exp(-2p\gamma a)$, it represents the disturbance due to the reflection of the first order crack opening displacement at the surface of the half-plane and coming back to the crack tip. Consequently, the second order solutions occur at the time $2sa$ after the first order solutions have been in effect.

$\tilde{\tilde{W}}_{d-}^{(0)}$ and $\tilde{\tilde{\Sigma}}_{d+}^{(0)}$

As mentioned above, (41) corresponds to the problem of a cracked infinite plane subjected to the moving dislocations. It, as one would expect, implies that in the short time range the Mode-III elastodynamic responses for a cracked half plane can be computed for a crack in an infinite plane, provided that the moving dislocations are the ones of the original half plane problem. The cracked infinite plane problem has been discussed by Brock and Wu [9] using of the linear superposition of the Green's function solutions of SH point force in the unbounded solid.

By the use of the Wiener-Hopf technique and the Cagniard-de Hoop method, we have

$$\frac{1}{2}\mu p\gamma_-\tilde{\bar{W}}_{d-}^{(0)} = \frac{\mu b}{2p}D_{d-}(\xi) \tag{47}$$

$$\tilde{\bar{\Sigma}}_{d+}^{(0)} = \frac{\mu b}{2p}\gamma_+ D_{d+}(\xi) \tag{48}$$

$$K_d^{(0)}(t) = \frac{\mu b\sin(\phi/2)}{(2\pi d)^{1/2}}\left\{1 - \left[\frac{s+s_1}{(t-t_d)/d+s_1}\right]^{1/2}\right\}H(t-t_d-sd) \tag{49}$$

with

$$D_{d+}(\xi) = -\frac{1}{2\pi i}\int_{Br}\frac{D_d(\eta)}{\eta-\xi}d\eta = D_d(\xi) - D_{d-}(\xi) \quad . \tag{50}$$

$\tilde{\bar{W}}_{d-}^{(1)}$ and $\tilde{\bar{\Sigma}}_{d+}^{(1)}$

From (42) and (47), the equation governing $\tilde{\bar{W}}_{d-}^{(1)}$ and $\tilde{\bar{\Sigma}}_{d+}^{(1)}$ is obtained as

$$-\frac{1}{2}\mu p\gamma_-\tilde{\bar{W}}_{d-}^{(1)} = \frac{\tilde{\bar{\Sigma}}_{d+}^{(1)}}{\gamma_+} - \frac{\mu b}{2p}(D_d - D_{d+})e^{-2p\gamma a} - \frac{\mu b}{2p}G_d(\xi) \quad . \tag{51}$$

As in the last section, the uses of the analytic continuation, the Liouville's theorem and the Cagniard-de Hoop method conclude the first order correction to the stress intensity factor is

$$K_d^{(1)}(t) = \frac{\mu b}{\pi(2\pi)^{1/2}}[K_{d1}(t) + K_{d2}(t) + K_{d3}(t)] \tag{52}$$

where

$$K_{d1}(t) = H(t - t_1 - t_d)\int_{t_1+t_d}^{t}(t-\tau)^{-1/2}\text{Im}\left[(s+\eta_1)^{-1/2}g(\eta_1)\frac{\partial\eta_1}{\partial\tau}\right]d\tau \tag{53}$$

$$K_{d2}(t) = -H(t - t_2 - t_d)\frac{1}{2\pi}\int_{s}^{(t-t_d-2sa)/d}\text{Im}\left[\frac{\eta_2 I(u)}{(s+\eta_2)^{1/2}(s_1+u)}\right]du \tag{54}$$

$$K_{d3}(t) = H(t - t_3 - t_d)\int_{t_3+t_d}^{t}(t-\tau)^{-1/2}\text{Im}\left[(s+\eta_3)^{-1/2}h(\eta_3)\frac{\partial\eta_3}{\partial\tau}\right]d\tau \tag{55}$$

with

$$I(v) = J\left(2a\sqrt{s^2-\eta_2^2}\right) - J\left(-2a\sqrt{s^2-\eta_2^2}\right) \tag{56}$$

$$J(q) = \frac{1}{(\bar{t}-q)^{1/2}}\log\left[-\frac{(\bar{t}-2sa)^{1/2}+(\bar{t}-q)^{1/2}}{(\bar{t}-2sa)^{1/2}-(\bar{t}-q)^{1/2}}\right], \quad \bar{t} = t-t_d-vd \tag{57}$$

$$\eta_1 = (s/t_1)\left[\tau d\cos\phi + i(2a + d\sin\phi)(\tau^2 - t_1^2)^{1/2}\right] \qquad (58)$$

$$\eta_2 = u\cos\phi + i\sin\phi(u^2 - s^2)^{1/2} \qquad (59)$$

$$\eta_3 = (s/t_3)\left[\tau d\cos\phi + i(2a - d\sin\phi)(\tau^2 - t_3^2)^{1/2}\right] \qquad (60)$$

$$t_1 = s(d^2 + 4ad\sin\phi + 4a^2)^{1/2} \qquad (61)$$

$$t_2 = s(d + 2a) \qquad (62)$$

$$t_3 = s(d^2 - 4ad\sin\phi + 4a^2)^{1/2} \qquad (63)$$

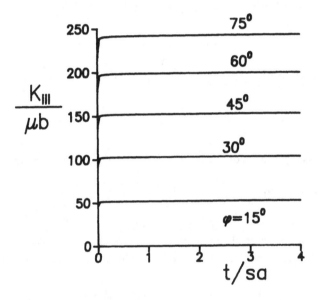

Figure 2. Mode-III stress intensity factor history for various emitting angle.

RESULTS AND DISCUSSIONS

There are numerous parameters in the present analysis. They may be subdivided into material parameters, geometric parameter, and loading parameters. The material parameters are the mass density ρ, the shear moduli μ of the half-plane. The geometric parameter is the characteristic length a. The loading parameters are the waveform f and the incident angle α of the incident wave, as well as the Burg's vector b, the initial emitting position d, the emitting angle ϕ, the moving speed v and the delay time t_d of the screw dislocation.

The elastodynamic Mode-III stress intensity factors of the sub-surface crack, which is parallel to the free surface of the half-plane, are the sum of (22) and (46). Numerical calculations have been carried out for various values of sv and angle ϕ. Only some typical results are presented in the paper. Figure 2 shows the dimensionless stress intensity factors versus the dimensionless time for various values of sv, ϕ with zero delay time, $d/a = 10^{-6}$, $b = d$, $\alpha = 30^0$, and $\sigma_0/\mu = 10^{-5}$.

For $sv = 0.1$, Fig. 2 shows $K_{III}/(\mu b)$ as a function of $t/(sa)$ for various values of ϕ. The result shows that the smaller the emitting angle is, the smaller the effect. A very small kink in K_{III} at time $t = 2sa$ is attributed to the arrival of the reflection of the diffracted wave. For a fixed emitting angle and various emitting speed, the results are almost identical, this suggests that the effect of dislocation emitting speed is negligible.

ACKNOWLEDGEMENT

This work was sponsored by the National Science Council of the Republic of China under Grant NSC80-0401-E002-10.

REFERENCES

1. Rice, J. R. and Thomson, R., Ductile versus brittle behavior of crystals. Phil. Mag., 1974, **29**, 73-97.

2. Ohr, S. M., An electron microscope study of crack tip deformation and its impact on the dislocation theory of fracture. Mater. Sci. Engng., 1985, **72**, 1-35.

3. Brock, L. M., An exact transient analysis of dislocation emission and fracture. J. Mech. Phys. Solids, 1989, **37**, 47-69.

4. Brock, L. M., Transient analyses of dislocation emission in the three modes of fracture. Int. J. Engng. Sci., 1989, **27**, 1479-1495.

5. Brock, L. M., Transient analysis of edge dislocation emission from a mode I crack, to appear 1990.

6. Brock, L. M. and Wu, J.-S., Transient generalized forces due to dislocation array-growing crack interaction. Int. J. Solids Structures, 1989, **25**, 393-405.

7. Brock, L. M. and Wu, J.-S., Transient analysis of dislocation emission from a growing crack. J. Mech. Phys. Solids, 1989, **37**, 583-601.

8. Brock, L. M. and Wu, J.-S., An exact transient study of dislocation emission and its effects on dynamic fracture initiation. J. Appl. Mech., 1989, **56**, 571-576.

9. Brock, L. M. and Wu, J.-S., An equation of motion for the emission of accelerating dislocations from a crack. J. Mech. Phys. Solids, 1990, **38**, 273-285.

10. Brock, L. M. and Wu, J.-S., An analysis incident wave behavior required for screw and edge dislocation emission from cracks. Int. J. Engng. Sci., 1989, **27**, 1223-1239.

11. Achenbach, J. D., Wave Propagation in Elastic Solids, North Holland, Amsterdam, 1973.

12. Knauss, W. G., J. Appl. Mech., 1966, **33**, 356-362.

13. Noble, B., Methods based on the Wiener-Hopf Technique, Pergamon Press, New York, 1958.

14. Burridge, L. and Knopoff, L., Body force equivalents for seismic dislocation. Bull. Seism. Am., 1964, **54**, 1875-1888.

MICROMECHANICS OF CRAZE-AIDED DUCTILITY OF RUBBER-TOUGHENED PLASTICS UNDER DYNAMIC LOADING

J. SHIOIRI, T. SEKI and M. HIRAI
College of Engineering, Hosei University
3-7-2 Kajinocho, Koganei-shi, Tokyo 184, Japan

ABSTRACT

Simultaneous measurements of the flow stress and stress whitening are conducted for HIPS and ABS, typical rubber-toughenend plastics, over wide ranges of the strain rate and temperature. The results indicate that the ductile elongation after yielding is of craze origin and the strain rate- and temperature-dependences of the flow stress can be expressed by Eyring's rate process equation. Theoretical analysis is made for the craze-aided ductile deformation using a micromechanical model of the craze in which the growth of the craze is controlled by the thermally assisted drawing out of the molecules from the matrix into the craze in the form of fibrils. The results show a fairly good agreement with the experimental results.

INTRODUCTION

Plastics for engineering use are required to have a sufficient impact resistance together with a reasonably high yield strength. In most cases, however, these two properties are incompatible with each other. One of the methods of improving the impact resistivity of the glassy brittle polymers such as polystyrene is the rubber-toughening. The remarkable increase in the impact resistivity by this method has been interpreted qualitatively in terms of the energy absorption by crazes or shear yielding originating from the finely dispersed rubber particles [1]. Quantitatively, however, it should be considered in terms of the ductile elongation after yielding under dynamic loading. In fact, rubber-toughened plastics such as HIPS and ABS show elongation more than several tens percent even under high speed loading. In case of the craze-aided flow, the deformation is dilatational, i.e. the elongation occurs by the volume change, since the density of the matter in the craze is lower than that of the matrix. On the other hand, in case of the flow caused by the shear yielding, the deformation is distortional and deformation occurs without volume change. Further, in case of the craze-aided flow, a remarkable decrease in the light transparency called "stress whitening" due to the scattering of light by crazes is observed. It has been pointed out that the ductile elongation in HIPS and ABS are the craze-aided flow except in ABS at high temperatures or/and at low strain rates.

In this work, firstly the simultaneous measurements of the flow stress and stress whitening were conducted over wide ranges of the strain rate and temperature, and secondly a micromechanical analysis was made by using a kinetic model of the craze. An emphasis was put on the thermally activated process controlling the rate of the growth of crazes.

EXPERIMENT

Simultaneous measurements of the flow stress and the decrease in the light transparency were made for commercial sheets of HIPS and ABS. Specimens were of dumbbell type. The thickness and the length and width of the parallel part were 1 mm, 15 mm and 10 mm, respectively. For HIPS the measurements were made at strain rates ranging from 5.5×10^{-4} to 2.8×10^{2} /sec and at temperatures from 295 to 353 K, and for ABS at strain rates from 2×10^{-2} to 2×10^{2} /sec and at temperatures from 253 to 353 K. In almost all the cases a large ductile elongation more than 20 percent was observed, but, except in cases of ABS at high temperatures and at very low strain rates, the lateral contraction was considerably small compared with the longitudinal elongation. Further, a remarkable whitening was observed at the yield point and the whitening grew with increase in the ductile flow. These experimental facts indicate that the large ductile elongation observed after yielding was essentially of craze origin. Typical stress strain relationships are shown in Figure 1 together with the transparency decreases due to the stress whitening.

Flow stresses at a strain of 0.20 are shown in Figure 2 for HIPS and ABS. As was pointed out by Truss and Chadwick [2] for the yield stress of ABS, the strain rate- and temperature-dependences of the flow stress shown in Figure 2 obey Eyring's rate process equation

$$\dot{\varepsilon} = \dot{\varepsilon}_{o} \exp[-(\Delta E - \alpha\sigma)/kT] \tag{1}$$

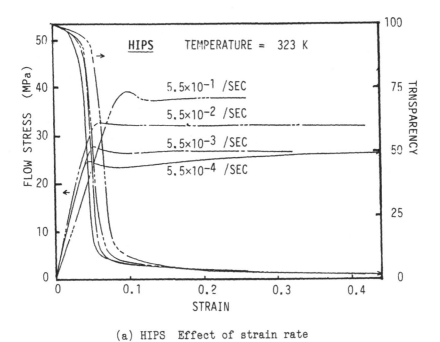

(a) HIPS Effect of strain rate

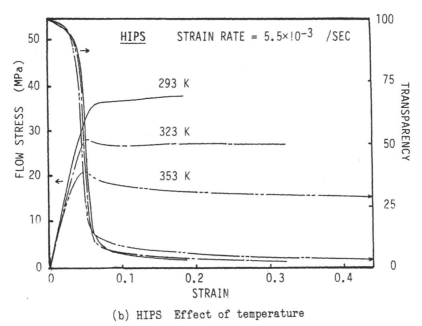

(b) HIPS Effect of temperature

Figure 1. Example of experimental stress-strain relationship. Change in
light transparency due to craze formation is also shown.
(Continue to next page)

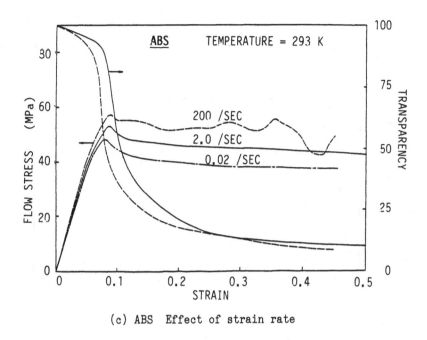

(c) ABS Effect of strain rate

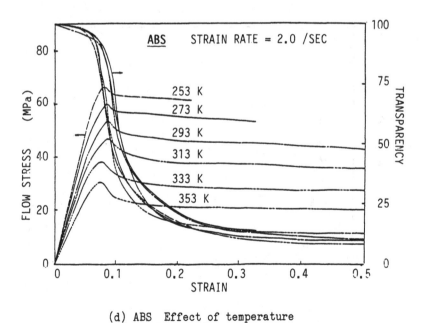

(d) ABS Effect of temperature

Figure 1. Example of experimental stress-strain relationship. Change in
light transparency due to craze formation is also shown.

(Continued from last page)

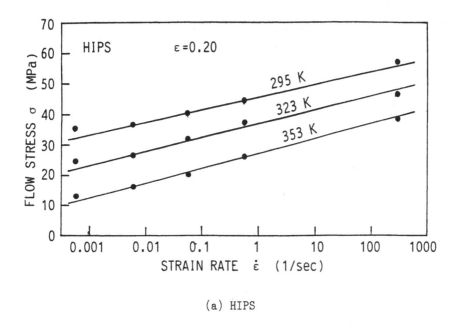

(a) HIPS

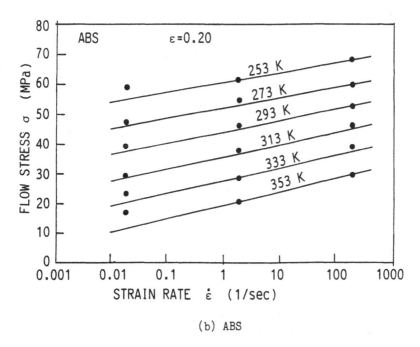

(b) ABS

Figure 2. Strain rate- and temperature-dependences of flow stress. Solid
lines are predicted using Eyring's rate process equation.

where $\dot{\varepsilon}$ is the strain rate, $\dot{\varepsilon}_0$ the pre-exponential term, ΔE the activation energy, α the activation volume, σ the stress, k Boltzmann's constant and T the temperature. The solid lines in Figure 2 are calculated using Equation (1). The values of ΔE, α and $\dot{\varepsilon}_0$ in Equation (1) which give a good fit with the experimental data are as follows:

for HIPS ΔE=190 KJ/mol, α=2275 $\overset{\circ}{A}^3$, $\dot{\varepsilon}_0$=4.07×10^{22} /sec

for ABS ΔE=245 KJ/mol α=2475 $\overset{\circ}{A}^3$, $\dot{\varepsilon}_0$=1.22×10^{32} /sec.

The values of ΔE and α for ABS are near to the values obtained by Truss and Chadwick for the yield stress.

MICROMECHANICAL ANALYSIS

Although the strain rate- and temperature-dependences of the experimental flow stress in the ductile flow region can be well described by the Eyring's rate process equation (1), the actual flow is caused by the initiation and growth of the crazes and accordingly the flow is heterogeneous in its nature. In this work, in order to interpret the ductile deformation in the rubber-toughened plastics, a micromechanical analysis is made. In the proposed model, the craze is simulated by a penny shaped-crack with oriented fibrils of molecules connecting the upper and lower surfaces. The growth of the craze is assumed to be controlled, in the thickness direction, by the rate of the thermally activated drawing out of the molecules from the matrix into the craze in the form of oriented fibrils, and, in the radial direction, by the stress intensity factor. In this model, the radial growth of the craze without increase in the thickness brings about decrease in the stress intensity factor, and accordingly the radial growth is also controlled by the thermally activated process controlling the thickness growth. Since the crazes spread along the planes normal to the tensile direction and since the mass density of the matter in the craze is smaller than that of the matrix, the growth of the crazes causes tensile elongation without the lateral contraction, in other words the elongation is dilatational.

Figure 3 shows the penny shaped-crack simulation of the craze. The penny shaped-crack is treated by using the distributed dislocation model. The tension of the fibrils per unit area, σ_t, caused by the z direction uniform stress applied at infinity, σ, is given by

$$\sigma_t(r) = \sigma + 2 \int_0^a [du(s)/ds] \, K[r,s] \, ds \qquad (2)$$

where $2u(r)$ is the thickness of the craze at the radius r determined by the length of the fibrils, a the radius of the craze and K]r,s] the z direction normal stress at r due to an edge dislocation ring of radius s and having unit Burgers vector in z direction. As was quoted by Mura [3], Kroupa [4] gave K[r,s] as

$$K[r,s] = [G/2(1-\nu)] \int_0^\infty sp\ J_0(pr)\ J_1(ps)\ dp \qquad (3)$$

where G is the modulus of rigidity, ν Poisson's ratio and J_0 and J_1 the Bessel functions of order zero and order 1, respectively. In the present model, the process of drawing out of molecules in the form of the oriented fibrils from the matrix into the craze is assumed to be a thermally acti-vated process. The rate of the increase in the craze thickness $2u(r)$ can be written as

$$\dot{u} = \dot{u}_0\ \exp[\ -\ (\Delta E\ -\ v\sigma_t)\ /kT] \qquad (4)$$

where $\dot{u}_0$ is the pre-exponential term and v the activation volume. The growth of the craze in the radial direction occurs when the stress intensi-ty factor K_I reaches K_{cr}, where K_I is given by

$$K_I = 2\sigma(a/\pi)^{1/2} - [2/(\pi a)^{1/2}] \int_0^a r\ \sigma_t(r)/(a^2 - r^2)^{1/2}\ dr\ . \qquad (5)$$

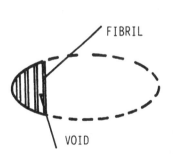

FIBRIL

VOID

(a) Modeling of craze

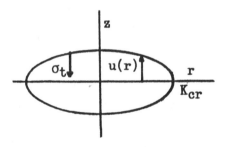

(b) Penny shaped-crack simulation

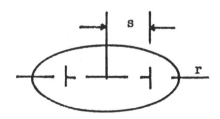

(c) Dislocation model

Figure 3. Modeling and penny-shaped crack simulation of craze.

The strain ε is

$$\varepsilon = (\sigma/E) + N \left[1 - (\rho_c/\rho) \right] V_c \qquad (6)$$

where E is the tensile modulus, N the number of the crazes in a unit volume, ρ and ρ_c the mass densities of the matrix and the matter in the craze, respectively, and V_c the volume of the craze given by

$$V_c = 4 \int_0^a \pi r \, u(r) \, dr \; . \qquad (7)$$

Using Equations (2) to (7), the stress strain relationship under constant strain rate can be followed numerically by taking a small time interval. In the present work, however, in order to obtain a wide perspective a simplified calculation was made. The shape of the penny shaped-crack of Mode I is ellipsoidal, but in the present model, the shape of the craze is not ellipsoidal owing to the normal traction exerted by fibrils unless the distribution of the traction is uniform. Although in the present case the traction is not uniform, the shape is assumed to be ellipsoidal. Under this assumption the shape of the craze can be given by two parameters, i.e. the mean thickness and the radius. Further, the interaction among the crazes is neglected. The results of the numerical calculation for ABS are shown in Figure 4 together with the experimental results. A fairly good agreement is seen between the results of the calculation and of the experiment over wide ranges of the strain rate and temperature. The numerical values of the parameters used in the calculation are as follows:

E=900 MPa, ν=0.35, N=6.0×10^{17} /m^3, K_{cr}=12000 N/m$^{3/2}$, ΔE=245 KJ/mol, v=3280 Å^3, $\dot{u}_0$=1×10^{24} m/sec, ρ_c/ρ=1/3.

For the initiation of the craze, the rubber particle was assumed to act as a zero thickness craze twice in radius. The distribution of the radius of the particle was assumed as follows:

radius (μm)	0.154	0.192	0.240	0.300	0.380	0.469	0.586
number ratio	1	6	15	20	15	6	1

The distribution of the logarithm of the radius is symmetrical around the center value 0.300 μm. Further, each elementary volume containing one particle was assumed to be subjected to the same strain rate, and the weighted mean of the flow stresses of the elementary volumes was taken as the flow stress σ.

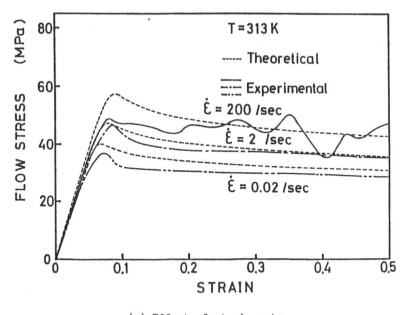

(a) Effect of strain rate

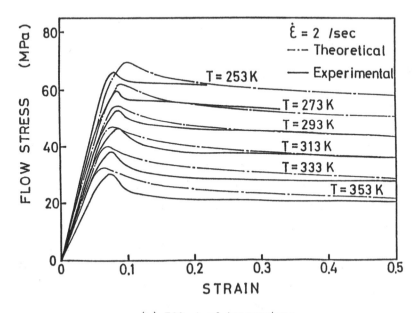

(b) Effect of temperature

Figure 4. Theoretical stress-strain relationship compared with experimental
results. ABS

DISCUSSION

As shown above, the proposed micromechanical model can describe the experimental stress-strain relationship, especially its strain rate- and temperature-dependences, fairly well. However, in order to prove the appropriateness of the model, the numerical values of the parameters used in the calculation should be carefully checked.

The value of E is the experimental one at 293 K but its temperature-dependence is neglected. The value of Poisson's ratio $\nu=0.35$ is a usual value. The number of the rubber particles in unit volume, N, is estimated from the rubber content and the mean radius of the dispersed rubber particles considering that ABS is a graft copolymers. In the present model the stress intensity factor at the tip of the growing craze should be kept at its critical value K_{cr}, and accordingly K_{cr} determine the shape of the craze during deformation. The value of $K_{cr}=12000$ N/m$^{3/2}$ was determined from the radius-thickness ratio of the craze usually seen in this type of polymers.

As was shown previously, macroscopically the strain rate- and temperature-dependences can be well predicted by the Eyring's rate process equation, Equation (1). This implies that, microscopically, the deformation rate is controlled by a thermal activation mechanism which has the same exponential term. Therefore, as the activation energy in Equation (4) the value obtained from the experiment, 245 KJ/mol, was used. Further, for the activation volume v of Equation (4), the same principle was applied; taking into account of the ratio of σ_t to σ the value of v was taken at 3280 Å^3 from the experimental value $\alpha=2475$ Å^3.

The matter in the craze is composed of the very fine fibrils and the interconnected voids. Therefore, the surface area of the fibrils per unit volume of the craze matter is very large. The mechanism of forming such a fine structure was explained by Argon and Salama [5] using Taylor's meniscus instability theory [6]. Kramer [7] pointed out that the work required for the growth of the craze is mostly the work for the fibril surface formation which is composed of the increase in the van der Waals surface energy and the scission of the entanglement network chains on the surface of the "phantom cylinder" of the fibril and that the latter plays more impotant role below the glass transition temperature. The value of the activation energy, 245 KJ/mol, obtained in the present experiment and used in the calculation is near to the bond energy of -C-C- bond in the chain molecules, and this agrees with the Kramer's mechanism. The value of the activation volume is also compatible with the Kramer's model.

On the other hand, the value of the pre-exponential term $\dot{u}_o$ in Equation (4) determined to give good agreement with experimental results is too large. As is often experienced in fitting the calculated results to experimental results in a thermally activated process, the value of the pre-exponential term is strongly affected by relatively small differences in the selected values of the activation energy and activation volume. However, even if the above experiences are taken into account, the value is too large. This means that the contribution of one thermal activation event is very large. Even if the exact calculation is made using Equations (2) to (7) without the assumption of the ellipsoidal shape, the above difficulty cannot be solved. One possible explanation may be that one thermally activated scission of a network chain triggers off a series of athermal scissions of neighbouring network chains due to taking over of the load. On this point, further investigation is left.

The condition of the initiation of the craze used in the present calculation was rather conventional one. Although this condition does not affect the flow stress after yielding so much, further research on the real mechanism is required, especially from the point of the materials design.

CONCLUSIONS

Experiments and a micromechanical analysis were made for the craze-aided ductile flow in HIPS and ABS. The conclusions are
(1) The strain rate- and temperature-dependences of the flow stress after yielding can be expressed by Eyring's rate process equation.
(2) Micromechanical analysis shows that the strain rate- and temperature-dependences can be interpreted in terms of the craze growth rate which is controlled by the thermally activated drawing out of the molecular fibrils from the matrix into the craze.
(3) The dominant thermally activated event is likely to be the scission of the entanglement network chain molecules, since the activation energy obtained from the experiment is nesr to -C-C- bond energy.
(4) As to the pre-exponential term in the Eyring's rate process equation for the drawing out rate of the fibrils uncertainties remain.
(5) Further research is required also for the treatment of the craze initiation process.

REFERENCES

1. Bucknall, C. B., Toughened Plastics, Applied Science Publishers LTD, London, 1977, pp. 182-242.

2. Truss, R. W. and Chadwick, G. A., Tensile deformation behaviour of ABS polymers. J. Mater. Sci., 1976, 11, 111-7.

3. Mura, T., Micromechanics of Defects in Solids, Martinus Nijhoff Publishers, Dordrecht, The Netherlands, 1987, pp. 280-97.

4. Kroupa, F., Circular edge dislocation loop. Czech. J. Phys., 1960, B10, 284-93.

5. Argon, A. S. and Salama, M. M., Growth of crazes in glassy polymers. Phil. Mag., 1977, 36, 1217-34.

6. Saffman, P. G. and Taylor, G. I., The penetration of a fluid into a porous medium or Hele-Shaw cell containing a more viscous liquid. Proc. Roy. Soc. Lond., 1958, A245, 312-29.

7. Kramer, E. J., Craze fibril formation and breakdown. Polym. Engg. Sci., 1984, 24, 761-9.

MICROSCOPIC FRACTURE MODES OF BRITTLE POLYMERS
IN DYNAMIC CRACK PROPAGATION

T. SHIOYA and R. ISHIDA
Department of Aeronautics, University of Tokyo
7-3-1 Hongo, Bunkyo-ku, Tokyo, Japan

ABSTRACT

Fracture process in the dynamic crack propagation is studied at microscopic level based on experiments of brittle fracture of PMMA plate in
fixed sided condition. The feature of fracture surface changes with the
crack propagation velocity; parabolic line pattern in low velocity
region, crazing and micro-branching surfaces in high velocity region, and
the transition to bifurcations of the crack path at high loading energy
case. The microscopic fracture surface patterns are caused by the growth
and coalescence of microcracks ahead of the running main crack. The
critical stress condition of the microcrack growth initiation is examined
from the fracture surface analysis and the dynamic stress analysis of the
crack. With use of the condition, the formation mechanism of parabolic
pattern is numerically simulated. The fracture surface formation energy
is calculated for the propagation velocity from the experimental result
and related with the mode transition mechanism.

INTRODUCTION

Fracture of brittle polymers is characterized by crack propagation with
high velocity. Microscopic feature of the crack surface as well as macroscopic crack path including branching or bifurcation depends on the
propagation velocity. Crack propagation path has been investigated in
dynamic continuum mechanics [1-3]. According to the analyses, in a fixed
sided plate, the crack tends to run towards the center line when the crack
propagation velocity is small, however it turns to deviate from the center
and goes toward the fixed side in case of high propagation velocity.
These analyses are within continuum mechanics and postulate no branching

of the path. In actual crack propagation branching is often observed at high velocity of propagation. There have been several theories of branching, mainly postulating a criterion based on the stress condition in continuum mechanics [4]. However, such phenomena depend much on material natures as well as loading conditions, and therefore should be examined by microscopic point of view together with macroscopic analysis.

In the present study, as a preliminary investigation, mechanism of crack propagation in brittle non-crystalline polymers is investigated based on experiments of running crack in PMMA plate. In the experiment, fracture surfaces are observed in detail by a scanning electron microscope and an optical microscope. The surface formation energy is also examined. Through the experiments and analyses, fracture mode transitions are discussed including microscopic fracture surface mode transition as well as macroscopic branching conditions.

EXPERIMENT

EXPERIMENTAL PROCEDURE

The fracture experiments were conducted in a brittle PMMA plate. Since the crack velocity should be refered to the elastic wave velocities, the dynamic elastic moduli were measured by dynamic methods, i. e., ultrasonic transmission method and vibration method. From the measured moduli, the wave velocities are calculated as v_l = 1945 m/s and v_s = 1205 m/s, where v_l and v_s are dilatational and distortional wave velocities, respectively. In order to obtain steady state crack propagation, slender rectangular shaped specimens were used with the longer sides rigidly fixed after tensile loading. The crack was triggered at the middle of a shorter side of the specimen plate after fixing the longer sides to specified tensile displacement. The crack propagation was along the longer sides direction (Mode I). The propagation velocities were calculated by measuring the time intervals between breaks of electrically conductive lines painted on the specimen (silver paint). In every case, the condition of constant crack velocity (terminal velocity) was achieved before the crack reaches the other end of the specimen. Examples of the velocity change with time are shown for different applied loading condition in Fig. 1.

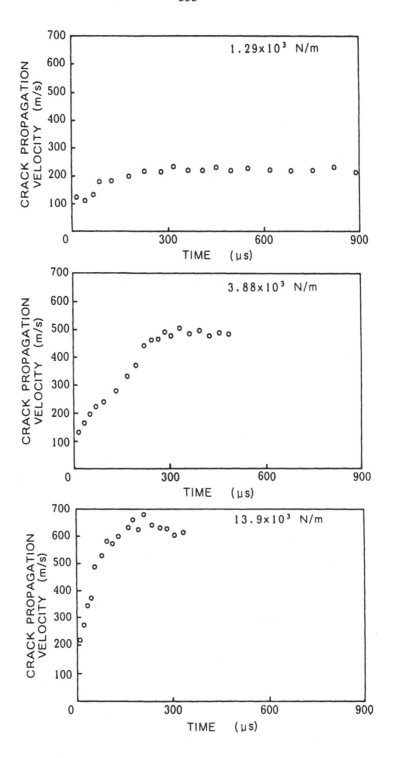

Figure 1. Measured Crack Propagation Velocity Vs. Time.

354

Fracture Surface Analysis

Parabolic Pattern Region: Generally the surface of brittle fracture is macroscopically flat in case of non-crystalline polymers, however, under the microscope field, the surface is revealed to have different patterns depending on the crack propagation velocity. The parabolic pattern appears at the terminal velocity when the applied load energy is relatively low, and also appears at low velocity in the accelerating stage in case of high applied loading energy. The features of the fracture surfaces are almost the same in both cases if the crack velocities are the same. These parabolic lines are thought to be intersection lines of the crack fronts of the main crack and the microcracks which nucleate, grow and coalesce into the main crack. The pattern changes in shape and density with the crack velocity. Typical fracture surfaces in the parabolic region with different crack velocities observed by a SEM are shown in Fig. 2 (a) and (b). The observed nuclei of micro-crack can be classified into two categories; large nuclei which are the foci of the main parabolic patterns and small foci which do not form patterns or form very shallow parabolic patterns. It becomes clear from the SEM photos of corresponding opposite fracture surfaces, that the nuclei sites are always indented implying that certain amount of permanent deformation has been caused forming void like shape before joining to the main crack. The depth of nuclei was measured by an optical microscope. The depth of the large nuclei increases with the crack velocity with their density almost unchanged, while, the density of small nuclei increases with the crack velocity without changing their depths. (Fig. 3 and 4) In either case, the depth is at most in the order of several micron meter.

Crazing Region: In case of intermediate loading energy, as the crack velocity is accelerated to a high velocity range, the smooth fracture surface turns to rough crazing surfaces. Often these crazing regions and the parabolic regions appear alternatively along the propagation direction [Fig. 2 (c)]. The depth or thickness of the craze is in the order of several 10s micron meter, which is much larger than that of parabolic patterns. Corresponding opposite fracture surfaces do not form cavity shape. These facts imply that the mechanism of craze formation is different from that of the parabolic pattern.

Micro- and Macro-Branching Region: As the applied loading energy becomes further higher, the craze size or thickness becomes deeper so that

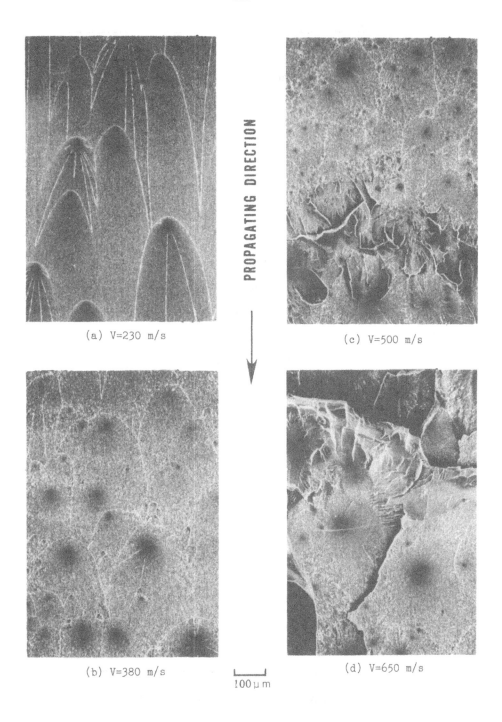

(a) V=230 m/s

PROPAGATING DIRECTION

(c) V=500 m/s

(b) V=380 m/s

100μm

(d) V=650 m/s

Figure 2. Fracture Surface of PMMA

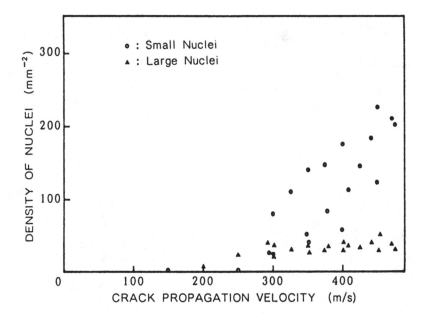

Figure 3. Density of Nuclei vs. Crack Propagation Velocity
in Parabolic Region.

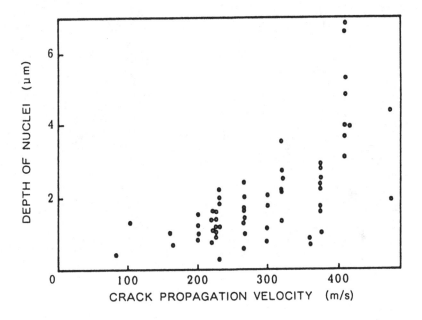

Figure 4. Depth of Nuclei vs. Crack Propagation Velocity.

the craze coalescence mode can be regarded as microscopic branching.
These microscopic branches can sometimes develop to macroscopic branches
or the macroscopic bifurcation of the crack path at high applied loading.
A typical example of the fracture surface in this region is shown in Fig.
2(d).

ANALYSIS

Parabolic Pattern Formation Mechanism and Simulation
The parabolic patterns are thought to be the intersections of the crack
fronts of the main and the secondary (micro-) cracks. The secondary
cracks are considered to start their propagation under the influence of
approaching cracks. The distance from the approaching crack at the instant of propagation initiation can be obtained from the configuration of
the parabolic patterns, and the stress at the instant can be calculated by
the dynamic stress intensity factor [5]. This gives the criterion of
micro-fracture initiation and the critical stress in case of the present
material is 140 MPa. A numerical simulation of the pattern formation is
carried out under the following assumptions: (a) The front lines of the
main and the secondary cracks spread in the same and constant velocity
concentrically from their nuclei which have some finite radius initially.
(b) The cracks start their propagation when the stress at the edge of the
nuclei reach to the critical value which is calculated from the dynamic
stress intensity factor of an approaching crack front. Figure 5
demonstrates the case when the main crack has started from the far
upstream position and under the influence of propagating main crack the
two nuclei situated in line initiate to spread micro cracks sequentially
forming intersection lines.

Surface Formation Energy and Simulation of Acceleration
In the steady state propagation (at the terminal velocity), the applied
strain energy per unit length (in the crack propagation direction) is in
equilibrium with the crack surface formation energy per unit length
(energy release rate) [5]. The surface formation energy is calculated
from this relation. Figure 6 shows the relation between the surface formation energy and the crack propagation velocity. The energy increases

358

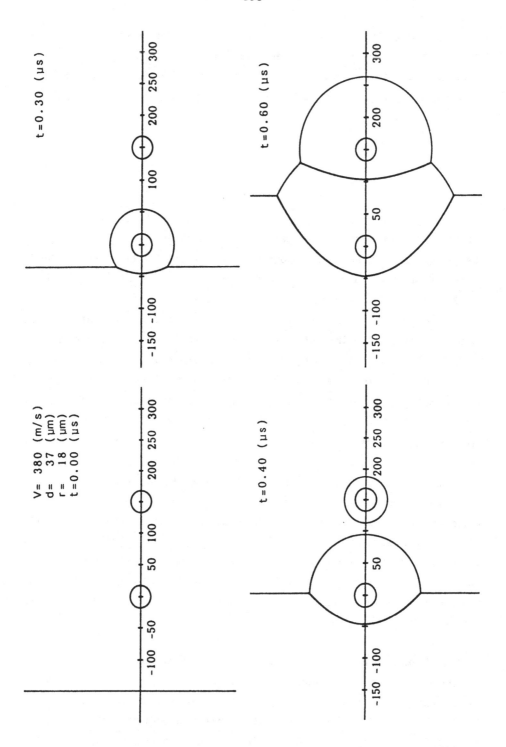

Figure 5. Simulation of Parabolic Pattern Formation.

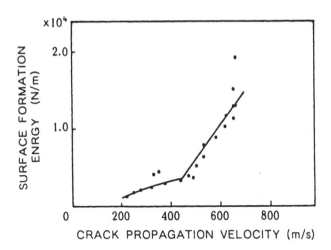

Figure 6. Surface Formation Energy vs. Crack Propagation Velocity.

with the velocity gradually in the low velocity region, however, the increase rate rises abruptly at about 450 m/s, which corresponds to the transition velocity from the parabolic region to crazing region. Numerical simulations are carried out for the acceleration stage of the propagation, assuming that the surface formation energy vs. crack propagation velocity relation is shown as the solid lines in Fig. 6 and the rest of applied energy is spent for the acceleration of the propagation. Figure 7 shows calculated examples which have good correspondence with Fig. 1. The further rise in the increase rate at about 650 m/s corresponds to the macroscopic bifurcation of the crack path. These facts together with the depth change of the patterns suggest that the surface formation energy is mainly the energy dissipation due to the parabolic pattern formation, the craze formation or branching in corresponding velocity regions.

DISCUSSIONS AND CONCLUSIONS

Dynamic crack propagation in brittle polymers is studied based on microscopic observation of the fracture surface. The fracture surface mode depends on the crack propagation velocity and changes at the velocity of about 450 m/s, from the parabolic pattern mode to rough crazing surfaces.

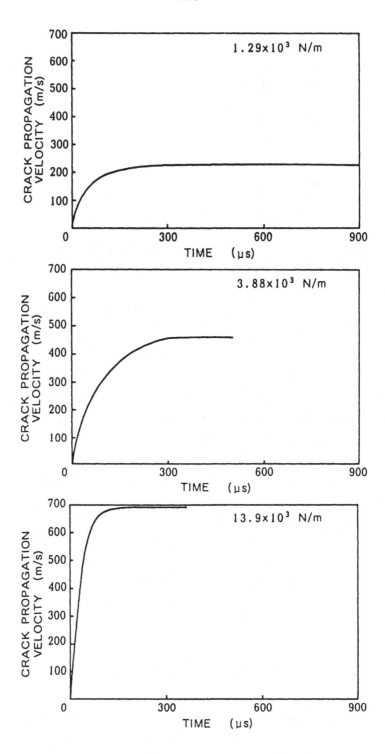

Figure 7. Simulation of Crack Acceleration. Crack Propagation
Velocity Vs. Time.

The change in surface formation energy increase rate corresponds to this mode change.

As to the macroscopic branching condition, the measured branching velocity of about 60 % of the shear wave velocity might be explained by the circumferential maximum stress in the singularity at the crack tip [4]. However, the experiment also claims that the actual branching is the joining of microcracks to the main crack so that the singularity in the continuum mechanics theory does not have definite meaning as the criterion of branching. A more reasonable explanation may be that the crazing and micro-branching is a necessary condition for macro-branching and some additional condition such as macroscopic energy criterion will be needed for actual branching of the path.

Acknowledgement

Important part of the present work has been done in the cooperation with Messrs. T. Kanazawa, and K. Yamamoto at Department of Aeronautics, University of Tokyo. The authors would like to express many thanks for their contributions.

REFERENCES

1. Shioya, T. and Fujimoto, K., A theoretical study on the running crack path by energy balance method. Trans. Japan Aeronau. Space Sci., 1983, 25,246-257.

2. Shioya, T. and Fujimoto, K., Application of Lagrange's Equation to Crack Propagation Problems. In Mathematical Modelling in Science and Technology, ed. X.J.R. Avula et al., Pergamon Press, New York, 1984, pp.513-518..

3. Fujimoto, K. and Shioya, T., A study on path of dynamic crack propagation. Int. Conf. Mechanical and Physical Behaviour of Materials under Dynamic Loading (DYMAT85), Journal de Physique, 1985, 46-C5,233-238.

4. Yoffe, E.H., The moving Griffith crack. Phil. Mag., 1951, 42, 739-750.

5. Shioya, T., Koga, Y, Fujimoto, K. and Ishida, R., Micro-mechanism of dynamic crack propagation in brittle materials. Int. Conf. Mechanical and Physical Behaviour of Materials under Dynamic Loading (DYMAT88), Journal de Physique, 1988, 49-C3,253-260.

SIMULATION OF A FAST DUGDALE CRACK MOTION
WITH VARYING VELOCITY

ANDRZEJ NEIMITZ, ZBIGNIEW LIS
Kielce University of Technology
Al.1000-Lecia P.P.7, 25-314 Kielce, POLAND

ABSTRACT

The fast non-steady-state, Mode III crack motion is analysed. The strip yield zone model of the crack has been adopted to simulate the crack motion within the elastic-plastic material. The "real" leading and trailing edge trajectories have been approximated by a piece-wise linear functions according to certain physically based arguments. The actual motion is selected with the help of two different equations of motion utilizing well known concept of CTOD as well as a new definition of the "driving force" on the strip yield zone. To predict the crack tip motion for various histories of loading and material parameters a numerical analysis has been utilized.

INTRODUCTION

The main purpose of the present paper is to reconsider the equations of motion of a crack that propagates within elastic-plastic material. The whole history of the crack motion, including: acceleration, steady-state, deceleration and arrest will be analysed.

Despite numerous articles published during the last three decades the problem of dynamic crack propagation still leaves many questions that should be discussed, reconsidered and finally answered to set up basic frames for further, deeper, more advanced analysis.

Similarly to quasi-static problems of crack propagation the dynamic analysis for linear-elastic materials provided us with first general picture of this very complex problem. The solutions presented in literature have been well defined and unquestionable from the mathematical as well as physical points of view. Both, mechanical fields and equations of crack motion that were postulated are generally accepted and understood. However, highly idealized linear-elastic models of materials should not be applied to most of the practical situations being met in engineering practice.

Real materials exhibit certain features that can not be explained by oversimplified elastic analysis. Intensive research concentrated on non-elastic materials provided an interesting, however very limited, information on the mechanical fields around moving crack tip. For non-elastic dynamic crack problems the stress intensity can not be defined precisely and related to the external loading. The energy release rate-type equation can be applied for non-steady-state motion for linear and non-linear elastic materials only. For these type of materials each motion is asymptotically steady-state. To overcome the above mentioned problems the simplified model of crack and loading mode have been adopted in order to obtain an unified picture of the complex fast crack motion problem. The strip yield zone model has been utilized to replace elastic-plastic analysis by elastic one with an additional boundary condition along the moving strip yield zone. Results and conclusions presented in this paper are strictly valid for Mode III loading. However, one can expect that certain features of crack motion are of more general nature and they can be helpful in solving more difficult (from mathematical point of view) Mode I and II problems. The steady-state, Mode III fast motion of Dugdale crack has been analyzed by several authors, among them by Glennie and Willis [1], Goodier and Field [2], Kanninen [3], and Achenbach and Neimitz [4]. The computational method utilized by Achenbach and Neimitz has been applied by Neimitz [5], [6] to analyse non-steady-state fast crack propagation. Present article is an extension of the last two [5], [6] and includes the computer simulation of the fast crack motion according to adopted model for various loading histories and hypothetical material properties - the resistance to crack motion.

From the two main differences distinguishing the fast crack motion from quasi-static one: the influence of inertial resistance of material to motion, and strain-rate dependent resistance of material to deformation the former one enters into analysis only. It has been assumed that the "yield stress" within the yield strip zone is independent of strain rate.

Before we proceed with an analysis the main assumptions, results and conclusions following from the computations [5] [6], that have been utilized in the present article will be listed:

a/ the tip of the strip yield zone will be called leading edge, and the "physical" tip of the crack will be called trailing edge,

b/ the scheme of both tips trajectories along with the notation used is shown in the Fig. 1,

c/ the general formulas to solve any kind of Mode III type of loading are listed below:

$$\tau\ (\xi,\eta) = \frac{1}{\pi}\ \frac{1}{\left[\eta - N(\xi)\right]^{1/2}} \int_{\xi}^{N(\xi)} \frac{f(\xi,\eta)\left[N(\xi) - u\right]^{1/2}}{(\eta - u)}\ du\ , \qquad (1a)$$

$$\tau(x,s) = \frac{1}{\pi} \frac{\left(1-\frac{dX}{ds}\right)^{1/2}}{\left[x-X(s)\right]^{1/2}} \int_{0}^{X(s)} \frac{f\left(\vartheta, s - X(s) + \vartheta\right)}{\left[X(s) - \vartheta\right]^{1/2}} d\vartheta , \qquad (1b)$$

$$w(\xi_1,\eta) = -\frac{1}{\sqrt{2}\,\pi\mu} \int_{K(\eta_1)}^{\xi_1} \frac{d\xi}{(\xi_1-\xi)^{1/2}} \int_{\xi}^{\eta_1} \frac{f(\xi,\eta)}{(\eta_1-\eta)^{1/2}} d\eta , \qquad (2)$$

where most of the symbols are defined in Fig. 1, $s_T = c_T t$, $c_T = \left(\frac{\mu}{\rho}\right)^{1/2}$ and $f(x,s)$ is given by relation (3). For strip yield zone model the function $f(-)$ should be assumed in the following form:

$$f(x,s) = -\tau_0(x_1) \, H[\,L(s)-x_1\,] \, H(s) +$$

$$+\tau_f \, H[\,x_1-T(s)\,] \, H[\,L(s)-x_1\,] \, H(s), \qquad (3)$$

where $\tau_0 = (2\pi r)^{-1/2} \, K_{III}(a)$, $K_{III}(a)$ is instantaneous, static SIF, $H(-)$ is a Heaviside function and τ_f is the shearing stress within strip yield zone.

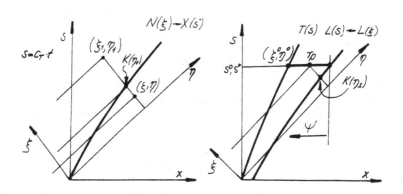

Figure 1. The scheme of the crack edge trajectories a/ elastic
case, b/ elastic-plastic case.

Rigorous evaluation of integrals (1) and (2) is not possible unless the crack tip trajectories are known a'priori. However, when trajectories are approximated by a piece-wise linear functions, the problem can be easily solved leading to the results:
d/ The length of the strip yield zone can be computed from (1)

and (3) yielding the following relation:

$$r_p = \frac{\pi}{8} \frac{K_{III}^2(a)}{\tau_f^2} (1 - \beta_T),$$ (4)

where $\beta_T = v_T/c_T$ and v_T is the speed of the trailing edge. An important feature of this relation is that the length of the strip yield zone depends on SIF-K, τ_f and the trailing edge speed only. Thus, it can be concluded, according to the model of crack used, that the steady-state can be reached when $\beta_T = \beta_L = $ const and $K_{III}(a) = $ const. If real trajectories of the crack edges are replaced by a piece-wise linear function the crack acceleration stage will take place if $\beta_T > \beta_L$ and deceleration stage if $\beta_T < \beta_L$.

e/ The crack tip opening displacement (CTOD) has a form:

$$\delta_T^d = \frac{K_{III}^2}{2\mu\tau_f} \left(\frac{1 - \beta_T}{1 + \beta_L} \right)^{1/2} \left[2 - \left(\frac{1 + \beta_T}{1 + \beta_L} \right)^{1/2} \right].$$ (5)

f/ The CTOD computed from Eq. (5) can be used directly to postulate the equation of motion in the form:

$$\delta_T^d = \delta_{Tc}^d,$$ (6)

where δ_{Tc}^d is considered as material property; $\delta_{Tc}^d = \delta_{Tc}^d(\beta_T)$

g/ The equation of motion following from the power conservation law can not be formulated in a usual way for non steady-state motion since the energy release rate is not uniquely determined for strip yield zone model. The contour of integration L in relation for energy flux is not shrunk in limit to the crack tip and has a definite - changing length. Instead of the energy release rate the driving force $\mathscr{G}_{III}^d$ on the strip yield zone has been defined in a form

$$\mathscr{G}_{III}^d = \int_0^{\delta_T} \tau_f d\delta + \frac{1}{v} \int_0^{r_p} \tau_f \frac{\partial \delta}{\partial t} \, d\zeta,$$ (7)

and utilizing Eq (2) can be computed yielding the following relations:

when $\beta_T = \beta_L = \beta$

$$\mathscr{G}_{III} = \frac{K_{III}^2}{2\mu} \left(\frac{1 - \beta}{1 + \beta} \right)^{1/2},$$ (8a)

for $\beta_T > \beta_L$

$$\mathcal{G}_{III} = \frac{K_{III}^2}{2\mu} \left(\frac{1 - \beta_T}{1 + \beta_L}\right)^{1/2} \left\{ 2 - \left(\frac{1 + \beta_T}{1 + \beta_L}\right)^{1/2} + \frac{2}{3}\frac{(\beta_L - \beta_T)}{\beta_T} + \right.$$

$$\left. + \left(\frac{\beta_T - \beta_L}{\beta_T}\right)^{1/2} \left(\frac{1 + \beta_L}{\beta_T}\right)^{1/2} \ln \left[\frac{(1 + \beta_T)^{1/2} + (\beta_T - \beta_L)^{1/2}}{(1 + \beta_L)^{1/2}}\right]\right\}, \qquad (8b)$$

for $\beta_T < \beta_L$

$$\mathcal{G}_{III} = \frac{K_{III}^2}{2\mu} \left(\frac{1 - \beta_T}{1 + \beta_L}\right)^{1/2} \left\{ 2 - \left(\frac{1 + \beta_T}{1 + \beta_L}\right)^{1/2} + \frac{2}{3}\frac{(\beta_L - \beta_T)}{\beta_T} + \right.$$

$$\left. + \left(\frac{\beta_L - \beta_T}{\beta_T}\right)^{1/2} \left(\frac{1 + \beta_L}{\beta_T}\right)^{1/2} \text{arc tg} \left(\frac{\beta_L - \beta_T}{1 + \beta_T}\right)^{1/2}\right\}. \qquad (8c)$$

In [6] the equation of motion has been postulated in the following form:

$$\mathcal{G}_{III}^d - \mathcal{G}_{IIIc}^d = \Psi \, a_T, \qquad (9)$$

where a_T is a crack tip acceleration, Ψ denotes an "equivalent" mass being proportional to density of material within the plastic zone and to $(r_p)^2$, $\mathcal{G}_{IIIc}^d$ is a material property that, in general, can be function of β_T, β_L and a_T. In the presented analysis equation of motion (9) will be modified and replaced by relation:

$$\mathcal{G}_{III}^d - \mathcal{G}_{IIIc}^d = \frac{d}{dt}(\Psi \, v_T). \qquad (9a)$$

BASIC RELATIONS TO BE UTILIZED IN THE COMPUTER ANALYSIS

The approximation of the real trajectories by piece-wise linear functions can not be introduced arbitrarily. The scheme is postulated in Fig. 2.

It is assumed that the crack starts propagation with a speed $\beta_T = \beta_T^1$ at $\beta_L = \beta_L^1 = 0$. The information that the trailing edge has already started propagation reaches the leading edge at time s^2 (see Fig.2). At this moment the leading edge starts motion with the speed $\beta_L^2 = \beta_T^1$ and sends information about that to the trailing edge. This information reaches trailing

edge at time $s = s^3$ and the leading edge corrects its speed
to a new value following from the equation of motion. Again
the signal is sent to the leading edge which at time $s = s^4$
adjusts its speed to the trailing edge, and so on |*.

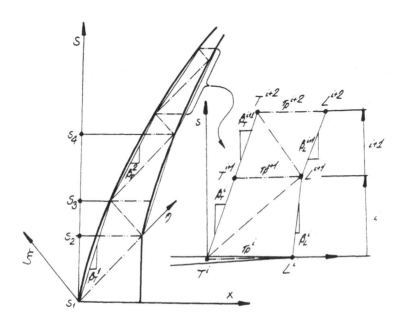

Figure 2. Approximation of the real trajectories by a piece-
-wise linear function.

        To perform a numerical analysis according to the proposed
model of crack kinetics the whole history of the crack motion
represented by crack tip trajectories should be divided on
geometrically similar elements following each other. According
to this procedure the computer will be able to repeat
computations for each consecutive elementary mesh.
The "representative" elementary mesh is shown in the Fig.(2b)
along with notation utilized in the computer code:

T^i, L^i - i-th points along T (trailing) and L (leading) edge
          trajectories,

r_p^i - the length of the plastic zone at the moment s^i,

β_T^i, β_L^i - trailing and leading edge speeds in s,x coordinate

---

*| This scheme is strictly true for $K_{III}(a)$ = const. For
$K_{III}(a)$ ≠ const β_L changes in phase with $K_{III}(a)$ according to
relation $\beta_L = \beta_L(\beta_T, K_{III}(a), \tau_\infty)$.

system,

λ_T^i, λ_L^i - trailing and leading edge speeds in ξ, η coordinate system,

The coordinates of characteristic points of the elementary mesh can be calculated from the relations:

$$s_\varepsilon^{i+k} = s_\varepsilon^{i+k-1} + \Delta^{i+k-1}, \tag{10}$$

$$x_\varepsilon^{i+k} = x_\varepsilon^{i+k-1} + \Delta^{i+k-1} \beta_\varepsilon^{i+k-1}, \quad i = 1,3\dots,\ k=1,2, \qquad \varepsilon=T,L,$$

where $\lambda_\varepsilon^\sigma = (1 + \beta_\varepsilon^\sigma)(1 - \beta_\varepsilon^\sigma)^{-1}, \qquad \sigma=1,2,\dots,\ \varepsilon=T,L,$

$$\Delta^\omega = r_p^{2m-1}(1-\beta_L^{2m-1})^{-1}(\lambda_T^{(\omega-2m+1)(2m-1)})^{-1};\ \omega=2m-1,2m,\ m=1,2,\dots . \tag{10a}$$

The lengths of the plastic zone at the moments s^{i+1}, s^{i+2} can be computed from the relations:

$$r_p^{i+k} = r_p^i(1 + \beta_T^i)^{-(k-2)}(1 - \beta_L^i)^{-1}(1 + \beta_L^{i+k-1})^{(k-1)}(\lambda_T^i)^{-1}, \tag{11}$$

where $i = 1,3,\dots,\quad k=1,2,\quad \lambda_T^i$ is defined by Eq. (10a).

Note also the following relations:

$$\beta_T^{i+1} = \beta_T^i, \qquad \beta_L^{i+2} = \beta_L^{i+1}, \qquad i=1,3,\dots,$$

$$\beta_L^{i+1} = \left[1+(1 + \gamma^{i+1})\beta_T^{i+1}\right](\varphi^{i+1})^2 - 1,\ i=1,3,\dots, \tag{12}$$

where $\gamma^m = r_p^m(a^m)^{-1}; \qquad \varphi^m = \tau_\infty^{m+1}(\tau_\infty^m)^{-1}; \qquad m=1,2,\dots . \tag{13}$

For $i= 1$ one obtains

$$\beta^1 = (1 + \gamma^1\beta^1) - (\varphi^1)^{-2}, \text{ where } \gamma^1, \varphi^1 \text{ see Eq. (13).} \tag{13a}$$

The crack tip opening displacement within the elementary mesh can be computed from the relation:

$$w^T(\xi^*,\eta^*) = \left(- \frac{1}{\sqrt{2\pi}\mu}\right)\left[\int_{\xi^{**}}^{\xi^*} \frac{d\xi}{(\xi^* - \xi)^{1/2}} \int_{\eta^T+\lambda_T^i(\xi-\xi^{T^i})}^{\eta^*} \frac{\tau_f d\eta}{(\eta^* - \eta)^{1/2}} + \right.$$

$$\left. + \int_{\xi^{***}}^{\xi^{**}} \frac{d\xi}{(\xi^* - \xi)^{1/2}} \int_{\eta^T+\lambda_T^{i-1}(\xi-\xi^{T^i})}^{\eta^*} \frac{\tau_f d\eta}{(\eta^* - \eta)^{1/2}}\right] + \left(\frac{2}{\pi}\right)^{1/2}\mu^{-1}\left(1+\beta_L^{i+1}\right)^{-1/2}K_{III}\left(a^* + r_p^*\right)r_p^* \tag{14}$$

with the help of the scheme of integration (Fig.3)

The crack tip opening displacement at an arbitrary point of the trailing edge trajectory T^{i+2} can be defined as follows:

$$\delta_T^{i+2} = \frac{K_{III}^2}{2\mu\tau_f}\left(\frac{1-\beta_T^{i+1}}{1+\beta_L^{i+1}}\right)\left[2-\left(\frac{1+\beta_T^{i+1}}{1+\beta_L^{i+1}}\right)^{1/2}\right], \quad i = 1,3,\ldots \quad .(15)$$

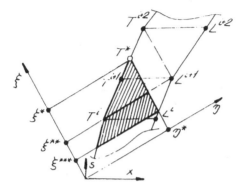

Figure 3. Scheme of integration to be utilized in Eq.(15).

To calculate the driving force $\mathcal{G}_{III}^d$ according to the relation (7) one must know the relation for stretch of the crack faces at an arbitrary point within strip yield zone. It can be computed with the help of the relation

$$\frac{1}{2}\delta_D = w_D(\xi^0,\eta^0) = -\frac{1}{\sqrt{2\pi}\mu}\left[\int_{\xi^{***}}^{\xi^0}\frac{d\xi}{(\xi^0-\xi)^{1/2}}\int_{\eta^{T^i}+\lambda_T^i(\xi-\xi^{T^i})}^{\eta^0}\frac{\tau_f d\eta}{(\eta^0-\eta)^{1/2}}+\right.$$
$$\left.+\left(\frac{2}{\pi}\right)^{1/2}\mu^{-1}\left(1+\beta_L^{i+1}\right)^{-1/2}K_{III}\left(a^{i+1}+r_p^{i+1}\right)(r_p^{i+1})^{1/2}\right. \quad (16)$$

and the scheme of integration shown in the Fig.4.

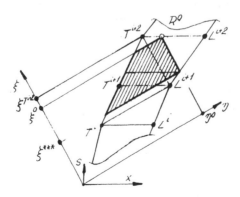

Figure 4. Scheme integration to be utilized in Eq.(16).

Evaluating integrals in (16) one obtains the final relation for stretch of the crack faces within the strip yield zone.

$$w_D^{T^{i+2}} = \frac{-\tau_f(1-\lambda_T^i\Gamma)}{\pi\mu(\lambda_T^i)^{1/2}} \left\{ \ln\left[\Omega^{1/2}+(\Omega-1)^{1/2}\right]+\left[\Omega(\Omega-1)\right]^{1/2}\right\}(\Delta s_T^i+\Delta x_T^i)$$

where $\quad \Omega = \left[1 - \lambda_T^i\left(\lambda_L^{i+1}\right)^{-1}\Sigma\right]\left[1 - \lambda_T^i\Gamma\right]^{-1}$, $\qquad\qquad$ (17)

$$\Sigma=\left(\Delta s_\omega+\Delta x_\omega\right)\left(\Delta s_\varphi+\Delta x_\varphi\right)^{-1}, \quad \Gamma=\left(\Delta s_\varphi-\Delta x_\varphi\right)\left(\Delta s_\varphi+\Delta x_\varphi\right)^{-1}, \quad \omega=T^{i+1}, \quad \varphi=T^i$$

$$\Delta s_\rho = s-s^\rho, \quad \Delta x_\rho = x-x^\rho, \quad \rho = T^i, T^{i+2}, \quad \lambda_T^i, \lambda_L^{i+1} - \text{see Eq.(10a).}$$

Above relation can be utilized to compute $\mathcal{G}_{III}^d$ according to the relation (7). It yields to the results for an arbitrary elementary mesh along the crack tip trajectories. The relation obtained is very similar in shape to the Eqs 8, but will not be shown here because of the limited space. It will be given in the forthcoming paper [7].

RESULTS OF NUMERICAL ANALYSIS

The results of numerical analysis presented in this paragraph are not based on experimental data, since such data do not exist in the literature for Mode III fast crack motion. Thus, all material parameters and loading histories have been assumed arbitrarily in order to test an introduced model of crack kinetics and equations of motion. Only a few examples of computations are included here in this article. More deep and extended analysis is being prepared at the moment and results will be published soon [7]. In this paragraph we would like to show that the model proposed can be used in dynamic analysis and it provides reasonable results. We would like also to test differencies in crack motion histories resulting from two different equations of motion adopted. From the physical point of view these two equations are equivalent for steady-state motion only. Within acceleration / deceleration stages they can generate different results.

The material parameters that have been utilized in the numerical analysis are as follows:

$K_{IIIc}= \quad 25.0 \quad \text{MPa} \; m^{1/2} \qquad\qquad \mu = 0.7923 \; 10^5 \; \text{MPa}$

$\tau_f \quad = 725.0 \quad \text{MPa} \qquad\qquad\qquad E = 2.06 \quad 10^5 \; \text{MPa} \qquad (18)$

$\delta_{Tc} \quad = K_{IIIc}^2(2\mu\tau_f)^{-1}= 5.44 \; 10^{-6}\text{m} \qquad r_p^1= 0.5 \quad 10^{-3}\text{m}$

To perform , calculation it has been assumed that $K_{III}(a) = \tau(\pi a)^{1/2}$ (semi-infinite specimen and finite crack of length a). Thus at $\tau = 50$ MPa, $a = 76.9 \; 10^{-3}$ m.

The fast crack motion will be analysed utilizing two
different equations of motion (6) and (9a). Computer performs
analysis for each consecutive elementary mesh according to
scheme:

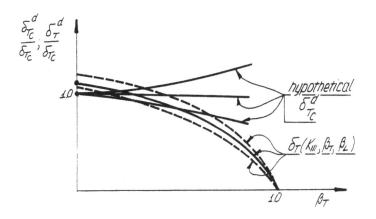

Figure 5. Scheme of the curves representing computed crack tip
opening displacement as a function of actual K_{III}, β_T and β_L
and hypothetical curves representing constitutive
properties of material.

The δ_T^d and δ_{Tc}^d curves are computed for actual external
loading, crack length , and crack tips velocities. The value
of β_T appropriate for the next elementary mesh follows from
the solution. Similar scheme of computation will be adopted
for the second equation of motion but the shapes of $\wp_{III}^d$ and
$\wp_{IIIc}^d$ curves will be different.
In the Figs 6a,6b the results of computation are shown for
the constant external loading τ and starting velocity
$\beta_T^1 = 0.209$. Both leading and trailing edge trajectories are
shown as functions of time (Fig.6a). Velocities of leading
and trailing edges are plotted as functions of time (Fig. 6b)
and the length of the plastic zone is presented in the Fig 6b.
Crack accelerates.

When the external loading decreases according to the relation
$\tau = (50 - 3.75 \; s)$ MPa but the starting speed β_T^1 is the same as
before the situation changes as is shown in the Figs 7a, 7b.
Crack decelerates.
If the rate of change of external loading is the same as in

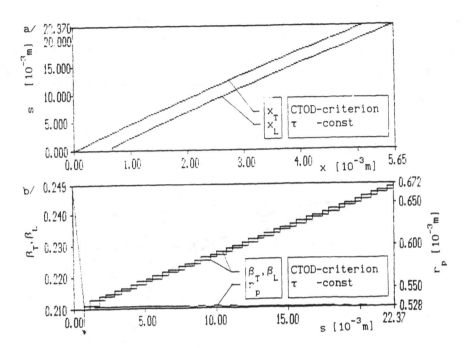

Figure 6. Leading and trailing edge trajectories vs. time (a),
leading and trailing edge velocities vs. time (b),
length of plastic zone vs. time (b).

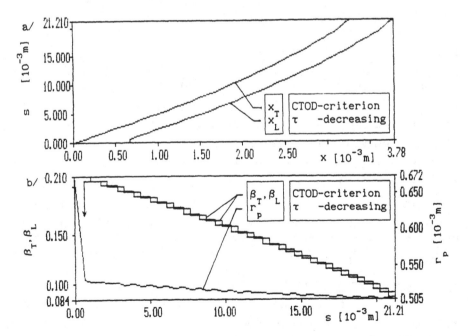

Figure 7. Leading and trailing edge trajectories vs. time (a),
leading and trailing edge velocities vs. time (b),
length of plastic zone vs. time (b).

the previous case but the starting speed β_T^1 is smaller and equals to 0.0506 the results of computations are shown in Figs 8a,8b. Before we proceed with the analysis based on the second equation of motion (Eq.9a) certain comments on an applicability of a CTOD equation of motion should be made in order to compare this criterion to the next one. First, the equation of motion based on a CTOD concept represents a part of an energy rate that is involved in a fracture process neglecting the rate of energy that is dissipated within the plastic zone (Freund [8]). The results obtained suggest that with the help of this criterion the whole history of crack motion can be predicted except the first short period (about 0.5 μsec). To perform analysis one must assume the value of the crack tip speed β_T^1 at the moment of the onset of crack propagation. It may be done with the help of various hypothesis more or less reasonable and results from the fact that the value of β_T^1 is a **mean** velocity within the first elementary mesh along crack tip trajectories. The best way to estimate the β_T^1 is to utilize experimental data if they are available. Both above disadvantages will be removed if the second equation of motion is adopted. The physical interpretation of this equation has been given by A.Neimitz [6]. It takes into account not only physical processes taking place at the crack tip but also effect of the changing length of the plastic zone.

When utilizing Eq.9a in our computer analysis the shape of the $\mathscr{G}_{IIIc}^d(\beta_T)$ function has been assumed arbitrarily basing on the physical intuition. We take a point of view that $\mathscr{G}_{IIIC}^d(\beta_T=0) = G_{IIIc}$ and decreases initially with increasing crack tip speed (there is less time for plastic deformation with increasing speed) reaching minimum at β= 0.25 (for this speed $\mathscr{G}_{IIIc}^d(\beta_T= 0.25\ c_T)=0.75\ G_{IIIc})$ and increasing beyond this point due to branching process. This constitutive function can be assumed arbitrarily within computer analysis but when analysing a real situation the $\mathscr{G}_{IIIc}^d(\beta_T)$ should be defined from an experiment.

In the Figs 9a, 9b the results of the computations are shown for external loading kept constant during crack motion. The propagation started at $K_{III}= K_{IIIc}$. At the beginning crack accelerates rapidly and after about 3 μsec very slowly. When we assume that external loading decreases from $K_{III}= K_{IIIc}$ according to the relation τ= (50 − 0.5 s) MPa $(K_{III}=\tau(\pi a)^{1/2})$ then the results are plotted in Figs 10a, 10b. One may now observe the steady-state motion at $\beta_T=0.272$ after 4 μsec. In Figs 11a,b results are plotted for external loading decreasing more rapidly (τ= (50 − 5.0 s) MPa) and now

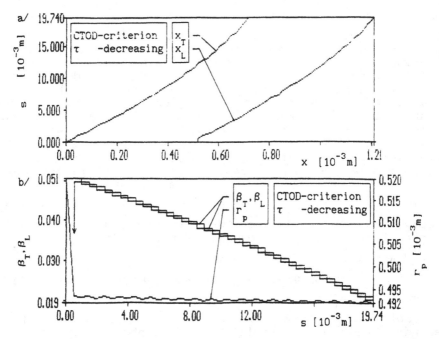

Figure 8. Leading and trailing edge trajectories vs. time (a),
leading and trailing edge velocities vs. time (b),
length of plastic zone vs. time (b).

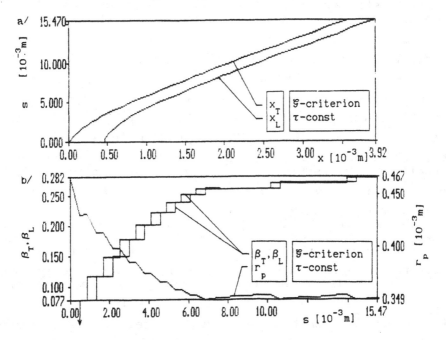

Figure 9. Leading and trailing edge trajectories vs. time (a),
leading and trailing edge velocities vs. time (b),
length of plastic zone vs. time (b).

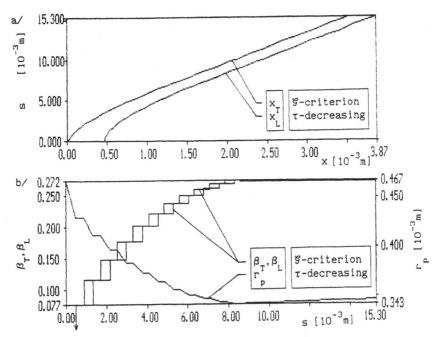

Figure 10. Leading and trailing edge trajectories vs. time (a),
leading and trailing edge velocities vs. time (b),
length of plastic zone vs. time (b).

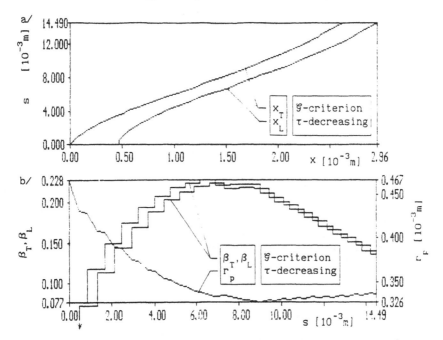

Figure 11. Leading and trailing edge trajectories vs. time (a),
leading and trailing edge velocities vs. time (b),
length of plastic zone vs. time (b).

we can see that crack first accelerates, then after a short steady-state stage decelerates.

If we load the specimen rapidly it can be assumed that $K_{III}(a=a_0)=1.25$ K_{IIIc} and later on decreases at the rate examined in Fig. 11. Results are presented in Figs 12a, 12b.

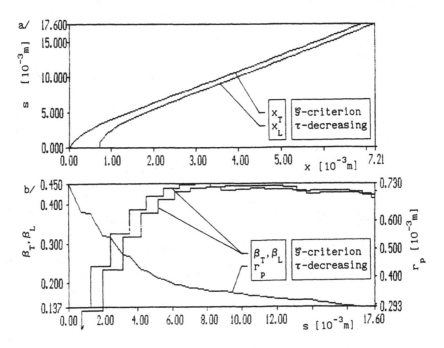

Figure 12. Leading and trailing edge trajectories vs. time (a), leading and trailing edge velocities vs. time (b), length of plastic zone vs. time (b).

CONCLUSIONS

Presented analysis has shown that the assumed model of crack kinetics can be utilized to the analysis of the fast crack motion.

The equation of motion in the form (9a) seems to be much more promising in a future analysis. More detailed results including various loading systems and constitutive properties will be published soon [7].

REFERENCES

1. Glennie, E.B. and Willis, J.R., An Examination of the effects of some idealized models of fracture on accelerating cracks. J.Mech.Phys.Solids., 1971, 19, 11-30.

2. Goodier, J.N. and Field, F.A., Plastic energy dissipation in crack propagation. In Fracture of Solids, ed. D.C. Drucer and J.J. Gilman, Wiley, New York, 1963, pp. 103-118.

3. Kanninen, M.F., An estimate of the limiting speed of a propagation ductile crack, J. Mech. Phys. Solids, 1968, **16,** 215-228.

4. Achenbach, J.D. and Neimitz. A., Fast fracture and crack arrest according to the Dugdale model, Eng. Frac. Mech., 1981, **14,** 385-395.

5. Neimitz, A., On fast crack motion in elastic-plastic materials. I-Mechanical fields around moving crack tip., 8th Europen Conference on Fracture Fracture Behaviour and Design of Materials and Structures., Italy, Torino, October 1-5, 1990, pp. 1119-1124.

6. Neimitz, A., On fast crack motion in elastic-plastic materials. II-Equations of motion.,8th European Conference on Fracture Fracture Behaviour and Design of Materials and Structures., Italy, Torino,October 1-5, 1990,pp. 1125-1130.

7. Neimitz, A. and Lis. Z., Computer simulation of the fast crack motion., To be published.

8. Freund, L.B., Dynamic crack propagation in the Mechanics of Fracture, In ADM, ed. F.Erdogan, ASME, 1976, **19,** pp.105-134.

MODELLING DYNAMIC FRACTURE IN POLYMERS
USING A LOCAL MODULUS CONCEPT

by

A. IVANKOVIC & J.G. WILLIAMS

Mechanical Engineering Department
Imperial College of Science, Technology and Medicine

Abstract

Experiments are described in which rapid crack propagation is measured in single edge notch tensile specimens of PMMA of various height to width ratios. The crack speed is found to be remarkably constant for given test and load dependent. These crack histories were analysed using a dynamic FE code employing a single modulus and shown to give large variations in fracture resistance. When a local, low modulus was used for a strip along the crack path the kinetic energy was increased and the resulting resistance was constant. The choice of modulus was justified by strain gauge measurements. A single mass-spring model using a low wave speed was shown to give the same value of resistance and to describe the variation of crack speed with geometry. This suggests that the expensive FE analysis may be avoided by using such a model.

1. Introduction

There has been a continuing interest in the analysis of dynamic fracture over many years. Recent improvements in experimental methods and equipment have made it possible to perform more informative experiments. However, the analysis of the resulting data has posed several questions. The presence of significant kinetic effects makes finite element solutions difficult and analytical solutions scarce. Much of the reported data casts doubt on the uniqueness of the measured or computed crack resistance and hence the utility of the analysis. The very large variations of crack resistance with crack length and speed which are implied seem to be particularly problematical.

The work described here attempts to overcome these difficulties by exploring the concept that the anomalies arise from large local strains around the crack tip resulting in the crack 'seeing' a low local modulus. This is done using a local strip along the crack path in an FE analysis and by using a low wave speed in a simple mass-spring model. These are shown to be equivalent and to resolve the issue of uniqueness.

2. Experimental Details

The material tested in this work is Polymethylmethacrylate (PMMA) supplied by ICI-UK. The mechanical properties of this viscoelastic material have been established experimentally by a number of authors (Crouch[1]) and are summarized in table 1:

Static Young's modulus E_s (GPa)	Dynamic modulus E_d (GPa)	Poisson's ratio v	Mass density ρ (kg/m^3)
3.0	6.0	0.33	1180.

Table 1: PMMA mechanical properties at room temperature.

The specimen type is chosen according to the following guide lines:
a) The force trace should be easily related to the crack history trace, especially to identify the instant of initiation.
b) The boundary conditions of loading and deflection should be very deterministic and easy to define (no slippage of supports, no unmeasurable dynamic contact stiffness, no bouncing of pins, no friction of a wedge etc.).
c) Again, in order to simplify the modelling and make it more accurate the crack should be encouraged to propagate with a straight front producing a planar surface.

A single edge notch tensile (SENT) sample, clamped rigidly between two heavy grips and loaded quasi-statically (QS) in tension until rapid crack propagation (RCP) occurs, seems to satisfy the above demands. Heavy clamps ensure 'fixed grip' conditions while the crack propagates (Dahlberg[2]).

The specimens are cut from a 13 mm thick sheet. Various width to height ratios (D (10, 30, 60 mm) / H (17, 40, 80 mm)) are employed in order to produce cracks of different speeds

(Williams & Ivankovic[3]) and also to investigate the influence of the specimen size on the dynamic fracture resistance (G_{ID}) results (Kobayashi et al [4]). As shown in figure 1, the specimens are prenotched by machining an initial notch (a_0), followed by sliding of a new razor blade to produce a sharp notch (a_1). The total length of the prenotch was $a_{0t} = 0.2D$.

The testing was performed at room temperature by using a conventional Instron tensile machine (model 1186). The straight crack was produced by clamping the SENT sample between two grips (the weight of each grip was 14.35 kg) which were free to rotate about their axes (no torsional load was applied on the specimen). The initial QS load and the cross head displacement (rate 10 mm/min) were recorded by the internal Instron chart recorder.

Two methods are employed to monitor the crack length histories, fast photography and a grid line technique. In the former method, a high speed camera (model 790 Imacon, Hadland Photonics Ltd) is operated in the streak mode creating a continuous picture of the crack history data (figure 2). The second method involves the application of conducting lines on the sample surface along the prospective crack path. The lines were connected in four groups as arms of a Wheatstone Bridge (figure 3) to improve the resolution of the resistance step reading. When the moving crack breaks the line it produces a step in the total resistance record (figure 4). Both methods require a triggering device to start recording and for this purpose a conducting triggering line applied near the tip of the prenotch was used.

Some important observations can be made regarding the test results:
a) In each test, the crack propagated at remarkably constant velocity.
b) Uniform fracture surfaces were associated with these constant speeds.
c) The crack speed was in the range of 200 to 420 m/s.
d) Higher crack speeds were found to correspond to higher breaking loads (P_b), the load prior to the crack initiation.
e) Details of the crack length at initiation could not be recorded and the initiation time was extrapolated.

3.   Analysis

The difficulties in dynamic fracture calculations are associated with the effects of the inertia and of stress wave propagation. Because of these effects the values of the stress intensity factor (K^{dyn}) and energy release rate (G^{dyn}) in dynamic fracture events cannot, generally, be derived from the crack length, loading and geometry as in the static cases.

An indirect method of characterizing RCP properties is used in this study. The method relies on the recorded crack length history and boundary conditions, which are then incorporated in a dynamic Finite Element (FE) code to generate the crack resistance. Employing the fixed mesh - 'node release' FE program (Keegstra[5]), the running crack is modelled by a sequential release of the nodes along the crack path progressively changing the boundary conditions. The simulation of the crack growth, between the two adjacent crack path nodes, is done by using a concept of a decaying 'holding back' force acting on the released node. Thus an energy sink near the moving crack tip is produced.

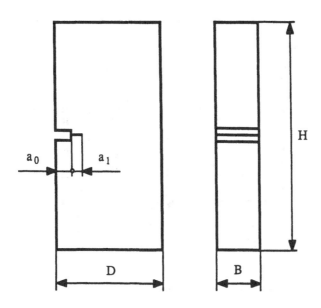

Figure 1: Single Edge Notch Tensile specimen configuration

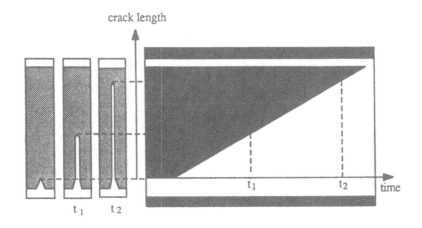

Figure 2: Streak image of a constant speed crack propagation

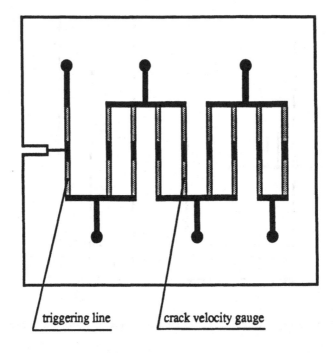

triggering line   crack velocity gauge

Figure 3: Timing line configuration (solid - silver paint, dotted - graphite)

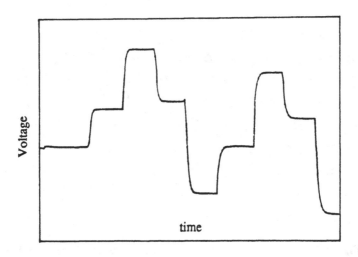

Figure 4: Voltage vs time record for a constant speed crack

The program is designed to analyse symmetrical bodies of isotropic elastic material. Because of symmetry, only half of the sample is modelled and the crack path becomes a boundary. Constant strain triangular elements are used in this 2-D code and the sample is represented by the FE mesh gradually refined towards the crack path. Prior to the crack initiation, an initial static solution, corresponding to the initial boundary loads and displacements, is calculated. Having obtained the initial static solution, a crack is propagated by the sequential release of boundary nodes at times specified by the experimental records. The dynamic solution is computed by an iterative step by step implicit procedure (Zienkiewicz[6]). Every time a node is released, the energy absorbed by the crack is calculated by both global energy balance (equation 1) and local work done by the 'holding back' force. When the solution is stable, both methods give the same result.

The calculation of RCP properties is based on the recorded crack length history, the magnitude of the breaking load (P_b) and dynamic modulus (E_d). It is assumed that as long as the crack propagates, the equality between the crack driving force (G^{dyn}) and the material resistance is satisfied, i.e. energy is conserved (equation 1):

$$G^{dyn} = \frac{dW}{dA} - \frac{dU_s}{dA} - \frac{dU_k}{dA} = G_{ID} \qquad (1)$$

dA - unit area of crack advance, W - external work, U_s - strain energy, U_k - kinetic energy. (G_{ID} data generated by this code have been compared to both theoretical and experimental results (Crouch[1]) and good agreement was found.)

Due to the uncertainties surrounding the crack initiation stage, the initiation time is arbitrarily extrapolated. Three different crack initiations are considered (figure 5). The crack is assumed to initiate with:
a) a speed equal to the recorded speed; the constant speed crack initiation case,
b) zero speed followed by a gradual crack acceleration to the measured speed; the accelerating crack initiation case, and
c) high speed and gradual deceleration to the true crack speed; the decelerating crack initiation case.

Figure 6 shows the G_{ID} vs a (a is the crack length advance) curves, generated by using the above assumptions. It indicates that the way the crack initiation is simulated is not critical in the analysis. Thus, in the rest of this work only the constant speed crack initiation case (a) is considered. The typical results of G_{ID} vs a, for various D/H ratios are presented in figures 7 and 8. After the crack is initiated, the highly tensile stressed region near the crack tip suddenly relaxes and a compressive stress wave is generated. During the crack propagation, compressive stress waves are continuously radiated from the wake of the moving crack. Until a reflected stress wave returns to the crack tip, the crack is 'unaware' of the specimen size and loading conditions and the failure is effectively at constant applied load. As the stress waves reach the crack tip the failure becomes governed by constant displacement conditions.

The main characteristics of the presented G_{ID} curves are summarized as follows:
a) Each figure demonstrates similar variation of the crack resistance with the crack length - the calculated G_{ID} gradually rises (up to 6 times) as the crack advances.

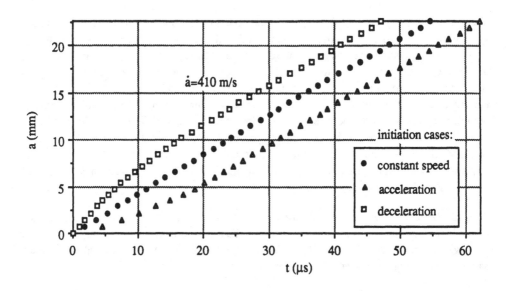

Figure 5: Crack length increment vs t - various initiation cases

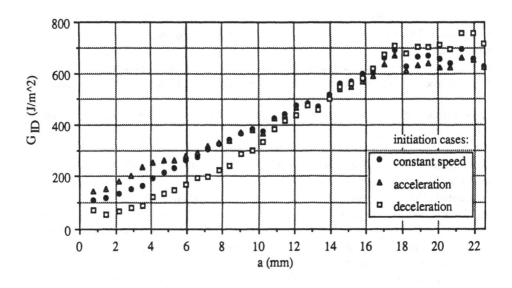

Figure 6: Crack resistance vs a - various initiation cases

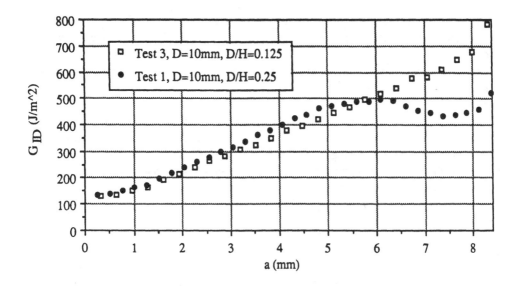

**Figure 7: Crack resistance vs a, ȧ=305m/s
influence of various D/H ratios**

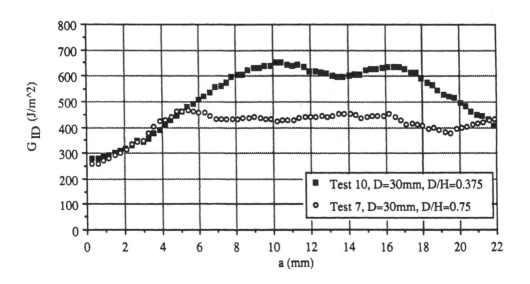

**Figure 8: Crack resistance vs a, ȧ=256m/s
influence of various D/H ratios**

b) Eventually, at the time the irrotational stress wave, travelling at speed of 2745 m/s, returns from the loaded boundary to the moving crack tip, G_{ID} stabilizes and remains almost constant.

c) Greater G_{ID} variations at a higher average value occur in the taller specimens (lower D/H ratio) at the same crack speed.

It is generally agreed that the fracture surface roughness is an indication of the energy absorbed in brittle fracture. Thus, the variation in G_{ID} is expected to correlate with the change in surface roughness. Unfortunately, during RCP at constant speed, the drastic increase in the crack resistance with the crack length (figures 7,8) does not correspond to any change in fracture surface appearance. Therefore, the roughness of the fracture surface does not indicate the large increase in G_{ID} suggested by the computed results.

4. Local Modulus Concept

An error in the analysis may be responsible for apparently improper characterization of G_{ID} vs a noted above. Possible errors may be generated by the unrealistic numerical simulation of RCP or by improper representation of the mechanical properties of the material.

FE analysis of the experimental RCP results relies on assumptions concerning:
a) the crack initiation,
b) the conditions of the loaded boundaries during RCP,
c) the choice of a single Young's modulus.
The examination of the influence of various crack initiation conditions on G_{ID} data (figure 6), showed no significant effect in G_{ID} vs a behaviour. An assessment of whether the loaded boundaries are likely to remain fixed during the crack propagation was also considered. The simulation was done by distributing the mass of the grip on the upper boundary nodes while the machine loading system is modelled by linear springs applied on each upper boundary node. Because of the clamp inertia, high machine stiffness ($1*10^8$ N/m) and short times to fracture a very small motion of these boundaries was found. However, the calculated G_{ID} cannot be affected by the way the clamped boundary is simulated until the stress waves return from that boundary to the crack tip (regardless of the grip mass and the stiffness of the machine). Thus, the crack resistance has enough time to rise, especially when the D/H ratio of the sample and time to fracture are small. Figure 12 (among other results) illustrates the effect of different moduli (E_s, E_d) on G_{ID} results. The crack resistance values obtained by using the static modulus are somewhat higher than the dynamic modulus results. Obviously, the amount of the strain energy available for the crack propagation was substantially larger in the static than in the dynamic modulus analysis. Both cases show a similar rising trend of the G_{ID} curves before the return of the stress wave.

Having examined the influences of the various crack initiation cases, single modulus and loaded boundary conditions on G_{ID} results, the question arises of whether the linear elastic analysis, with the single constant modulus, can be used to model RCP adequately. A remaining possible source of error in the analysis, which ought to be examined, could be the improper

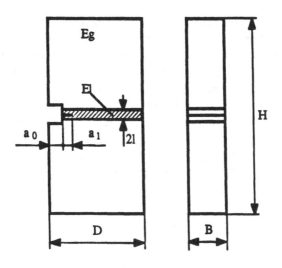

Figure 9: Local strip configuration (SENT)

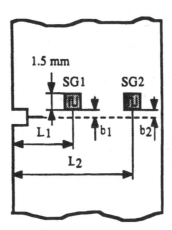

Figure 10: Strain measurement configuration

representation of the material behaviour around the rapidly moving crack tip. It has been recognized that any fracture event is dominated by the stress-strain-strain rate relation of the material. This relation governs both the local process at the crack tip and the capability of the remote stress field to transmit its energy into the region where the crack propagation takes place. Broberg[7,8] pointed out that the neighbourhood of the crack tip is always a region of material instability. During dynamic fracture this implies that the propagation velocities of irrotational and equivoluminal waves could be very low, irrespective of plastic flow being present or not. As a consequence, the energy flow to the crack tip is greatly reduced compared to the homogeneous case. A number of authors have attempted to allow for non-linear effects in their dynamic solutions (Kanninen[9], Rosakis[10]).

Unfortunately, the parameters of the tested-material behaviour near the rapidly moving crack tip are uncertain because of large strains and high strain rates. Also, the incremental modelling of non-linearity increases the complexity of the analysis as well as the computational cost. Considering the above arguments, it was decided to approximate the local non-linear effects by a local strip along the prospective crack path (figure 9). Both the thin strip layer and the surrounding material are taken as isotropic and linearly elastic (Broberg[8]). The crack is propagated in the plane of symmetry of the strip. The large strain around the moving crack tip would lead to a locally small modulus, so the modulus of the local strip (E_l) is assumed lower than of the base surrounding material (E_g). The energy dissipation due to eventual plastic deformation is not considered by the local strip model.

In order to support the local strip analysis and define the local zone parameters (the thickness of the layer-$2l$ and E_l), strain history measurements were performed. As shown in figure 10, two Micro Measurement Strain Gages (gauge type is EA-06-06027-120, gauge length 1.5 mm) are placed on the sample side close to the prospective crack path. During the crack propagation, the strain-ε_{yy} (ε_{yy} is a normal strain in respect to the crack plane) is recorded as a function of the time at two positions. As the moving crack tip approaches the strain gauge (SG), the strain reaches a maximum value. The average crack speed can be obtained from the distance between SGs and the time difference between the peaks. The details about the typical specimen dimensions, SGs positions, crack velocity (à) and breaking load are listed in table 2:

B	D	H	a_0+a_1	L_1	L_2	b_1	b_2	à (m/s)	P_b (KN)
9.93	25.75	104.	3.3	10.	19.6	0.55	0.5	348.	2.824

Table 2: Sample dimensions and SGs positions (dimensions in mm).

The experimentally obtained ε_{yy} vs t results are presented in the figure 11. Figure 11 also shows the ε_{yy} vs t as predicted numerically using the single constant modulus $E_g = E_d$. The line joining adjacent peak strains, termed the peak strain line, is assumed to be straight. The peak strain lines, shown by figure 11 are:
line 1 - experimentally recorded line,
lines 2,3,4 - peak strain lines generated numerically using single constant E_d, single constant E_g and local strip with $E_l = 0.1$ GPa, $E_g = E_d$, $l = 0.56$ mm analysis, respectively.

The corresponding, computed, crack resistance curves are presented in figure 12.

This analysis (figures 11 and 12) reveals several important conclusions:

a) The strain, calculated using a single constant E_s, does not agree with the experimentally determined strain. Neither the strains measured when the crack tip is far from the gauge, nor the measured peak strains are reproduced.

b) ε_{yy} vs t data, obtained from the numerical simulation with a single constant E_d compare well with the strain measured when the crack tip is not near the gauge. Thus, for the region away from the crack tip, the choice of the short term modulus (E_d) for RCP analysis is justified.

c) The measured peak strains history cannot be reproduced by computed peak strains from the numerical analysis with the single constant modulus (either E_s or E_d). The slopes of the numerically generated peak strain lines (lines 2 and 3) are considerably larger than the slope of the corresponding experimental line (line 1).

d) The variation in the crack resistance results, over the distance from L_1 to L_2, correlates with the square of the variation in calculated strain over the same region, i.e.:

$\varepsilon_{yy2} / \varepsilon_{yy1} = 1.545$ and $(G_{D2} / G_{D1})^{1/2} = 1.549$ for the single constant E_s analysis,

$\varepsilon_{yy2} / \varepsilon_{yy1} = 1.489$ and $(G_{D2} / G_{D1})^{1/2} = 1.483$ for the single constant E_d analysis,

subscripts 1 and 2 refer to the crack lengths L_1 and L_2 respectively.

e) A very good fit, between the measured peak strain line (line 1) and numerically generated peak strain line (line 4), is obtained by using the strip zone with local parameters $E_l = 0.1$ GPa, l=0.56 mm and the modulus of the surrounding material equal E_d. The local strip zone along the entire crack path cannot properly model the material behaviour far ahead of the moving crack tip. The calculated strain of course does not agree with the strain recorded long before the crack approached the SGs. However, this is not essential for G_{ID} computation. Generated crack resistance data are nearly independent on the crack length $(G_{D2} / G_{D1})^{1/2} = 1.13$, figure 12), as suggested by measured strains $(\varepsilon_{yy2} / \varepsilon_{yy1} = 1.125)$ and fracture surface roughness.

Note that similar G_{ID} curves can be obtained by employing a different set of the local strip parameters (E_l, l) and matching the slope of the experimental peak strain line, e.g. $E_l = 0.08$ GPa and l=0.43 mm; $E_l = 0.06$ GPa and l=0.28 mm, but these do not satisfy the strain magnitudes. Only one combination of E_l and l will give a constant resistance and model the strain histories.

Figure 13 shows the crack resistance vs crack speed results obtained by the single constant E_d approach and local strip analysis. In the later case, E_d is chosen for the base material while the local properties are defined according to the measured peak strain lines (as described above). The elastic-single constant modulus interpretation of G_{ID} vs à results, shows highly scattered and multivalued G_{ID} data (in a good agreement with Rosenfield & Kanninen[11], Crouch[1] results etc.). By including the local strip with a low modulus (E_l), these are essentially eliminated and much of the speed-dependence is also removed. The crack resistance generated on this way is around 220 J/m^2, and is close to the reported G_{ID} lower bound data at lower speeds.

The changes in the energy components of both single constant E_d and local strip approaches (test 10, figure 8) are shown in figure 14. The local strip model generated a substantially larger amount of the kinetic energy in the sample than the single modulus approach.

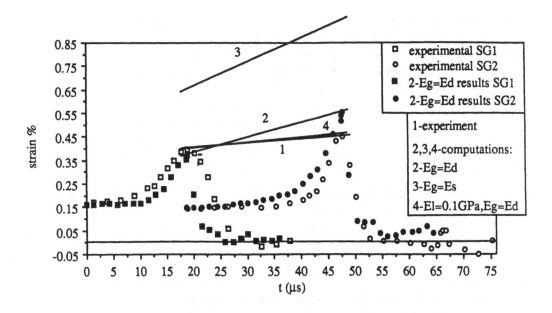

Figure 11: Strain vs time histories at SG1 and SG2 numerical-experimental data comparison

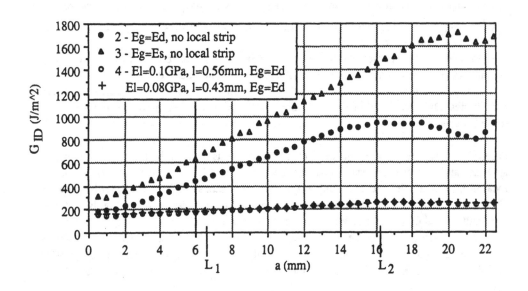

Figure 12: Crack resistance vs a Ed, Es, El+Ed results

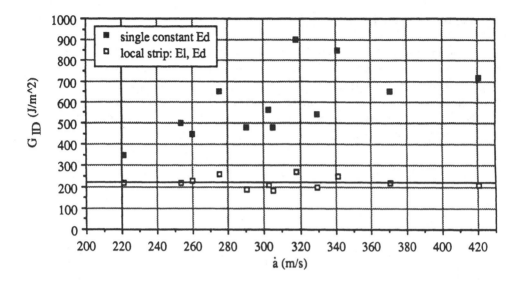

Figure 13: Crack resistance vs crack speed

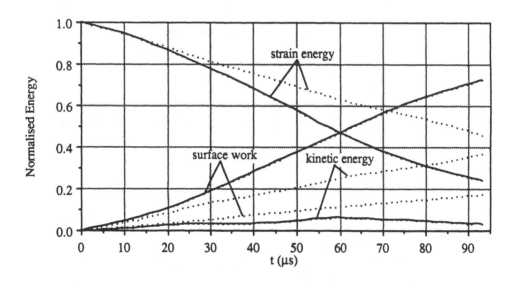

Figure 14: Energies vs time, test 10
single constant Ed results-lines
local strip (El, Ed) results-dots

5. Mass-Spring Modelling

The simple tension test may be modelled by a two spring-single mass model shown in figure 15. k_1 is the contact stiffness comprises the grips and machine stiffness, and $k_2(a)$ is the crack length dependent specimen stiffness. m is the equivalent mass of the specimen. The mass of the clamps is not included here since measurements of the machine and clamp stiffness showed that they are essentially immobile and all the kinetic energy arises in the specimen. The applied displacement (u_0) gives rise to a displacement in the specimen (u) of:

$$u = \frac{k_1}{k_1 + k_2} u_0 \qquad (2)$$

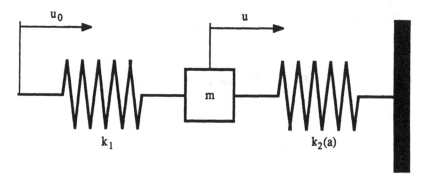

Figure 15: The mass-spring model

In the analysis, it is assumed that all motion arises from the crack growth and that u_0 is constant. The energy release rate (G), for the short crack lengths, is given by Williams[12] as:

$$\frac{G}{G_s} = 1 - \frac{1}{\bar{C}^2} \frac{d(a\,\dot{a})}{dt} \qquad (3)$$

where:

$$G_s = \left[\frac{\left(\frac{k_1}{BE}\right)H}{D + \left(\frac{k_1}{BE}\right)H} \right]^2 \left(\frac{u_0}{H}\right)^2 E\pi a \qquad (4)$$

the static energy release rate and $\bar{C}$ is a characteristic speed given by:

$$\bar{C} = C \sqrt{\frac{3}{4\pi}} \left[\sqrt{\frac{D}{H}} + \left(\frac{k_1}{BE}\right)\sqrt{\frac{H}{D}} \right] \qquad (5)$$

and:

$$C = \sqrt{\frac{E}{\rho}}$$
(6)

the tensile wave speed.

$\bar{C}$ depends on the material properties mostly via C but is also geometry dependent via D, H and B, and is system dependent via k_1. It is significant that a minimum occurs in $\bar{C}$ when $D/H=k_1/BE$ and that $\bar{C}$ varies rather slowly with D/H.

If we define a G_s value for $a = a_{0t}$ as G_0 and assume G = R, a constant crack resistance, then equation (3) becomes an equation of motion for a:

$$\frac{R}{G_0} \frac{a_\alpha}{a} = 1 - \frac{1}{\bar{C}^2} \frac{d(a\,\dot{a})}{dt}$$
(7)

which may be solved for the boundary conditions $\dot{a} = 0$ at $a = a_{0t}$ to give:

$$\left(\frac{\dot{a}}{\bar{C}}\right)^2 = \left(1 - \frac{a_{0t}}{a}\right)^2 + 2\frac{a_{0t}}{a}\left(1 - \frac{a_{0t}}{a}\right)\left(1 - \frac{R}{G_0}\right)$$
(8)

In the real experiment an apparently almost constant speed over much of the crack growth was observed. For simplicity, we shall assume that the speed was observed at, say, half way across the ligament, i.e. at:

$$a = \frac{D + a_{0t}}{2}$$
(9)

from which:

$$\frac{a_{0t}}{a} = \frac{2\frac{a_{0t}}{D}}{1 + \frac{a_{0t}}{D}}$$
(10)

In all tests (section 2), $a_{0t} = 0.2D$ so $a_{0t}/a = 1/3$, and from the equation (8):

$$\frac{1}{EG_0} = \frac{2}{ER} - \frac{9}{4ER} \frac{\dot{a}^2}{\bar{C}^2}$$
(11)

Further, G_0 can be expressed as:

$$G_0 = \frac{\pi\sigma^2 a_{0t}}{E}$$

where, the nominal stress σ is:

$$\sigma = \frac{P_b}{BD}$$

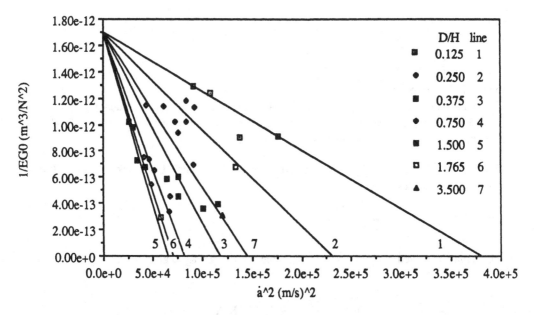

Figure 16: 1/EG0 vs à^2 experimental results

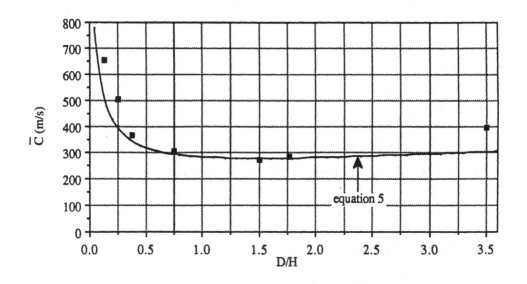

Figure 17: C̄ vs D/H

and P_b is the experimentally recorded breaking load. Thus, the left hand side term of equation (11) takes the form:

$$\frac{1}{EG_0} = \frac{1}{\pi \left(\dfrac{P_b}{BD}\right)^2 a_{0t}}$$

and is calculated for each test from section 2.

Figure 16 shows $1/EG_0$ as a function of the square of measured crack speed ($\dot{a}^2$) for various D/H values. The lines (figure 16) are fitted through each set of points, i.e. for each D/H ratio. From the intercepts between these lines and the horizontal axes, $\bar{C}$ data are obtained for different D/H values using equation (11). A crack resistance (R) of 200 J/m² is generated from the common intercept of the lines and vertical axes. This value is very close to that of 220 J/m² computed by dynamic FE with the local strip zone. Figure 17 presents the $\bar{C}$ data as a function of D/H ratio. The line, drawn by employing the equation (5) and using measured $k_1 = 1*10^8$ N/m with the global modulus E = Ed, is also shown in figure 17. The minimum in $\bar{C}$ occurs for D/H = 1.5. The most important observation is the very good agreement between the experimental $\bar{C}$ data and theoretical prediction (equation 5) with C = 230 m/s. This is much lower value of C than would be expected (2745 m/s) and it gives the modulus of $7*10^7$ N/m², which is close to E_1 used in the FE analysis. Lower bound G_{ID} data (figure 13) were obtained using the local strip model with $E_1 = 5*10^7$ to $1.1*10^8$ N/m².

6.   Conclusions

The dynamic FE method was used to analyse RCP in PMMA-SENT specimens. The most important observation is that the systematic analysis, based on a single constant modulus, could not reproduce the measured strain histories along the crack path. Consequently, this linear elastic analysis generated highly scattered and sample size dependent G_{ID} data. Thus, the experimental methods based on linear elastic solutions might not be as accurate as believed. It is worth mentioning that at least two, near crack path, strain histories have to be recorded and some kind of full dynamic analysis employed if the validity of predicted strain histories (and so G_{ID}) are to be assessed. The local strip, low modulus, model generated crack resistance data nearly independent of the speed and sample size. This local modulus concept is thought to be very important in understanding dynamic fracture problems. It was observed that most of the features detected by the FE analysis and reported above could be described with a mass-spring model. G_{ID} and E_1 values obtained from the mass-spring model were in a very good agreement with those generated by the FE analysis. Hence, the authors are encouraged to believe that the simple mass-spring model could be a useful alternative to the more complicated and expensive FE analysis.

Acknowledgements

The authors wish to thank ICI for their financial support of this work.

References

[1] B. Crouch, Ph.D. Thesis, University of London, Imperial College, (1986)

[2] L. Dahlberg et al, Crack Arrest Methodology and Applications, ASTM STP 711 (eds. M.F.Kanninen and G.T.Hahn), pp.89, (1980)

[3] J.G. Williams and A. Ivankovic, Int. J. of Fracture, (in press 1990)

[4] A.S. Kobayashi et al, Int. J. of Fracture 30, pp.275, (1986)

[5] P.N.R. Keegstra, Ph.D. Thesis, University of London, Imperial College, (1977)

[6] O.C. Zienkiewicz, The Finite Element Method, McGraw-Hill, London, (1971)

[7] B.K. Broberg, J. Mech. Phys. Solids 19, pp.407, (1971)

[8] B.K. Broberg, Dynamic Crack Propagation, (ed. G.C. Sih), pp.461, Noordhoff (1973)

[9] M.F. Kanninen, Int. J. of Fracture 27, pp.299, (1985)

[10] A.J. Rosakis and A.T. Zehnder, Int. J. of Fracture 27, pp.169, (1985)

[11] A.R. Rosenfield and M.F. Kanninen, J. Macromolecular Sci., Vol. B7, pp.609, (1973)

[12] J.G. Williams, Int. J. of Fracture 33, pp.11, (1987)

AUTHOR INDEX

Printed in the United States
By Bookmasters